2/24

Introductory
POLYMER
CHEMISTRY

G.S. MISRA

Former Professor & Head
Department of Chemistry, Universities of Jabalpur and Jammu
and
Ex-Director, Indian Lac Research Institute
Namkum, Ranchi

30 YEARS OF PUBLISHING FOR ONE WORLD

NEW AGE INTERNATIONAL (P) LIMITED, PUBLISHERS
New Delhi • Bangalore • Bombay • Calcutta • Guwahati • Hyderabad
Lucknow • Madras • Pune • London • Bangkok

NEW AGE INTERNATIONAL (P) LIMITED, PUBLISHERS

NEW DELHI : 4835/24 Ansari Road, Daryaganj, New Delhi 110 002
BANGALORE : No. 35, Annapoorna Complex, South End Road,
Basavangudi, Bangalore 560 004
BOMBAY : Room No. 3, 1st Floor, 128 Noorani Building,
L.J. Road, Opp. Mahim Bazar P.O., Bombay 400 016
CALCUTTA : 40/8, Ballygunge Circular Road, Calcutta 700 019
GUWAHATI : Pan Bazar, Rani Bari, Guwahati 781 001
HYDERABAD : 1-2-412/9, Gaganmahal, Near A V College, Domalguda,
Hyderabad 500 029
LUCKNOW : 18, Pandit Madan Mohan Malviya Marg, Lucknow 226 001
MADRAS : 20, IInd Main Road, Kasthuribai Nagar, Adyar, Madras 600 020
PUNE : Flat No. 2, Building No. 7, Indira Co-op Housing
Society Ltd. (Indira Height), Paud Fatta, Erandawane,
Karve Road, Pune 411 038
LONDON : Wishwa Prakashan Ltd., Spantech House, Lagham Road,
South Godstone, Surrey, RH9 8HB, U.K.

ISBN 81-224-0471-5

Published by H.S. Poplai for New Age International (P) Limited, 4835/24, Ansari Road,
Daryaganj, New Delhi 110 002. Typeset by Printek India, C4F/27IB Janakpuri, New Delhi 110
058 Ph. 5554775 and printed at Taj Press, Mayapuri, New Delhi 110 064.   Printed in India.

In Kind Remembrance

Dedicated to the Memory of
(Late) Prof. T.R. Seshadri, FRS

PREFACE

The present book is an outcome of the teaching and research experience of the author in the realm of polymer chemistry for over 35 years. Polymer chemistry is now taught as an independent subject in a number of Indian Universities and Engineering Institutes. With the opening of manufacturing units for plastics, elastomeric materials and fibers, and establishment of various defence and space research organizations, the demand for properly trained personnel in polymer science and technology has increased. Many chemistry students, have very little mathematical background and yet wish to seek a career in the polymer field. It is considered that the elementary treatment provided in this book will meet their requirement, if they have a background of organic chemistry. The arrangement of topics in this book is rather unorthodox. The synthetic approach is emphasized in the first four chapters in addition to outlining some elementary physical chemistry necessary for the understanding of the mechanisms and theoretical principles involved. This is followed by an account of morphology and molecular weight determinations of polymers in the next two chapters. Chapter seven discusses some of the technological aspects of elastomers, fiber forming materials and plastics.

A description of the characterization of polymers, specially by spectroscopy and thermal analysis is followed by a discussion of important polymer reactions and polymer reactants. The last three chapters are concerned with solubility, flow properties of polymers and naturally occurring polymers.

The subjects covered conform roughly to American Chemical Society's model syllabus for Introductory Polymer Chemistry (Appendix IV). A special feature of the book is the incorporation of questions in the text, followed by their answers at the end of the chapters. To add to the usefulness of the book, ten specimen laboratory exercises have been provided (Appendix II).

I would like to take this opportunity to thank the many publishers who cooperated by allowing the generous use of data and illustrations from their publications. I am also thankful to my publishers, Wiley Eastern, New Delhi for undertaking the publication of this book.

I shall be failing in my duty if I do not thank a host of my former co-workers who helped me in various ways. Special thanks are due to Drs. J.S. Shukla, U.D.N. Bajpai, B.C. Srivastava, B.D. Arya, Prema and Lakshmi M. Subramanian. My daughter Ms. Manjvi Chaturvedi also helped me in preparing the first few chapters of the manuscript, for which I thank her.

Lucknow   G.S. MISRA

ACKNOWLEDGEMENTS

I. Permission granted by the following copyright holders to reproduce their copyright materials is gratefully acknowledged.

Copyright holder	Journal/Book reference	Copyright year	Figure/Table in this book (Material reproduced)
1	2	3	4
Academic Press Inc., New York	Bovey, F.A. *Chain Structure and conformation of macromolecules*	1982	Fig. 13.2 and Fig. 13.3
Akademische Verlaggesellschaft, Leipzig	Schulz, G.V. and Blaschke, F.H., *Z, Physik Chem*, B51, (1942) 75	1942	Fig. 3.2
American Chemical Society, Washington D.C.	Mayo, F.R., Gregg, R.A. and Matheson, M.S. *J. Amer. Chem. Soc.*, 73 (1951) 1691.	1951	Fig. 3.2
	Arnett, L.M. *J. Amer. Chem. Soc.*, 74 (1952) 2027	1952	Fig. 3.2
	Billmeyer Jr., F.W. and de Than, C.B. *Ibid.*, 77 (1955) 4764, Table 1.	1955	Fig. 6.14
	Henrici-olive, G and olive, s. chem tech 11 (1981) 746	1981	Fig. on Page 69
American Chemical Society (Division of Chemical Education), Journal of Chemical Education, Easton, Pa	Neckers, D.C., *J. Chem. Education*, 52, (1975) 692-697.	1975	Fig. 14.1
	Harris, F.W., *Ibid.*, 58, (1981) 840	1981	Fig. 7.1
	McGrath, J.E., *Ibid.*, 58, (1981) 856	1981	Fig. 4.2 and Fig. 4.3
	Cloutler, H. and Prudhomme, R.E., *Ibid*, 62, (1985) 818	1985	Fig. 13.1

1	*2*	*3*	*4*
Applied Science Publishers, U.K. (Editor)	Grassie, N., Editor, Developments in polymer degradation	1977	Fig. 4.4
Cambridge University Press, Cambridge	Grassie, N. and Scott, G., *Polymer degradation and stabilisation*, 1985	1985	Scheme11.3 Scheme11.9 Scheme11.12
Dow Chemicals (Dr. John C. Moore)	Altigelt, K.H. and Moore, J.C. *in Polymer Fractionations* Ed. M.J.R. Cantow, Academic Press, 1967	1967	Fig. 6.7
Dow Chemicals, Mid. Michigan, U.S.A.	Nyquist, R.A. *Infra-red spectra of plastics and resins*, 2nd Edition, 1961. IR spectra of poly (ethylene), poly (propylene), 1, 4-poly (isoprene), poly (styrene) and poly (vinyl chloride).	1961	Fig. 13.4
Elsevier Science Publications, Amsterdam	Sudol, R.S., *Anal. Chim, Acta*, 46 (1969) 231.	1969	Fig. 13.6
Federation of Societies for Coating Technology. Philadelphia, U.S.A.	Hoy, K. *J. Paint Tech.* 42 (1970) 118.	1970	Table 15.1
Helvetica Chimica Acta (Redaktien)	Signer, R. And Gross, H. *Helvetica Chimica Acta*, 17, (1934) 59, 335, 726	1934	Table 6.1 & Figs.6.1, 6.2, 6.3 & 6.4
Huthig and Wepf Verlag, Basel	Schulz, G.V. Harborth, G., *Makromolekulare Chemie*, 1, (1947) 106.	1947	Table 3.2
John Wiley & Sons New York	Billmeyer Jr., F.W. *Text book of polymer science*, Third Edition (1984)	1984	Fig. 6.14, Fig. 7.5, Fig 7.6 Fig 13.6 Fig. 16.4, Table 15.1 and Table 15.2

1	*2*	*3*	*4*
John Wiley & Sons, New York	Odian, G. *Principles of polymerization*, Second Edition (1981)	1981	Fig. 3.6 Fig. 3.7 Fig. 3.8 Fig. 5.3 Fig. 5.5 and Fig. 7.7
	Stille, J.K. *Introduction to polymer chemistry* (1962)	1962	Table 3.6
John Wiley & Sons, New York	*Journal of Polymer Science* Maley, L.R., C8, (1965); 253	1965	Fig. 6.10
	Grubisie, Z. *et al.*, B5, (1967) 755	1967	Fig. 6.9
	Natta, G., 48, (1960) 219;	1960	Fig. 3.6
Marcel Dekker Inc., New York	Marvel, C.S. in Segal, C.L. (Ed.) *High temperature polymers*, 1967	1967	Fig. 11.1
McGraw-Hill Book Company, New York	Seymour, R.B. *Introduction to polymer chemistry*, 1971	1971	Fig. 16.3 and Table 12.2
	Schmidt, D.L., *Modern plastics*, Dec. 1960, 38 4, 147.	1960	Fig. 11.3
Mir Publishers, Moscow	Tager, A. *Physical chemistry of polymers*, 1972	1972	Fig. 5.4 and Fig. 6.13
Royal Society of Chemistry, London	Gregg, R.A. and Mayo, F.R., *Discussion of Faraday Soc.* 2, 328 (1947)	1947	Fig. 3.3
	Pepper, D.C., *Quarterly Reviews*, 8, (1954) 88	1954	Fig. 4.1
Scientific American, New York	Natta, G. *How giant molecules are made* Scientific American, 197, (1957) 98-99	1957	Fig. 5.6 & Fig. 5.7
Springer Verlag, Heidelberg	Vollmer, Bruno, *Polymer chemistry* (Springer Verlag, New York)	1973	Fig. 3.2 Fig. 6.5 Fig. 6.15 & Fig. 16.5

1	2	3	4
VCH Verlagsgesell-Schaft Weinheim, West Germany	Schulz, G.V., *Chem. Ber.* 80, (1947) 232.	1947	Fig. 3.4
	Patat, F. and Sinn. H., *Ange-wandte Chemie*, 70, (1958) 496.	1958	Fig. 3.6

II. The Following materials in the public domain are also reproduced. The sources as indicated below are acknowledged.

(i) Flory, P.J., *J. Amer. Chem. Soc.*, 61, (1939) 3334
For Figs. 2.1 and 2.2 in the book.

(ii) Flory, P.J., *Ibid.* 58 (1936) 1877. For Figs. 2.3 and 2.4 in this book.

Addition (chain) polymerization: This occurs when small molecules join together under the stimulus of a catalyst, heat or radiation to form a linear polymer usually without the elimination of a small molecule. This can be of the following three types (a)–(c):

(a) Free radical addition polymerization: In this type, chains are initiated by a free-radical such as phenyl.

(b) Cationic addition polymerization: The active species which initiates the addition polymerization is a cation such as a proton.

(c) Anionic addition polymerization: The initiating species in this case is an anion such as $NH2^{(-)}$.

Coordination polymerization: There are a number of coordination catalysts such as a combination of aluminium trialkyl and titanium or vanadium chloride, which will polymerize olefenic compounds to yield a stereospecific polymer, e.g., isotactic polypropylene from propylene.

Adhesive: Material that binds and holds the surfaces together.

Amorphous: A non-crystalline polymer or non-crystalline areas in a polymer.

Atactic: Polymer in which there is a random arrangement of pendant groups on each side of the chain.

Biopolymer: A naturally occurring polymer such as cellulose.

Block co-polymer: The repeating unit consists of segments or blocks of similar monomers tied together along the macro-molecular chain.

Branched chain polymer: A polymer having extensions of polymer chain attached to the polymer backbone.

Calendering: A process of making polymeric sheets by means of a machine containing counter-rotating rolls.

Compression moulding: A fabrication technique of moulding a thermosetting polymer by means of heating and applying pressure.

Chain-transfer: A reaction in which a free-radical abstracts an atom or group of atoms from a solvent, initiator, monomer or polymer.

Chain-polymerization: See addition polymerization

Colligative properties: Properties of a solution which are dependent on the number of solute molecules present.

Co-polymers: A long chain polymer composed of at least two different monomers, joined together in an irregular sequence.

Critical chain length: The minimum chain-length required for the entanglement of the polymer chains.

Cross-links: Covalent bonds between two or more polymer chains.

Crystalline polymers: A polymer with an ordered structure, which has been allowed to disentangle and form a crystal.

Condensation (step) polymerization: The polymers are formed by various organic condensation reactions with the elimination of small molecules such as water.

Crystalline melting point (T_m): This is the range of melting temperature of the crystalline domain of a polymer sample and is accompanied by change in polymer properties. It is also the first-phase transition when the solid and liquid phases are in equilibrium.

Degree of polymerization ($\overline{DP}$ or $\overline{P}$): It is the average number of repeating units in a macromolecule. The degree of polymerization is obtained by dividing the (average) molecular weight by the molecular weight of the momomer.

Elastomer: These are the non-crystalline high polymers or rubbers that have three-dimensional space network structure (e.g. that produced by vulcanization), which improves stability or resistance to plastic deformation. Normally, elastomers exhibit long range elasticity at room temperature.

Extrusion moulding: A fabrication technique by which a heat softened polymer is forced continuously by a screw through a die.

Fibers: A fiber is a thread or thread like structure composed of strings or filaments of linear macromolecules that are cross-linked in such a manner as to give rise to an assemblage of molecules having a high ratio of length to width.

Functionality: The number of reactive groups in a molecule.

Glass transition temperature (T_g): This is the temperature at which an amorphous polymer starts exhibiting the characteristic properties of the glassy state, (because of the onset of segmental motion) stiffness, brittleness and rigidity.

Graft co-polymer: When a monomer is polymerized onto the primary high polymer chain obtained by the polymerization of another kind of monomer, a graft polymer results.

Inhibitor: An additive which reacts with a chain-forming radical to produce non-radical products or radicals of low reactivity, incapable of adding fresh monomer units.

Injection moulding: A fabrication process in which a heat-softened polymer is forced continuously through a die by means of a piston.

Intermolecular forces: Secondary valency forces among different molecules.

Intramolecular forces: Secondary valency forces within the same molecule.

Isotactic polymers: A polymer in which all the pendant groups are arranged on the same side of the polymer backbone.

Kinetic chain-length: It is defined as the average number of monomer molecules contained per radical which initiates a polymer chain.

Linear chain polymer: It consists of a linear polymer chain wihtout any branching.

Macromolecules: See polymers

Mer: The repeating unit in a polymer chain.

Micelle: It is an aggregation of crystallites of colloidal dimensions and exists either in solid state or in solution.

Molecular weight: Most polymer are poly-disperse or mixtures containing polymer molecules of different molecular weights. Different measures of molecular weight are defined as:

Number average molecular weight ($\overline{M_n}$): The arithmetical mean value obtained by dividing the sum of molecular weights by the number of molecules.

Weight average molecular weight ($\overline{M_w}$): The second power average of molecular weight in a polydisperse polymer.

Z-average molecular weight ($\overline{Mz}$): The third power average of molecular weight in a polydisperse polymer.

Monomer: All high polymers are formed by the joining together of many molecular units of groups of molecular units. The number of units or mer of a polymer is the unit of the molecule which contains the same kind of and number of atoms as the real or hypothetical repeating unit.

Oligomer: A polymer containing very few repeating units, usually between 2 and 10.

Osmotic pressure: The pressure that, a solute would exert in solution if it were an ideal gas at the same volume.

Pendant groups: Groups attached to the main polymer chain or backbone. An example is the methyl groups in polypropylene.

Plastics: A group of artificially prepared substances usually of organic origin which sometimes during their stage of manufacture have passed through a plastics condition.

Plasticizer: An additive which reduces the inter-molecular forces between polymer chains and thus acts as an internal lubricant.

Polymer or macromolecule: A giant molecule made up of a large number of repeating units such as polyethylene which may contain 100 or more ethylene monomer units.

Polymerization: It is the process of formation of large molecules from small molecules with or without the simultaneous formation of byproducts such as water. A classical example is the formation of polystyrene from styrene molecules.

Polydispersed: A polymer containing molecules of different molecular weights.

Rayon: Regenerated cellulose in the form of a filament; used as a fiber.

Retarder: An agent which acts as a chain-transfer agent to produce less reactive free-radicals.

Rheology: The science of flow.

Ring opening polymerization: Formation of polymers by the opening of rings such as those of ethers or lactams. The formation of Nylon-6 from caprolactam is an example.

Spinneret: A metal plate with many small holes of uniform size used for spinning.

Step polymerization: See condensation polymerization.

Stereoselective polymerization: In this type of polymerization one type of ordered structure is preferentially formed in contrast to the other.

Stereospecific polymerization: Polymerizations which yield ordered structures (isotactic or syndiotactic).

Syndiotactic: A polymer in which pendant groups are arranged alternately on each side of the polymer backbone.

Tacticity: The arrangement of pendant groups in space.

Thermoplastic: These soften in a reversible physical process under the influence of heat and sometimes of pressure and can be moulded into different shapes under this condition. They retain their shapes on cooling.

Thermosetting resins: These soften under the influence of heat and pressure and can be moulded into different shapes. They become hard and infusible on account of chemical change and cannot be remoulded.

Theta temperature: A temperature at which a polymer of infinite molecular weight starts to precipitate from a solution.

Viscosity: Resistance to flow.

Viscosity, intrinsic [η]: The limiting viscosity number obtained by the extrapolation of relative viscosity to zero concentration.

Viscosity, reduced: The specific viscosity divided by concentration.

Viscosity, relative: The ratio of the viscosities of the solution and the solvent.

Viscosity, specific: The difference between the relative viscosity and one.

Vulcanization: This is a process by which cross-links between linear elastomer chains are introduced. An example is the introduction of sulphur cross-link in natural rubber hy heating it with sulphur.

Ziegler-Natta catalyst: A catalyst with the composition $TiCl_3$-AlR_3, obtained from titanium tetrachloride and aluminium trialkyl.

CONTENTS

CHAPTER 1

INTRODUCTION

1.1. HISTORICAL INTRODUCTION

Early in the history of modern chemistry, those organic compounds which could not be crystallised were considered to be tars and were thrown into the sink. Polymer also came in this category. In 1920 Herman Staudinger published his famous paper on the structure of macromolecules. In this and subsequent papers, he argued that, "in high polymers many single molecules are held together by normal valency bonds. The tendency to form such compounds is observed in particular in organic chemistry due to the special nature of carbon." This view met with a great deal of opposition from famous chemists like the Nobel Prize winner Heinrich Wieland who declared:

> "Dear Colleague: please leave aside the idea of big molecules. There are no organic molecules with a molecular weight greater than 5000. Purify your products e.g. rubber, then it could be crystallised and it would be found that it is a low molecular weight substance."

Some others gradually began to support Staudinger's views. K.H. Meyer and H.F. Mark were his early supporters. In 1929, W.H. Carothers published his work on synthetic polymers, and thus Staudinger and Carothers were the two men who placed polymer chemistry on a firm footing.

1.2. NATURAL AND SYNTHETIC POLYMERS

Polymers can be divided into two classes—natural and synthetic. An example of natural polymer is wool, which is a protein fiber. Cotton, another natural polymer, is an example of vegetable fiber which is composed of cellulose. Natural rubber is another example of a natural polymer. It is obtained from the trees of *Hevea* species.

Synthetic polymers can be divided into three broad based divisions:

(i)   *Fibers* The most well known examples are nylon and terylene.
(ii)   *Synthetic rubbers*—These are not produced merely as substitutes for natural rubber. Some of these synthetic rubbers have properties superior to those of the natural product and are thus better suited for specific purposes, e.g., Neoprene rubber.
(iii) *Plastics*—These were originally produced as substitutes for wood and metals. Now they are important materials in their own right and tailor-made molecules are synthesized for a particular use, e.g., heat resistant polymers which are used in rocketry.

1.3. POLYMERIZATION

The fundamental process by which low molecular weight compounds are converted into high molecular weight compounds is called polymerization. This process is illustrated below:

| Low molecular weight material (possessing two or more reactive groups) | High temp. and/or pressure ──────────────→ and/or catalyst | High molecular weight material |

The monomers must have two or more reactive groups. These may, for example, be an amino group and a carboxyl group as in the case of a poly (amide):

$$n \, NH_2 - R - NH_2 + n \, HOOC - R' - COOH \longrightarrow$$
(Diamine)   (Dicarboxylic acid)

$$(2n - 1) \, H_2O + NH_2 \{R - NH.CO - R'CO\}_n \, OH \qquad (1.1)$$
[Poly(amide)]

Alternatively, reactive groups may be double bond as in the case of vinyl polymerization:

$$n \, CH_2 = CHY \xrightarrow{\hspace{2cm}} \{CH_2 - CHY\}_n \qquad (1.2)$$
(Vinyl monomer)   (Polymer)

The polymerization process may be divided into three categories on the basis of the type of reaction taking place: (i) condensation polymerization, (ii) addition polymerization and (iii) ring opening polymerization.

1.3.1 Condensation Polymerization

Condensation polymerization is a process of formation of polymers from polyfunctional monomers of organic molecules with the elemination of some small molecules such as water, hydrochloric acid, ammonia etc. An example has already been given earlier (Eq. 1.1).

In this type of polymerization also known as step-growth polymerization, the molecular weight of the polymer chain builds up slowly and there is only one reaction mechanism for the formation of the polymer. The distinct initiation, propagation and termination steps of chain growth are meaningless in step-growth polymerization (see section 1.3.2). The polymerization process proceeds by individual reactions of the functional groups of the monomers. Thus two monomers react to form a dimer. The dimer may now react with another dimer to produce a tetramer, or the dimer may react with more monomer to form a trimer. This process continues, each reaction of the functional groups proceeding essentially at the same reaction rate. The reaction proceeds for a relatively long period of time until a high molecular weight polymer is obtained.

1.3.2 Addition Polymerization

Addition polymerization is the process of formation of addition polymers from monomers without the loss of small molecules. Unlike condensation polymers, the repeating unit of an addition polymer has the same composition as the original monomer. The polymerization of ethylene to give poly (ethylene) is an example of this type of reaction. Addition polymers are prepared from olefines by a chain polymerization reaction, which usually leads to high molecular weight products. In contrast to condensation polymerization, there seems to be no intermediate low molecular weight units which slowly produce the high molecular weight materials.

It is generally believed that addition polymerization involves three distinct steps called initiation, propagation and termination. These steps apply to all types of addition polymerizations such as those initiated by free-radicals, cations, anions, y-rays etc.

We may represent these steps as following:

(1) Initiation

$$\dot{R} + CH_2 = CHY \longrightarrow R - CH_2 - \dot{C}HY$$

Here a free-radical $\dot{R}$ attaches itself to the olefin and produces a new radical consisting of radical and the monomer unit.

(2) Propagation

$$R - CH_2 \dot{C}HY + nCH_2 = CHY \longrightarrow R(CH_2 - \dot{C}HY)_n + 1$$

This step is very rapid and leads to high molecular weight products. No low polymers are produced.

(3) Termination

Termination may occur by several processes. For example, coupling is termination by combination of two growing chain radicals.

$$2R - (CH_2 - \dot{C}HY)_n \longrightarrow R(CH_2 - CHY)_n - (CHY - CH_2)_n - R$$

Termination can also take place by disproportionation which involves the transfer of a hydrogen atom from one growing chain radical to another. Other mechanism of termination such as due to transfer are also known.

1.3.3 Ring Opening Polymerization

A third type of polymerization is ring opening polymerization. An example of this type is the polymerization or epsilon caprolactam:

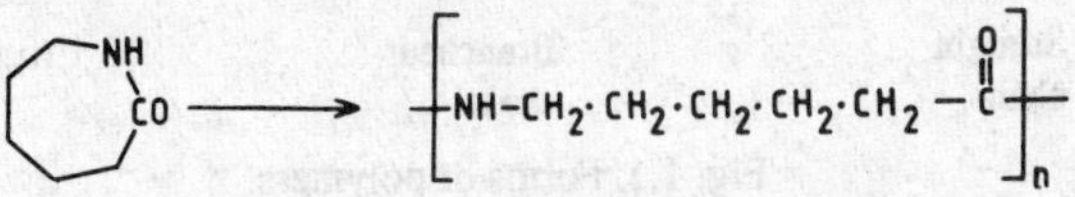

This type of polymerization has some of the features of both condensation and addition polymerization as far as kinetics and mechanism are concerned. It resembles addition or chain-polymerization in that it proceeds by the addition of monomer (but never of larger units) to growing chain molecules. However, the chain-initiating and subsequent addition proceeds at similar rates. If so, these are not chain reactions in the kinetic sense. As in stepwise polymerization the polymer molecules continue to increase in molecular weight throughout the reaction.

Most of cyclic compounds like lactans, cyclic ethers, lactones etc. polymerize by ionic mechanisms in the presence of strong acids or bases, when water and alcohols are excluded. The polymerizations are often very rapid.

Q. 1.1. Which of the following polymers are natural and which are synthetic?
   (i) Cellulose, (ii) Poly(styrene), (ii) Terylene, (iv) Starch, (v) Proteins, (vi) Silicones, (vii) Orlon [poly(acrylonitrile)], (viii) Phenol-formaldehyde resins.
Q. 1.2. Identify which of the following polymers are condensation polymers and which are addition polymers.

(i) $+CO\,(CH_2)_4\,CO.NH\,(CH_2)_6\,NH\,+_n$

(ii) $+CH_2 - \underset{\underset{C_6H_5}{|}}{CH}+_n$

(iii) $+CH_2 - CH = CH - CH_2+_n$

(iv) $\left[-OCH_2\cdot CH_2O\cdot CO -\bigcirc- CO-\right]_n$

1.4. FORMS OF POLYMERS

Polymer materials can exist in at least three general forms. First basic arrangement is the straight chain where the monomer molecules are joined together in a linear arrangement. The second arrangement is that of a branched chain. These branches may arise as a result of a secondary process or the presence of three reactive groups in a monomer molecule. In other cases branches might have been produced on purpose. A special category in branched chain polymers, graft polymers, is obtained when the branch consists of a chian(s) composed of a variety of monomer species. Yet another, a third arrangement is possible. Here cross-links are produced between linear chains (Fig. 1.1).

Straight chain   Branched chain   Cross-linked

Fig. 1.1. Forms of polymers.

Q. 1.3. Classify the following polymers into straight chain, branched and cross-linked polymers:

(i) $+CH_2 - \underset{\underset{CN}{|}}{CH}+_n$

(ii)

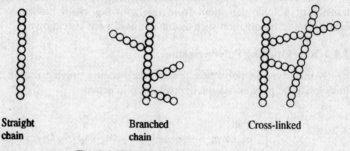

(iii) 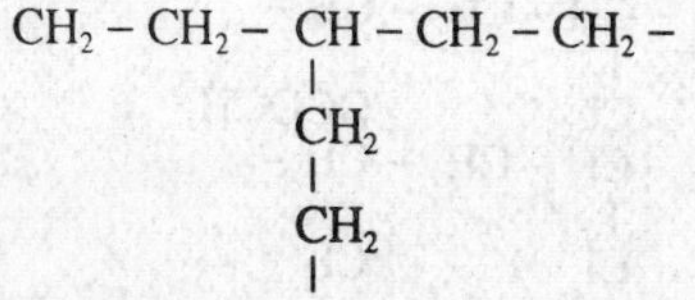 $- CH_2 - CH_2 - CH - CH_2 - CH_2 -$
$|$
CH_2
$|$
CH_2
$|$

1.5. COPOLYMERIZATION

It is possible to polymerize not only one monomer but a mixture of two or three different monomers to get copolymers. Copolymers can be random copolymers (Fig. 1.2. (a)) where we have a random arrangement of the two monomers A and B along the chain or alternating copolymers as shown in Fig. 1.2. (b).

$$- A - A - B - A - A - B - A - B - A - A - B - A -$$

Fig. 1.2. (a) Random copolymers.

$$- A - B - A - B - A - B - A - B - A - B -$$

Fig. 1.2. (b) Alternating copolymers.

1.5.1. Block and Graft Polymers

These are copolymers produced by special techniques to give certain specific properties to the products (Figs. 1.3 and 1.4).

$$- A - A - A - B - B - B - B - A - A - A - B - B - B. - B -$$

Fig. 1.3. Block copolymers.

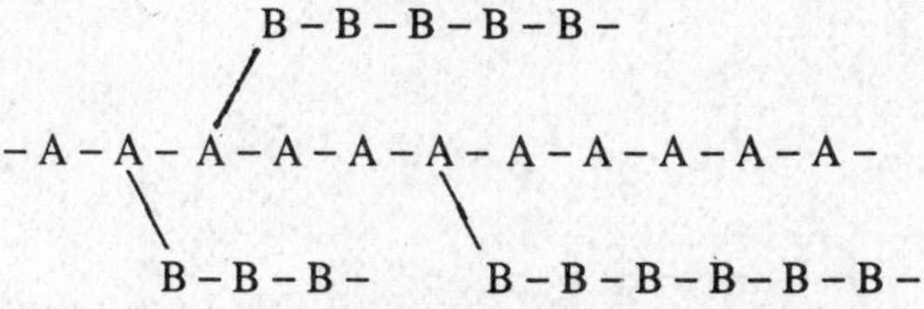

$$B - B - B - B - B -$$
$$/$$
$$- A - A - A - A - A - A - A - A - A - A - A -$$
$$\backslash \qquad\qquad \backslash$$
$$B - B - B - \qquad B - B - B - B - B - B -$$

Fig. 1.4. Graft copolymers.

Q. 1.4. Classify the following as random, alternating, block or graft copolymers.

(i) 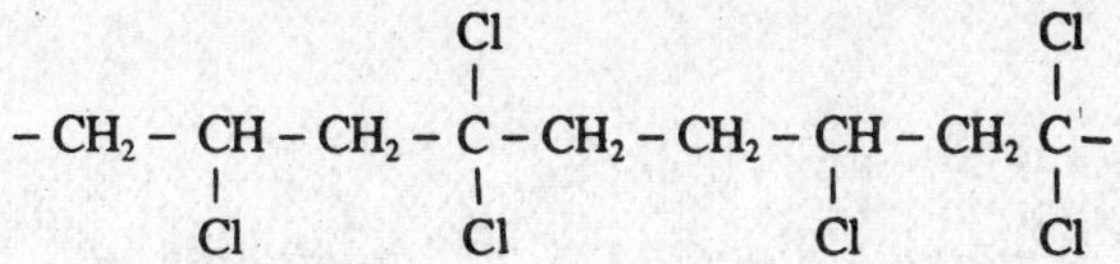
$$Cl Cl$$
$$| |$$
$$(i) \quad - CH_2 - CH - CH_2 - C - CH_2 - CH_2 - CH - CH_2 \; C -$$
$$| | | |$$
$$Cl Cl Cl Cl$$

(ii) 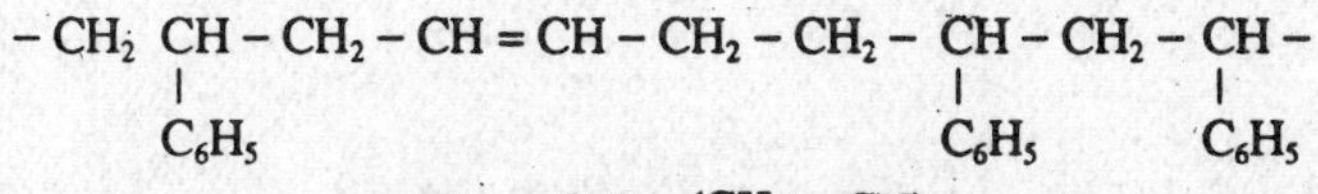
$$(ii) \quad - CH_2 \; CH - CH_2 - CH = CH - CH_2 - CH_2 - CH - CH_2 - CH -$$
$$| | |$$
$$C_6H_5 C_6H_5 C_6H_5$$

$$CH_2 - CH - (CH_2 - CH)_n$$
$$| |$$
$$CN CN$$

(iii)
$$(iii) - CH_2 \; CH - CH_2 - \; C - CH_2 - CH - CH_2 - CH -$$
$$| | | |$$
$$C_6H_5 C_6H_5 C_6H_5 C_6H_5$$

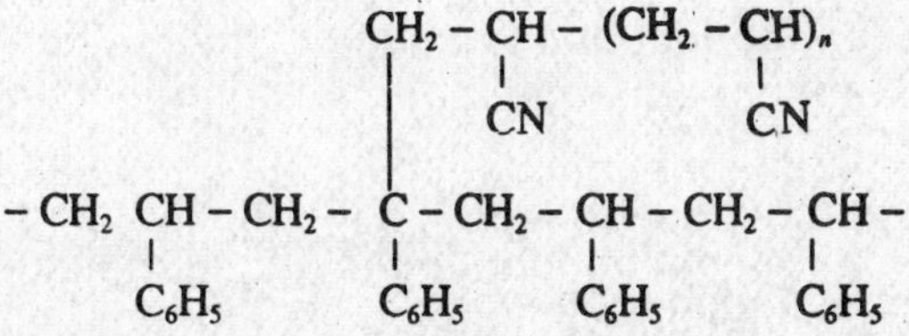

(iv) $-CH_2-CH-CH_2-CH-CH_2-CH-CH_2-CH-$
$||||$
$ClClClOCOCH_3$
$-CH_2-CH-CH_2-CH-CH_2-CH-CH_2-CH-$
$||||$
$OCOCH_3OCOCH_3ClCl$

ANSWERS TO QUESTIONS

A.1.1   (i) Natural, (ii) Synthetic, (iii) Synthetic, (iv) Natural, (v) Natural, (vi) Synthetic, (vii) Synthetic, (viii) Synthetic.

A.1.2   (i) Condensation, (ii) Addition, (iii) Addition, (iv) Condensation.

A.1.3   (i) Straight chain, (ii) Cross-linked, (iii) Branched-chain.

A.1.4   (i) Alternating, (ii) Random, (iii) Graft, (iv) Block.

CHAPTER 2

STEP (CONDENSATION) POLYMERIZATION

2.1. INTRODUCTION

Condensation polymerization involves the reaction of a polyfunctional molecule or molecules to give a macro-molecule (polymer) with the loss of some simple molecules such as water or hydrochloric acid. When a bi-functional molecule is used, a linear condensation product is formed. The following example shows the reaction between a bi-functional dihydric alcohol and a dicarboxylic acid:

$$HO\,(CH_2)_x\,OH + HOOC\,(CH_2)_y\,COOH \longrightarrow$$
(Dihydric   (Dicarboxylic
alcohol)   acid)

$$HO\,(CH_2)_x\,O.\,OC(CH_2)_y\,COOH + H_2O \xrightarrow{\ HO\,(CH_2)_x\,OH\ }$$

$$HO\,(CH_2)_x O.\,OC(CH_2)_y\,CO.O\,(CH_2)_x\,OH \xrightarrow{\ 2HOOC(CH_2)_y.COOH\ }$$

$$HOOC\,(CH_2)_y\,CO.O\,(CH_2)_x\,O.OC -$$

$$-\,(CH_2)_y\,CO.O\,(CH_2)_x\,O.CO\,(CH_2)_y\,COOH \longrightarrow$$

$$HO\,{+\!CO\,(CH_2)_y\,CO.O\,(CH_2)_x.O\,}_{\,n}^{}\,H \qquad\qquad (2.1)$$
[Poly(ester)]

In the above reaction a polymer molecule is gradually formed until all the functional groups have reacted. It is necessary to have equimolecular proportions of the two monomers and these should be extremely pure; otherwise, the chains would end in the functional group in excess and a low molecular weight polymer would be produced. A third requirement is that the reaction responsible for the polymerization must be a very high yield reaction and the side reactions should be absent. In the following example a hydroxy acid can condense with itself to form a polymer as follows:

$$HO\,R - COOH + HO - R - COOH \longrightarrow$$
(Hydroxy acid)

$$HO - R - CO.\,O - R - COOH \xrightarrow{\ HORCOOH\ }$$

$$HO - R - CO.\,O - R.\,CO.O - R - COOH \longrightarrow$$

$$H\,{+\!O - R - CO.O - R - CO\,}_{\,n}^{}\,OH \qquad\qquad (2.2)$$

2.2. TYPES OF CONDENSATION POLYMERS

There are many types of condensation polymers. Some important ones are illustrated in Table 2.1. Examples of four types are given below:

(i) *Poly(ester):* – A poly(ester) is obtained either by the self-condensation of a hydroxy carboxylic acid or by the reaction of a dicarboxylic acid with a glycol.

$$n \, HO - R - COOH \; + n \, HO - R - COOH \longrightarrow$$

(Hydroxy acid)

$$H \{ O - R - CO . O - R - CO \}_n \, OH \qquad (2.2)$$

[Poly(ester)]

$$n \, HOOC - R - COOH + n \, HO - R' - OH \longrightarrow$$

(Dicarboxylic acid)   (Glycol)

$$HO \{ CO - R - CO.O - R' - O \}_n \, H \qquad (2.3)$$

[Poly(ester)]

(ii) *Poly(amides)*: – These are obtained by the condensation of a diamine with a dicarboxylic acid.

$$n \, H_2N - R - NH_2 + n \, HOOC - R' - COOH \longrightarrow$$

(Diamine)   (Dicarboxylic acid)

$$H \{ HN - R - NH.CO - R' - CO \}_n \, OH \qquad (2.4)$$

[Poly(Amide)]

(iii) *Poly(ethers)*: – When a compound containing an epoxide group reacts with a glycol, a poly(ether) results:

$$\overset{\displaystyle O}{\overbrace{CH_2 - CH}} - (CH_2)_n \, Cl + HO - R - OH \longrightarrow$$

(Epoxy derivative)   (Glycol)

$$- O - CH_2 - \underset{\underset{OH}{|}}{CH} - (CH_2)_n \; - O - R - O - \qquad (2.5)$$

(Polyether)

(iv) *Poly(urethanes)*: – A poly(urethane) is obtained by the condensation of a glycol with a diisocyanate:

$$HO - R - OH + OCN - R' - NOO \longrightarrow$$

(Glycol)   (Diisocyanate)

$$- \{ O - R - O - \underset{\underset{O}{||}}{C} . NH - R' - NH - \underset{\underset{O}{||}}{C} \}_n \qquad (2.6)$$

[Poly(urethane)]

2.3. KINETICS OF LINEAR CONDENSATION POLYMERIZATION

It is now fully understood that the reaction between two functional groups is independent of the chain length of the molecules attached to the two functional groups except in the initial stages of the reaction. Let us first of all consider the case of poly(esters), in which the water formed is continuously removed from the vicinity of the reaction.

Thus if N_0 is the initial number of reactive groups and N the number of unreacted groups remaining at time t, then the extent of the chemical reaction p is given by,

$$p = \frac{N_0 - N}{N_0} \qquad (2.7)$$

Table 2.1. Typical Condensation Polymers

Type	Characteristic linkage	Monomer	Trade name	Uses
Poly(amide)	$-NH-CO-$	Dibasic acid + Diamine or ε-Aminocarboxylic acid	Nylon Perlon	Moulded objects. fibers
Poly(ester)	$-CO-O-$	Poly (hydroxy glycol) + Dicarboxylic acid	Mylar (film). Alkyd, Terylene	Used as surface coating material or films, fibers
Poly (urethane)	$-O-CO-NH-$	Glycol + Diisocyanate		Fibers, rubbers, foams used as packaging materials
Poly (siloxane)	$-Si-O-$	$HO-Si-OH + OH-Si-OH$		Temperature resistant lubricants, rubbers and water repelling coatings
Phenol-formaldehyde	$-Ar-CH_2-$	Phenol + Formaldehyde	Bakelite, PF	Reinforced moulded objects, lacquers, adhesives
Urea-formaldehyde	$-NH-CH_2-$	Urea + Formaldehyde	Beetle, scarab	Moulded objects; textile coatings, adhesives
Melamine-formaldehyde	$-NH-CH_2-$	Melamine + Formaldehyde		Moulded objects
Poly(ether)	$-R-O-$	e.g. Epichlorohydrin + Bisphenol–A or other dihydric alcohols	Araldite, Epoxy	Adhesives, surface coatings

Here p is defined as the fraction of the functional groups, which has been esterified at the time when a sample was removed for estimation. If we take the well known example of the reaction between ethylene glycol and adipic acid, then in the absence of added strong acid, the rate of reaction will be given by the expression,

$$-\frac{dC}{dt} = k \, [\text{OH}] \, [\text{COOH}]^2 = kC^3 \qquad (2.8)$$

Here C denotes the concentration of the reacting groups. The rate of reaction is thus proportional to the first power of hydroxyl concentration and the square of the carboxyl concentration as the latter acts as a catalyst as well. k is the rate constant. On integration, Eq. (2.8) gives (with the condition that at time $t = 0, C = C_0$),

$$2kt = \frac{1}{C^2} - \frac{1}{C_0^2} \qquad (2.9)$$

If it is assumed that the water formed during the reaction is negligible, then $C = C_0 (1 - p)$ and the expression can be written as,

$$2kt \, C_0^2 = \frac{1}{(1-p)^2} - 1 \qquad (2.10)$$

It is seen from Fig. 2.1 that the plot of $(1 - p)^2$ against time t gives a straight line. In the case of an added catalyst the rate of reaction is proportional to the hydroxyl, carboxyl and the catalyst concentration and it can be written as,

$$-\frac{dC}{dt} = k' C_{\text{cat}} C^2 \qquad (2.11)$$

Since the catalyst concentration is constant, on integration the following expression is obtained

$$\frac{1}{C} = k''t + \frac{1}{C_0}$$

where $k'' = k' C_{\text{cat}}$ and $C = C_0 (1 - p)$ $\qquad (2.12)$

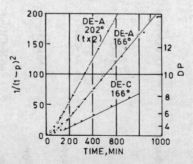

Fig. 2.1. Reaction of diethylene glycol with adipic acid (DE-A) and of diethylene glycol with caproic acid (DE-C). Time values at 202°C have been multiplied by two.

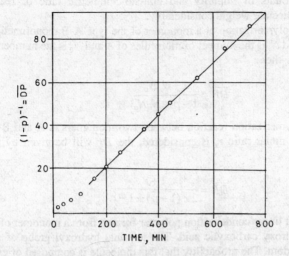

Fig. 2.2. Reaction of diethylene glycol with adipic acid at 109°C catalysed by 0.4 mol per cent of p-toluene sulphonic acid.

$$k''t\,C_0 = \frac{1}{1-p} - 1 \tag{2.13}$$

The linearity of the plot of $(1-p)^{-1}$ against t has been established for a variety of polyesterification reactions and for polymers of molecular weights up to 10,000. Fig. 2.2 Shows this.

The extent of reaction at a given time has already been divided by p, which is the ratio of the number of molecules undergoing reaction to the initial number of molecules.

$$p = \frac{N_0 - N}{N_0} \tag{2.7}$$

and

$$N = N_0(1-p) \tag{2.7a}$$

$\overline{DP}$, the degree of polymerization, is given by the original number of molecules divided by the number of molecules remaining unreacted,

$$\overline{DP} = \frac{N_0}{N} = \frac{N_0}{N_0(1-p)} = \frac{1}{(1-p)} \tag{2.14}$$

As stated earlier, to get a high molecular weight polymer, the extent of conversion must be high. Thus, to get $\overline{DP} = 50$, so that a polymer of desirable mechanical strength is obtained, it is necessary to take the reaction to the extent of 98% as the following calculation shows. If $p = 0.98$,

$$\overline{DP} = \frac{1}{1 - 0.98} = 50$$

Small amounts of impurity and non-stoichiometric ratio of reactants would depress the molecular weight considerably.

If in the polymerization of a monomer of the type A-B monofunctional impurity A is present and N_i is the number of molecules of A and N_0 is the number of molecules of A-B present, then,

$$\overline{DP} = \frac{1 + N_i/N_0}{1 - p + N_i/N_0} \tag{2.15}$$

If the polymerization reaction between two monomers A-A and B-B, which are present in the molar ratio r, is considered, the $\overline{DP}$ will be given by the following expression,

$$\overline{DP} = \frac{1 + r}{2r(1 - p) + 1 - r} \tag{2.15a}$$

Consider a linear condensation polymer formed from a monomer of the A-B type such as ω–hydroxy carboxylic acid. The terminal hydroxyl group of a molecule is selected at random. The probability that this molecule is composed of exactly x units is to be ascertained.

$$\text{H} - \text{ORC} - \text{OR.CO} -\text{.....} - \text{ORCO} - \text{OH}$$
$$1 \qquad\qquad 2 \qquad x-1 \qquad x$$

The probability that the carboxyl group of the first unit is esterified is equal to p. The probability that the carboxyl of the second unit is esterified is likewise equal to p. This is in accordance with the principle of equal reactivity at every stage of the polymerization, irrespective of the size of the molecule (as already stated). The probability that this sequence continues for $x - 1$ linkages is the product of these separate probabilities or p^{x-1}. *This is the probability that the molecule contains at least $x - 1$ ester groups, or units.* The probability that the xth carboxyl groups is unreacted, thus limiting the chain to exactly x units, is $1 - p$. Hence, the probability that the molecule in question is composed of exactly x units is given by

$$N_x = NP^{x-1}(1 - p)$$

if the total number of units present is N_0 and $N = N_0(1 - p)$, then,

$$N_x = N_0(1 - p)^2 p^{x-1} \tag{2.16}$$

This is the number distribution function for linear stepwise polymerization with the extent of reaction equal to p.

The weight fraction of x-mers is given by the expression,

$$W_x = \frac{x N_x}{N_0} = x(1 - p)^2 p^{x-1} \tag{2.16a}$$

This is the weight distribution function of linear stepwise polymerization for the extent of reaction p.

The same derivation holds for linear condensation polymers formed from A–A and B–B reactions.

Equations 2.16 and 2.16a are illustrated in Figs. 2.3 and 2.4 respectively.

The curves in Figs. 2.3 and 2.4 show that if we consider the number fraction, an appreciable quantity of very low molecular weight species will be present at all stages of conversion, as compared to high molecular weight species. On the other hand, if we consider the weight fraction, the very low molecular weight species will be present in a negligible quantity as compared to high molecular weight species.

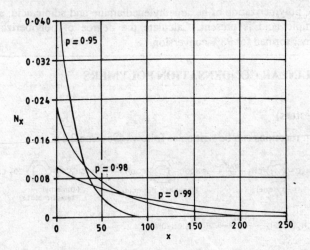

Fig. 2.3. Molecular weight distribution for linear condensation polymers on a number basis for several extents of reaction, p. N_x = mole fractions; x = number of structural units.

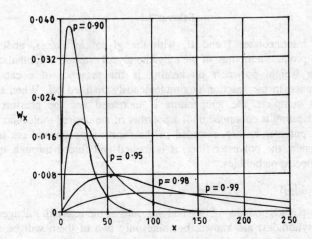

Fig. 2.4. Molecular weight distribution for linear condensation polymers on a weight basis for several extents of reaction, p.

Q. 2.1 Calculate the number average degree of polymerization of an equimolecular mixture of hexamethylenediamine and adipic acid for the extents of reaction 0.500, 0.800, 0.900, 0.950, 0.970, 0.990 and 0.995.

Q. 2.2. In the polymerization of ω–hydroxycaproic acid, $HO(CH_2)_5 COOH$, a 2% impurity is present. Calculate the degree of polymerization of the polymer formed.

Q. 2.3. In the polymerization of hexamethylenediamine and adipic acid, a 2% excess of adipic acid is present. Calculate the degree of polymerization of the polymer formed for 98% conversion.

2.4 SOME LINEAR CONDENSATION POLYMERS

2.4.1. Poly(esters)

The main reactions can be written as follows (Scheme 2.1):

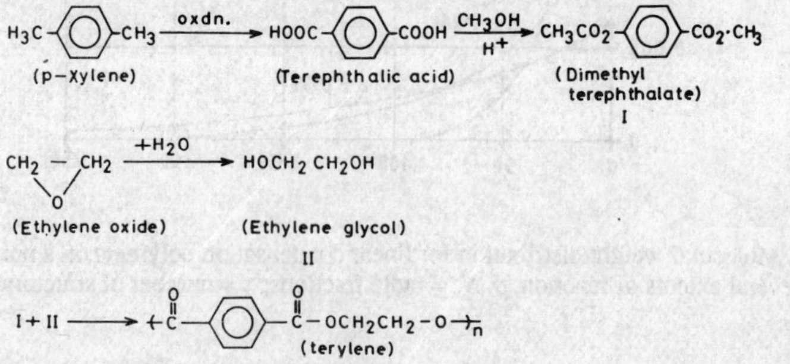

Scheme 2.1

The two intermediates I and II, with the glycol in excess, undergo ester interchange to produce a mixture of the ethylene glycol ester of terephthalic acid and low molecular weight polymers on heating in the presence of a catalyst. The methanol obtained in the reaction is simultaneously distilled off. When the ester–interchange is complete, the temperature is increased and the pressure reduced considerably. Heating is continued until a polymer of the desired molecular weight is obtained. The polymer is then extruded in the form of ribbon and cut into small pieces. To prepare the polyester fiber, it is melted and forced through spinnerets, cooled and collected on bobbins.

2.4.2. Poly(amides)

These compounds contain – CO – NH – groups as the interunit linkages. A large number of poly(amides) are known, but here only two of them will be discussed Nylon 6,6 is obtained as shown in Scheme 2.2.

To obtain Nylon of desired chain-length, equivalent amounts of hexamethylene diamine and adipic acid are taken. To purify the 1:1 salt, called the Nylon salt, it is dissolved in methanol in which it is completely soluble. The salt is then polymerized in a pressure vessel, called autoclave. When the reaction is complete, the water vapour is allowed to escape and the temperature is raised. The resin is extruded in the form of a ribbon and cut into small bits. The main use of Nylon is in the preparation of Nylon fibers.

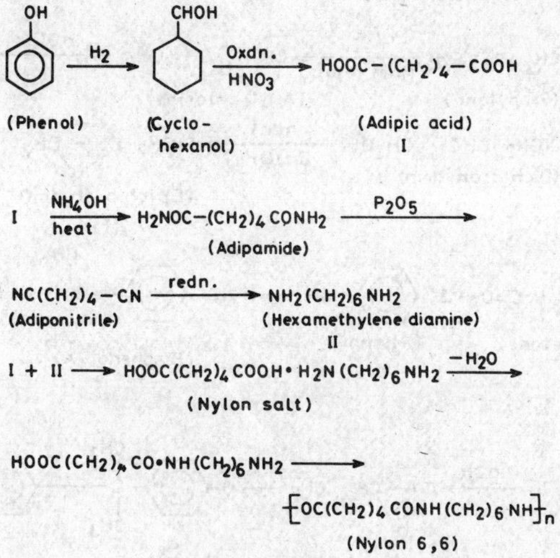

Scheme 2.2

Scheme 2.3 outlines the preparation of Nylon, 6 or perlon

Poly(amides) containing aromatic diamines and aromatic diacids can be prepared indirectly. Scheme 2.4 gives the preparation of one such polyamide.

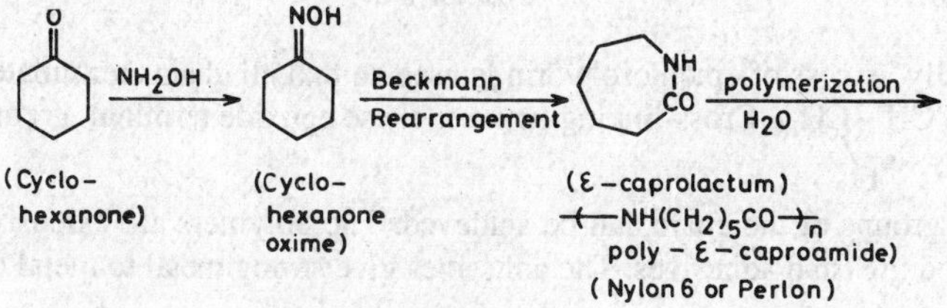

Scheme 2.3

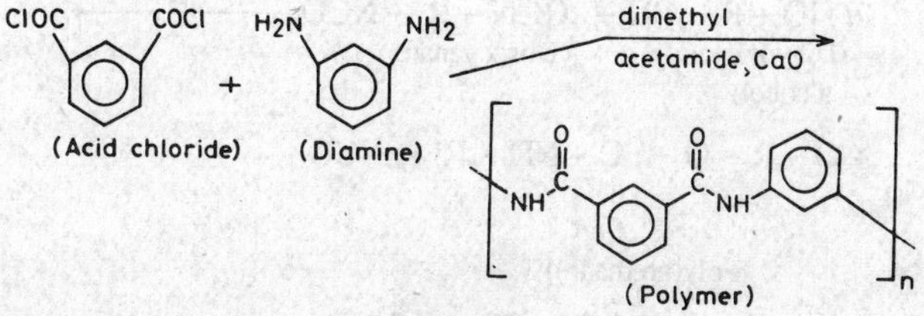

Scheme 2.4

2.4.3. Poly(ethers)

The well known epoxy resins belong to the class of polyethers. A typical preparation is shown in Scheme 2.5.

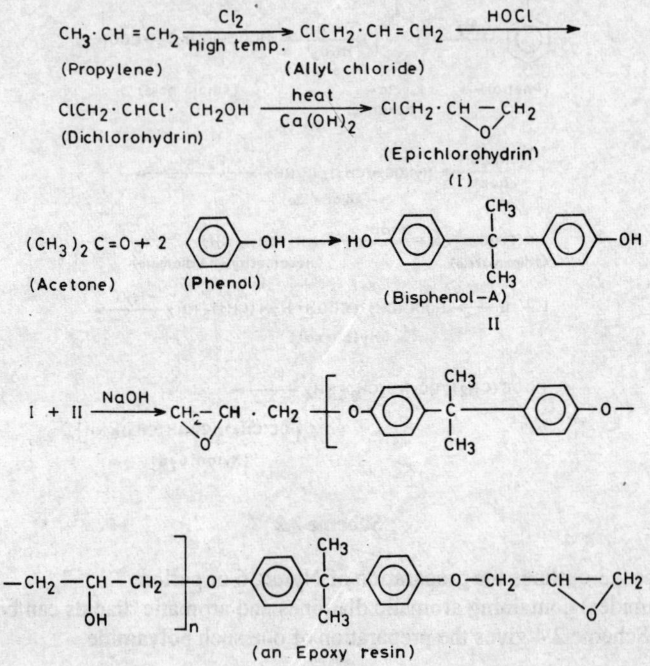

$$CH_3 \cdot CH = CH_2 \xrightarrow[\text{High temp.}]{Cl_2} ClCH_2 \cdot CH = CH_2 \xrightarrow{HOCl}$$
(Propylene)   (Allyl chloride)

$$ClCH_2 \cdot CHCl \cdot CH_2OH \xrightarrow[\text{Ca(OH)}_2]{\text{heat}} ClCH_2 \cdot CH \overset{\text{O}}{-} CH_2$$
(Dichlorohydrin)   (Epichlorohydrin)
(I)

$$(CH_3)_2 C = O + 2 \underset{\text{(Phenol)}}{\bigcirc\!-OH} \longrightarrow HO-\bigcirc-\overset{CH_3}{\underset{CH_3}{C}}-\bigcirc-OH$$
(Acetone)   (Bisphenol-A)
II

I + II $\xrightarrow{NaOH}$ an Epoxy resin (structures as shown)

(an Epoxy resin)

Scheme 2.5

Usually, excess of epichlorohydrin is used so that all chains terminate in epoxide groups $(-CH-CH_2)$. Cross-linking between these epoxide terminal groups with the hydroxyl groups of the chain can be achieved. The polymers are extensively used in surface coatings and adhesives. The adhesives give strong metal to metal bonds.

2.4.4. Poly(urethanes)

The basic reactions in the preparation of poly(urethanes) are as shown in Scheme 2.6:

$$n \, HO - R - OH + OCN - R' - NCO \longrightarrow$$
(Dihydric   (Diisocyanate)
alcohol)

$$\{O - R - O - \underset{O}{\overset{\|}{C}} - NH - R'\}_n NCO$$
[Poly(urethane)]

Scheme 2.6

The linking unit here is $- NH - CO - O -$ unlike $- NH - CO -$ group in Nylons. The extra oxygen gives increased flexibility to the chains. Perlon–U is a typical poly(urethane) fiber. The preparation of a poly(urethane) foam is given in Scheme 2.7.

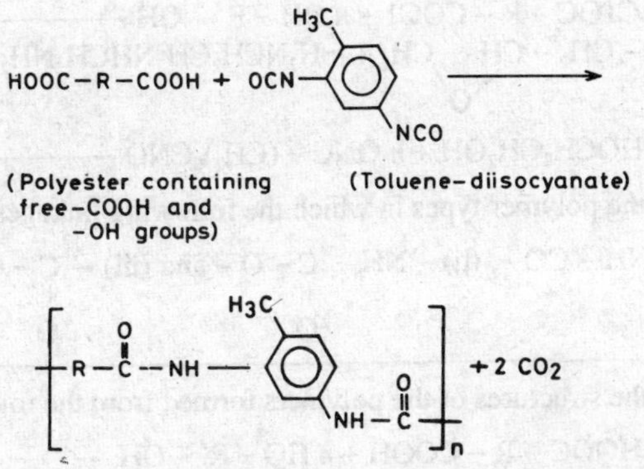

$$
HOOC-R-COOH + OCN-\underset{NCO}{\overset{H_3C}{\bigcirc}} \longrightarrow
$$

(Polyester containing   (Toluene–diisocyanate)
free –COOH and
–OH groups)

$$
\left[-R-\overset{O}{\underset{||}{C}}--NH-\underset{}{\overset{H_3C}{\bigcirc}}-NH-\overset{O}{\underset{||}{C}}- \right]_n + 2\ CO_2
$$

Scheme 2.7

Poly(urethanes) in the form of either rigid or flexible foams find a great deal of use. Solid foams are obtained by making a poly(ester), containing free hydroxyl and carboxyl groups, react with diisocyanates. The carbon dioxide liberated in the reaction blows the foam.

2.4.5. Poly (carbonates)

A common method for the synthesis of poly(carbonates) is to react Bisphenol– A with phosgene in the presence of a base such as pyridine (Scheme 2.8). The same reaction can take place in a two-phase system consisting of methylene chloride and aqueous sodium hydroxide.

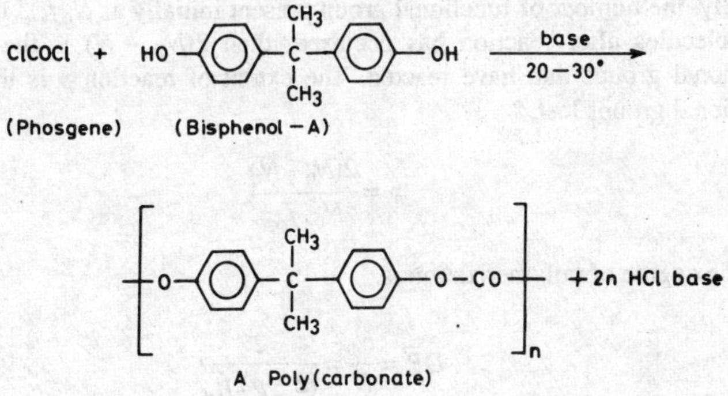

$$
ClCOCl + HO-\bigcirc-\underset{CH_3}{\overset{CH_3}{C}}-\bigcirc-OH \xrightarrow[20-30°]{base}
$$

(Phosgene)   (Bisphenol – A)

$$
\left[-O-\bigcirc-\underset{CH_3}{\overset{CH_3}{C}}-\bigcirc-O-CO- \right]_n + 2n\ HCl\ base
$$

A   Poly(carbonate)

Scheme 2.8

In poly(carbonates), $-CO-O-$ linkage is present unlike the $-CO-NH-$ linkage in poly(amides) or $-NH-CO-O-$ linkage in poly(urethanes). This resin is high melting. It has good impact strength and good electrical resistance. It is a moulding type of resin.

Q.2.4. Complete the following equations:—

    (i) n ClOC – R – COCl + n H$_2$ N –'R' – NH$_2$ ⟶

    (ii) n ClOC – R – COCl + n OH – R' – OH ⟶

    (iii) n – CH$_2$ – CH – CH$_2$ + n H$_2$NCH$_2$CH$_2$NHCH$_2$NH$_2$ ⟶
$$\diagdown \text{O} \diagup$$

    (iv) n HOCH$_2$CH$_2$OH + n ONC – (CH$_2$)$_6$CNO ⟶

Q.2.5. Name the polymer types in which the following linkages are present:—

    (i) – NH – CO –, (ii) – NH – C – O – and (iii) – C – O –
$$\qquad\qquad\qquad\quad \overset{\|}{\underset{O}{}}\qquad\qquad\qquad \overset{\|}{\underset{O}{}}$$

Q.2.6. Draw the structures of the polymers formed from the following monomers:—

    (i) n HOOC – R – COOH + n HO – R' – OH

    (ii) n HOOC – R – COOH + n HO – R' – OH
$$\qquad\qquad\qquad\qquad\qquad\qquad\quad |$$
$$\qquad\qquad\qquad\qquad\qquad\qquad\text{OH}$$

    (iii) n HO – R – COOH

2.5. THREE-DIMENSIONAL POLYMERS

2.5.1. Carother's Equation

In order to control the cross-linking of a polymer to the desired stage, it is important to know the relationship between gelation (the stage at which a resin is converted into a rubbery infusible mass) and the extent of polymerization. Thus, for a system containing two moles of glycerol (a triol) and three moles of phthalic acid (a dibasic acid), there is a total of twelve functional groups per five moles of the mixure and f_{ave} (average number of functional groups) is 12/5 or 2.4. In a system containing equivalent number of two reactants A and B, if N_0 is the number of molecules present initially, the number of functional group present initially as $N_0 f_{ave}$. If N is the number of molecules after reaction has occurred, then $2(N_0 - N)$ is the total number of functional groups that have reacted. The extent of reaction p is the fraction of the functional groups lost.

$$p = \frac{2(N_0 - N)}{N_0 \cdot f_{ave}} \qquad (2.17)$$

and the degree of polymerization is,

$$\overline{DP} = \frac{N_0}{N} = \frac{2}{(2 - p) \cdot f_{ave}} \qquad (2.18)$$

which can be rearranged to,

$$p = \frac{2}{f_{ave}} - \frac{2}{\overline{DP} \cdot f_{ave}} \qquad (2.19)$$

Equation 2.19 is called the Carother's equation. At the gel point (at this point one observes the visible formation of a gel or insoluble polymer fraction), when the number average degree of polymerization becomes infinite, the second term can be dropped and the equation reduces to,

$$P_c = \frac{2}{f_{ave}} \qquad (2.20)$$

Thus, the glycerol-phthalic acid (2:3 molar ratio) system mentioned above has a calculated critical extent of reaction equal to 0.883.

Q.2.7. Calculate the extent of reaction when phthalic anhydride and glycerol react in the stoichiometric amounts.

Q.2.8. Calculate the extent of reaction when phthalic anhydride and glycerol react in the molar ratio 1.500 : 0.980.

2.5.2 Poly(ester)resins (Alkyd resins)

Poly(ester) resins obtained from a triol and a dibasic acid are called alkyd resins. Thus, glycerol reacts with phthalic acid to give an alkyd resin.

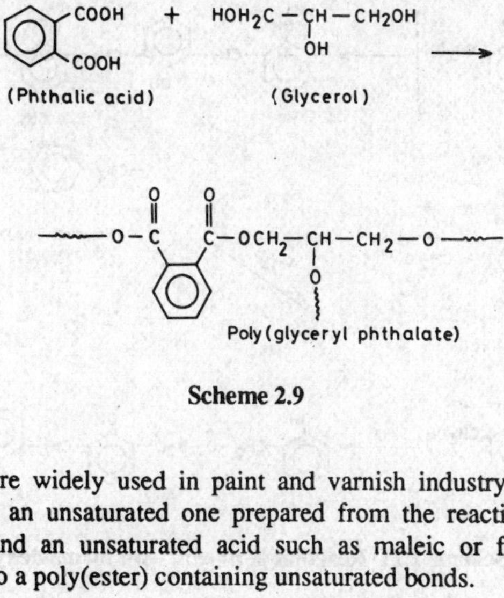

Scheme 2.9

Alkyd resins are widely used in paint and varnish industry. Another type of poly(ester) resin is an unsaturated one prepared from the reaction of ethylene or propylene glycol and an unsaturated acid such as maleic or fumaric acid. This reaction gives rise to a poly(ester) containing unsaturated bonds.

HO CH_2CH_2OH + HOOC. CH = CH. COOH ⟶

(Ethylene glycol)   (Maleic/Fumaric acid)

$\{OCH_2CH_2OCO.CH = CH.CO\}_n$

[Unsaturated poly(ester)]

The cross linking of this poly(ester) can be achieved by copolymerizing the unsaturated poly(ester) with a monomer like styrene or methyl methacrylate.

$$- O_2C.CH = CHCO.O.CH_2CH_2.O - + nCH_2 = CHX \longrightarrow$$

$$- O_2C.\overset{|}{C}H - CH. \, CO.O.CH_2CH_2 - O -$$

$$(\overset{|}{C}H_2$$

$$\overset{|}{C}HX)_n$$

$$- O_2C.\overset{|}{C}H - CH.CO.O.CH_2CH_2 - O -$$

Scheme 2.10

Usually, a reinforcing filler such as fiber-glass is added. This gives rise to a very strong yet light product useful in a wide variety of applications such as boat hulls, suitcases and car bodies.

2.5.3. Phenol-formaldehyde Resins

The reaction between phenol ($f = 3$) and formaldehyde ($f = 2$) is not a simple reaction. The reaction can take place in the presence of an acid or a base (Scheme 2.11):

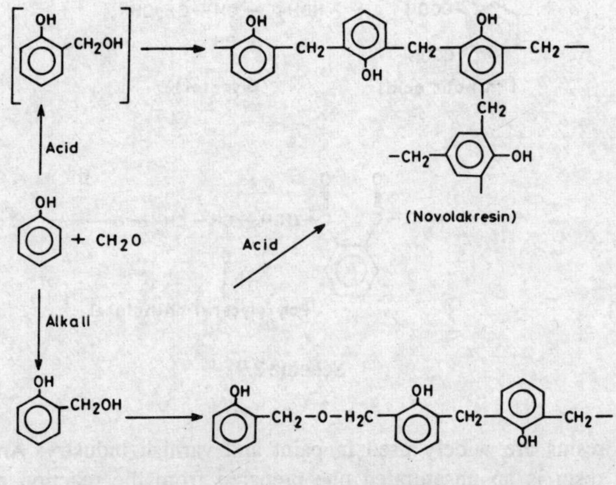

Scheme 2.11 Reaction of phenol with formaldehyde

The reaction can be conducted in the presence of an acid like hydrochloric or oxalic acid and formaldehyde is used in insufficient amount. The resin obtained is called the Novolak resin. At this stage it is soluble in alkali and many organic solvents. To bring it as near as possible to the final stage, when it can be moulded in a compression moulding machine, a calculated amount of hexamethylene tetramine, a filler (wood flour or cellulose) and other additives are mixed together. The mixture is given a heat treatment and finally powdered. This is called a two-stage process and is the only one now used industrially.

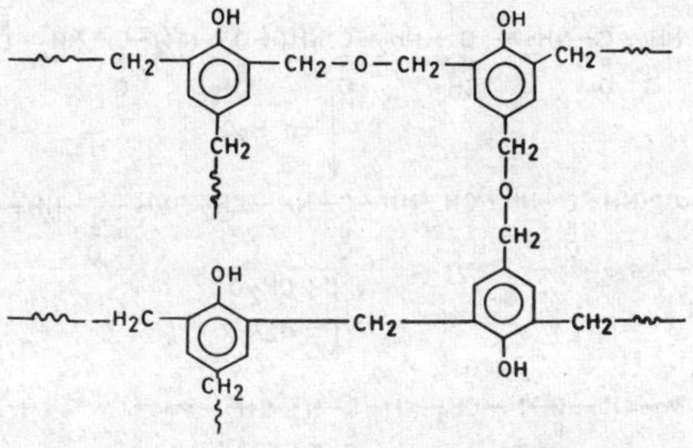

Scheme 2.12 Phenol-formaldehyde polymer

In the one-stage process, the reaction is conducted taking calculated amounts of phenol and formaldehyde (1:1.5 molar ratio) and the resin formed has ether linkages in addition to methylene bridges. The resin obtained is dissolved in methylated spirit, mixed with a filler and finally dried. After mixing with other additives, it is given a heat treatment and powdered to get the moulding powder. After themosetting in the mould, the polymer has the following tentative structure (Scheme 2.12).

2.5.3.1. *Preparation of a Novolak Resin (Two-stage process)*

Phenol and formaldehyde (40% w/v) in the molar ratio (1 : 0.9) are heated cautiously under reflux in the presence of oxalic acid (2% of the weight of phenol) as the reaction is exothermic. On heating for about 45 minutes, the reaction mixture clouds out into two layers, the upper being the resin layer. After the reaction is complete the mixture is heated on a water bath under reduced pressure until all the water is distilled off and the resin attains a brittle bead condition (liquid while hot and solid when cold). It is then poured into shallow trays. If solidifies on cooling and is powdered before using it for the preparation of moulding powder.

2.5.4. Urea-formaldehydehydre Resins

Urea and formaldehyde, in weakly alkaline solution, condense as shown in Scheme 2.13:

The reaction of urea with formaldehyde to give condensation products can also be formulated as given below:

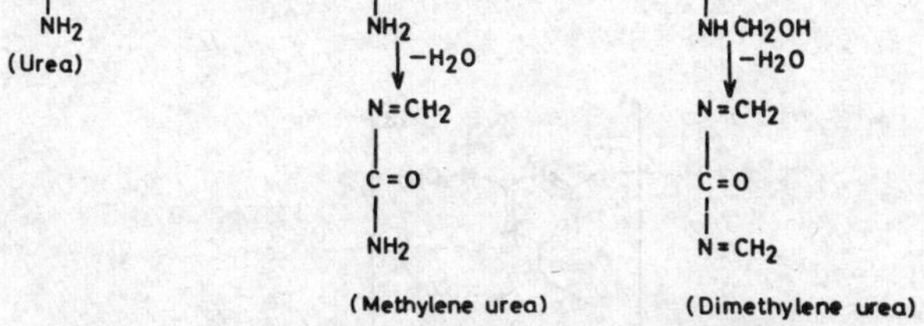

Scheme 2.13 Reaction of urea with formaldehyde

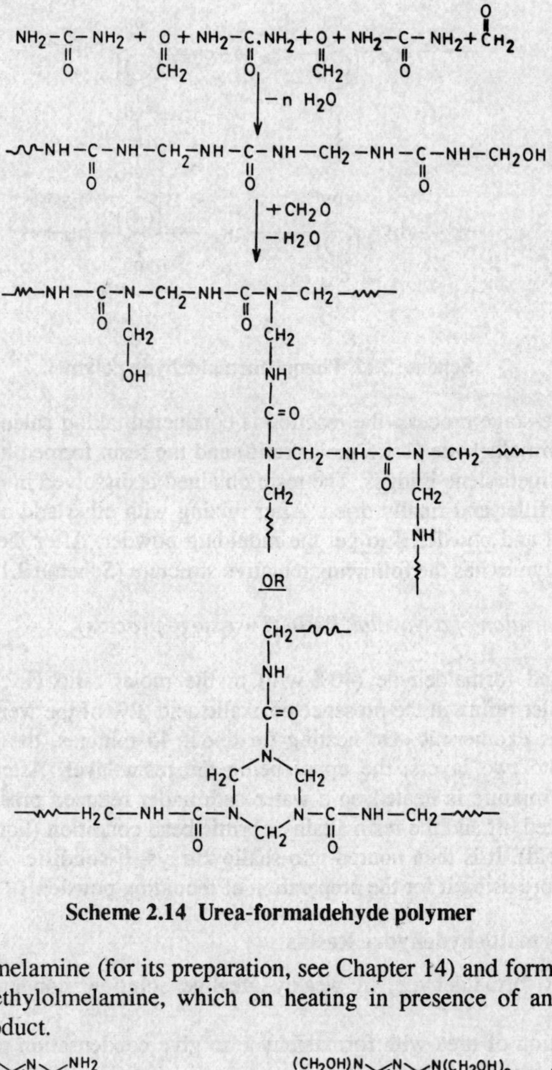

$$NH_2-C-NH_2 + O + NH_2-C.NH_2 + O + NH_2-C-NH_2 + CH_2$$

(chemical scheme with urea and formaldehyde condensation)

$$\downarrow -n\ H_2O$$

$$\text{---}NH-C-NH-CH_2-NH-C-NH-CH_2-NH-C-NH-CH_2OH$$

$$\downarrow \begin{array}{c}+CH_2O\\-H_2O\end{array}$$

Scheme 2.14 Urea-formaldehyde polymer

Similarly, melamine (for its preparation, see Chapter 14) and formaldehyde react to give hexamethylolmelamine, which on heating in presence of an acid yields a cross-linked product.

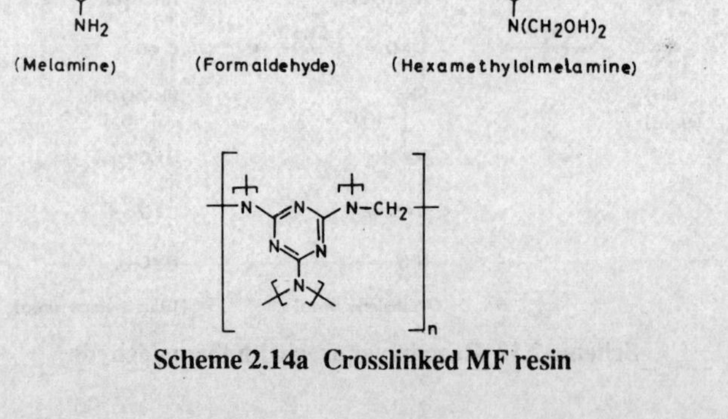

(Melamine)   (Formaldehyde)   (Hexamethylolmelamine)

Scheme 2.14a Crosslinked MF resin

Melamine-formaldehyde resins are more resistant to heat and moisture than urea-formaldehyde resins and are, therefore, preferred for making dinner-ware and decorative table tops (Formica).

2.5.4.1. *Preparation of Urea-Formaldehyde Resin*

Urea and formaldehyde are used in the molar ratio 1:1.45. Formaldehyde (40%, w/v) is taken in a jacketed vessel (which can be cooled and steam passed through it) and is stirred with magnesium carbonate under cooling until the mixture reacts to neutral to universal indicator. It is allowed to stand at 15–20°C and ammonia (0.880 mole) and urea (1 mole) are added to it. It is then stirred and the temperature allowed to rise to 35–40°C. After allowing to stand for 2 hours, the syrup is mixed with a filler like sulphite wood pulp and sodium hydroxide solution added so that on thoroughly mixing, pH value of 7.75 is attained. It is then dried and mixed with a little tricresyl phosphate (plasticizer). A plasticizer acts as a lubricant between polymer molecules. Chloroacetamide (catalyst), other additives like a dye and a mould lubricant are added and the material given a heat treatment. It is then cooled, powdered and stored in tins with lid (moisture free) until required for use.

2.5.5. Aromatic Poly(imides)

These are formed by the reaction of an aromatic dianhydride and an aromatic diamine. The reaction is usually carried out in two stages (Scheme 2.15). This is done to avoid the production of insoluble and infusible polymers.

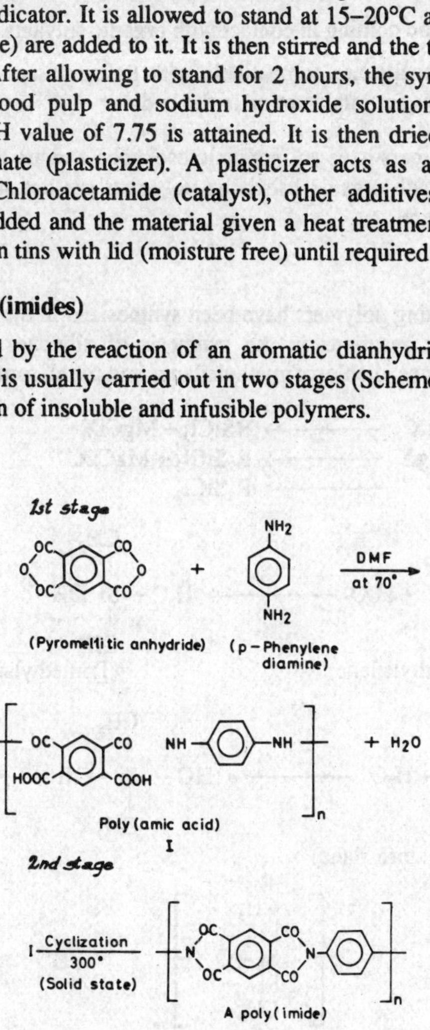

Scheme 2.15

Q.2.9. Give the mechanism for the formation of o– and *p*– hydroxybenzyl alcohol from phenol and formaldehyde in the presence of alkali.

2.6. INORGANIC POLYMERS

The organic polymers which have been discussed in this chapter have a carbon backbone; thus, they have poor flame resistance and are also unstable at higher temperatures. Polymers like polyacrylonitrile emit a poisonous gas, hydrocyanic acid, on heating. Some other defects of organic polymers are:

   (i) They have poor weather resistance. They become inflexible and brittle at very low temperatures. They also deteriorate by slow oxidation in the tropics.

   (ii) They swell on coming in contact with organic solvents.

   (iii) They are usually not compatiable with living tissues and cannot be used to make devices for implantation in the body.

Inorganic polymers have an inorganic backbone and are thus useful for a variety of purposes for which organic backbone polymers are not suitable. A few of them will be discussed here.

2.6.1. Silicones

Silicon containing polymers have been synthesized in the form of oils, gums and rubbers. The basic reactions in the synthesis of silicones are the hydrolysis of chlorotrimethylsilanes, dichlorodimethylsilanes and trichloromethylsilanes.

$$SiCl_4 + R\,MgX \longrightarrow RSiCl_3 + MgClX$$
$$RSiCl_3 + RMgX \longrightarrow R_2SiCl_2 + MgClX$$
$$Si + 2\,R\,Cl \longrightarrow R_2SiCl_2$$

(i)
$$H_3C - \underset{\underset{CH_3}{|}}{\overset{\overset{CH_3}{|}}{Si}} - Cl + H_2O \longrightarrow H_3C - \underset{\underset{CH_3}{|}}{\overset{\overset{CH_3}{|}}{Si}}\ OH$$

(Chlorotrimethylsilane)   (Trimethylsilol)

(ii)
$$Cl - \underset{\underset{CH_3}{|}}{\overset{\overset{CH_3}{|}}{Si}} - Cl + H_2O \longrightarrow HO - \underset{\underset{CH_3}{|}}{\overset{\overset{CH_3}{|}}{Si}} - OH \longrightarrow$$

(Dimethyldichlorosilane)

$$\left[\overset{\overset{CH_3}{|}}{\underset{\underset{CH_3}{|}}{Si}} - O \right]_n$$

After it attains a certain chain-length, the polymer obtained in step (II) can be reacted with chlorotrimethylsilane to get a silicone oil type of product.

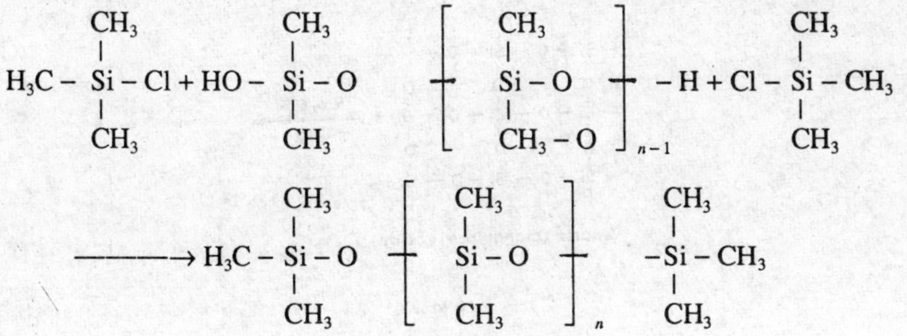

$$\begin{array}{c} CH_3 \\ | \\ H_3C - Si - Cl \\ | \\ CH_3 \end{array} + \begin{array}{c} CH_3 \\ | \\ HO - Si - O \\ | \\ CH_3 \end{array} \quad \left[\begin{array}{c} CH_3 \\ | \\ Si - O \\ | \\ CH_3 - O \end{array} \right]_{n-1} - H + Cl - \begin{array}{c} CH_3 \\ | \\ Si - CH_3 \\ | \\ CH_3 \end{array}$$

$$\longrightarrow H_3C - \begin{array}{c} CH_3 \\ | \\ Si - O \\ | \\ CH_3 \end{array} \quad \left[\begin{array}{c} CH_3 \\ | \\ Si - O \\ | \\ CH_3 \end{array} \right]_{n} \begin{array}{c} CH_3 \\ | \\ - Si - CH_3 \\ | \\ CH_3 \end{array}$$

(Silicone oil)

Scheme 2.16

(iii)   If one adds trichlorsoilane to the mixture, one obtains a crosslinked polymer,

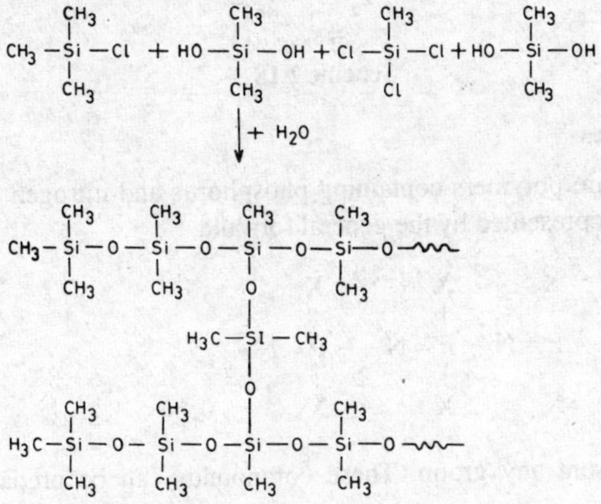

Scheme 2.17

Silic acid $Si(OH)_4$ can polymerize to give cross-linked polymers.

2.6.2. Glass

The formation of inorganic glasses can be described in terms of the *trans*-condensation of SiO_2 (quartz sand) with the aid of Na_2CO_3 and CaO (Scheme 2.18).

According to their structures and properties, silicones lie between the organic and inorganic polymers. As in the case of the glasses, some of the silicon atoms in the silicone chain can be replaced by other atoms such as Al, B or Pb. The silicones do not react with water and are actually water repelling. Water on a silicone treated fabric or surface appears as beads and will not go into the interstices between the threads. Yet a silicone treated fabric will breathe.

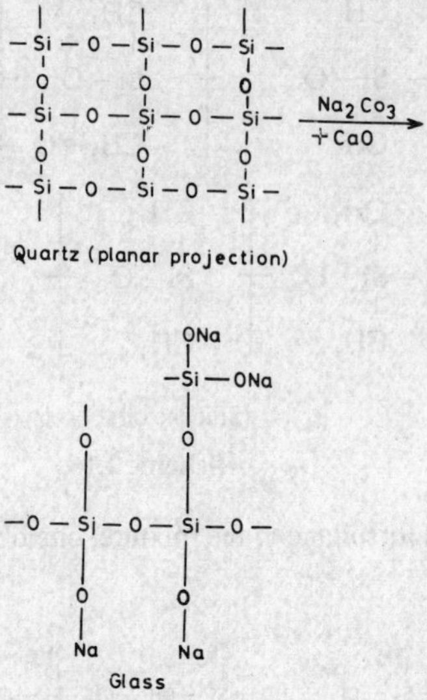

Quartz (planar projection)

Glass

Scheme 2.18

2.6.3. Phosphazenes

Phosphazenes are polymers containing phosphorus and nitrogen atoms arranged alternately and are represented by the general formula,

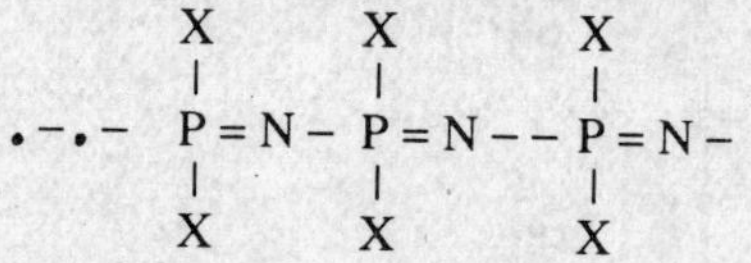

where X can represent any group. These compounds can be prepared as follows (Scheme 2.19):

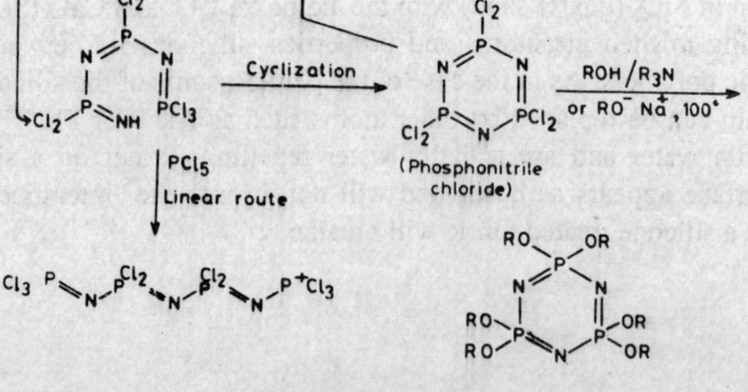

Scheme 2.19

When the group $X = OCH_2CF_3$, the polymers obtained are strongly water resistant, even more than silicones. They do not react with animal tissues, remain flexible even at very low temperatures and are non-inflammable. They find use in replacing blood vessels. They are used for a variety of purposes and unlike rubber do not swell on coming in contact with organic solvents.

2.6.4. Poly(phosphates)

The poly(phosphates) cannot really be considered to be polymers as their chain lengths are very short. They mainly occur in the following forms:

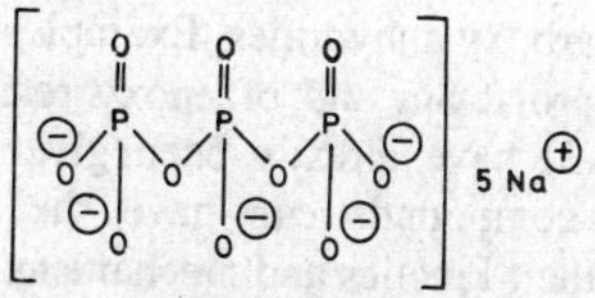

$Na_5P_3O_{10}$
(Sodium triphosphate)

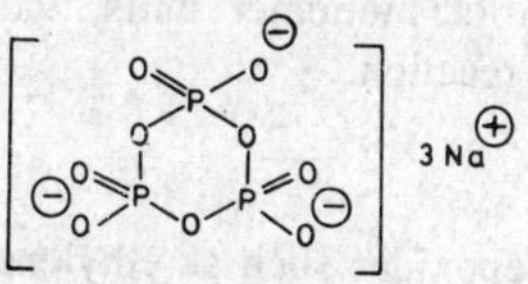

$Na_3P_3O_9$
(Sodium metaphosphate)
(Soluble form)

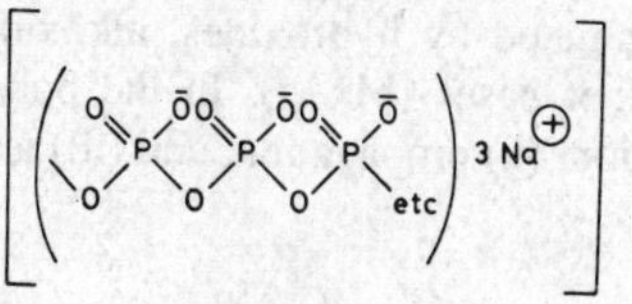

$(Na_3P_3O_9)_n$
Sodium metaphosphate
(insoluble form)

Scheme 2.20

Of the above mentioned three types, sodium triphosphate is used as a water softner.

Q. 2.10. Give the structure of the polymer/product formed in the following reactions

(i)
$$CH_3 - \underset{\underset{CH_3}{|}}{\overset{\overset{CH_3}{|}}{Si}} - Cl + H_2O \longrightarrow$$

(ii)
$$CH_3 - \underset{\underset{Cl}{|}}{\overset{\overset{Cl}{|}}{Si}} - Cl + H_2O \longrightarrow$$

$$\text{(iii)} \quad Cl - \underset{\underset{CH_3}{|}}{\overset{\overset{CH_3}{|}}{Si}} - Cl + H_2O \longrightarrow$$

2.7. RING OPENING POLYMERIZATION

Polymers can be prepared by reactions involving ring opening. Some typical examples of ring-scission polymerization involve lactams, cyclic ethers, lactones, cyclic anhydrides and N-carboxy anhydrides. Examples of the preparation of Nylon 6 (Section 2.4.2) from ε-caprolactam and of epoxy resins from epichlorohydrin and bisphenol-A (Section 2.4.3) have already been given earlier in this chapter. The polymerization of these compounds can have the features of both chain and condensation reactions in their kinetics and mechanism. Ring opening polymerization resembles chain-polymerization in the sense that it proceeds by the addition of monomer units; it also resembles step polymerization as the polymer growth takes place by stepwise addition of monomer units, the molecular weight increasing throughout the course of the reaction.

2.7.1. Cyclic Ethers

The polymerization of epoxides such as ethylene and propylene oxides to high molecular weight compounds can be accomplished in a variety of ways. Lewis acid catalysts, e.g., boron trifluouride (along with a co-catalyst such as water) or stannic chloride generally give low molecular weight polymers. The anionic polymerization of these epoxides can be initiated by hydroxides, alkoxides, metal oxides, organo-metallic compounds and other bases (M^+A^-). In the polymerization reaction three main stages occur: (i) initiation, (ii) propagation, and (iii) termination.

Initiation

$$\underset{\text{(Ethylene oxide)}}{H_2C - CH_2} + \underset{\text{(Initiator)}}{M^+A^-} \longrightarrow A - CH_2CH_2O^-\, M^+ \tag{2.21}$$

Propagation

$$A - CH_2CH_2O^-\, M^+ + H_2C - CH_2 \longrightarrow ACH_2CH_2OCH_2CH_2O^-\, M^+$$

$$A \overset{}{\leftarrow} CH_2CH_2O \overset{}{\underset{n-1}{\rightarrow}} CH_2CH_2O^-\, M^+ + CH_2 - CH_2 \longrightarrow$$

$$A \overset{}{\leftarrow} CH_2CH_2O \overset{}{\underset{n}{\rightarrow}} CH_2CH_2O^-\, M^+ \tag{2.22}$$

Many of these polymerization reactions have the characteristics of living polymers (see Section 3.9.2.2) in that there is no termination in the absence of an added terminating agent.

In the case of an unsymmetrical epoxide, like propylene oxide, it may appear that different polymers can form depending upon the mode of opening of the ring (Eq. 2.23).

$$
\text{CH}_3 \cdot \underset{\underset{\text{O}}{\diagup\diagdown}}{\text{CH}-\text{CH}_2}
\quad
\begin{array}{c}
\text{CH}_3 \\
| \\
\text{-w-CH} - \text{CH}_2\,\text{O}^-\text{K}^+ \\
\diagdown \\
\text{CH}_3 \\
| \\
\text{-w-CH}_2\text{-CH} - \text{O}^-\text{K}^+
\end{array}
\tag{2.23}
$$

(Propylene oxide)

However, it is found in practice that the same polymer is obtained except with slight differences in the nature of the end groups. Several epoxide polymerizations initiated by alkoxides are carried out in the presence of alcohols (usually the alcohol whose alkoxide is used). This provides a homogeneous system by solubilization of the initial ion; the rate of polymerization also registers an increase, probably due to an increase in the concentration of free ions.

2.7.2. Cyclic amides(lactams)

The polymerization of cyclic amides or lactams is initiated by water and anionic reagents. Cations are not usually used as they lead to polymers of low molecular weights. The hydrolytic polymerization of ε-caprolactam can be summarised as follows:

$$
n \ \overline{\text{Co}-\text{NH}-(\text{CH}_2)_5} \xrightarrow[\substack{250-270^\circ \\ 12-24\,\text{hrs.}}]{\text{H}_2\text{O}} \left[\text{NH}-(\text{CH}_2)_5 - \text{CO} \right]_n
\tag{2.24}
$$

In this type of polymerization three main equilibria are involved.

Hydrolysis

$$
\underset{(\,\varepsilon-\text{Caprolactam}\,)}{(\text{CH}_2)_5 - \overset{\overset{\text{O}}{\parallel}}{\underset{|}{\text{C}}} - \text{NH}} + \text{H}_2\text{O} \rightleftharpoons \underset{(\text{Amino acid})(\,\varepsilon-\text{amino Caproic acid})}{\text{HOOC}-(\text{CH}_2)_5-\text{NH}_2}
\tag{2.25}
$$

Step polymerization

$$
\text{HOOC}-(\text{CH}_2\text{)}_5\,\text{NH}_2 \ + \ \text{HOOC}(\text{CH}_2)_5\,\text{NH}_2 \rightleftharpoons \text{H}_2\text{O} +
$$
$$
\text{HOOC}(\text{CH}_2)_5\text{-NHCO}(\text{CH}_2)_5\text{NH}_2
\tag{2.26}
$$

Ring opening polymerization

Initiation

$$
\text{HOOC}(\text{CH}_2)_5\,\text{NH}_2 \ +(\text{CH}_2)_5 - \overset{\overset{\text{O}}{\parallel}}{\text{C}} - \text{NH} \rightleftharpoons \text{HOOC}(\text{CH}_2)_5\,\text{NH}\overset{\overset{\text{O}}{\parallel}}{\text{C}}(\text{CH}_2)_5\text{NH}_2
\tag{2.27}
$$

Propagation

$$\text{---}\text{NH}_2 + (\text{CH}_2)_5\text{-NH} \rightleftharpoons \text{---}\text{NHCO}(\text{CH}_2)_5\text{-NH}_2 \qquad (2\cdot28)$$

The predominant propagating reaction is the ring opening rather than the step polymerization of the amino acid.

Strong bases (B⁻M⁺) like alkali metals or sodium methylate can initiate the polymerization of cyclic amides by forming the lactam anion. A tentative scheme is given below:

Initiation

$$(\text{CH}_2)_5\text{-NH} + \text{B}^-\text{M}^+ \rightleftharpoons (\text{CH}_2)_5\text{-N}^-\text{M}^+ + \text{BH}$$

$$(\text{CH}_2)_5\text{-N}^-\text{M}^+ + \text{HN}-(\text{CH}_2)_5 \underset{\text{fast, Caprolactam}}{\overset{\text{slow}}{\rightleftharpoons}} (\text{CH}_2)_5-\text{N}-\text{CO}(\text{CH}_2)_5-\overset{\text{H}}{\text{N}}^-\text{M}^+ \qquad (2.29)$$

$$(\text{CH}_2)_5-\text{N}^-\text{M}^+ + (\text{CH}_2)_5-\text{N}-\text{CO}(\text{CH}_2)_5\text{NH}_2$$

N-(ε-aminocaproyl) caprolactam

Propagation

$$(\text{CH}_2)_5-\text{N}^- + (\text{CH}_2)_5-\text{N}-\text{CO}(\text{CH}_2)_5\text{NH}_2 \longrightarrow$$

$$(\text{CH}_2)_5-\text{N}-\text{CO}(\text{CH}_2)_5-\text{N}^-\text{-CO}(\text{CH}_2)_5\text{NH}_2$$

+ Caprolactam

$$(\text{CH}_2)_5-\text{N}^- + (\text{CH}_2)_5-\text{N}-\text{CO}(\text{CH}_2)_5-\text{NH}-\text{CO}(\text{CH}_2)_5\text{NH}_2 \qquad (2.30)$$

$$(\text{CH}_2)_5-\text{N}-\text{CO}(\text{CH}_2)_5\ \text{N}^-\text{-CO}(\text{CH}_2)_5-\text{NH}-\text{CO}(\text{CH}_2)_5\text{NH}_2$$

+ Caprolactam

etc.

Termination

$$(\text{CH}_2)_5-\text{N}^- + \text{H}_2\text{O} \longrightarrow (\text{CH}_2)_5-\text{NH} + \overset{-}{\text{O}}\text{H} \qquad (2.31)$$

According to this mechanism, initiation of the reaction occurs with the formation of N- (ε-caproyl) caprolactam. The chain-growth occurs with the opening of the caproylcaprolactam ring by the addition of caprolactam anion. There is always a caproylcaprolactam ring at the chain end.

Even the five-membered pyrrolidone, which does not get polymerized with water, can be polymerized to a poly(amide) with alkali and acetic anhydride.

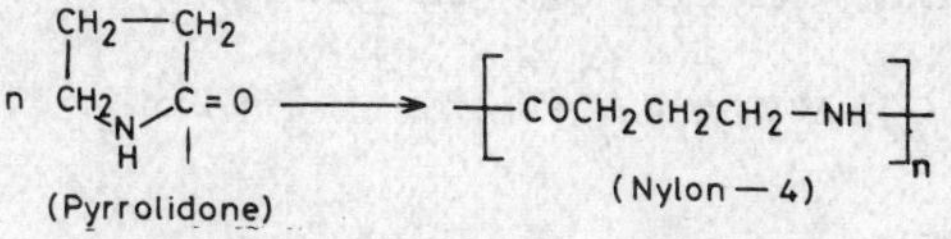

$$n \quad \underset{\text{(Pyrrolidone)}}{\begin{array}{c} CH_2-CH_2 \\ | \quad\quad | \\ CH_2 \quad C=O \\ \underset{H}{N} \end{array}} \longrightarrow \left[COCH_2CH_2CH_2-NH \right]_n \quad\quad (2.32)$$

(Nylon — 4)

ANSWERS TO QUESTIONS

A.2.1 $\overline{DP} = \dfrac{1}{1-p}$

p	0.500	0.800	0.900	0.950	0.970	0.980	0.990	0.995
$\overline{DP}$	2	5	10	20	33.3	50	100	200

A.2.2. $\overline{DP} = \dfrac{1+N_1/N_0}{1-p+N_1/N_0} = \dfrac{1+\frac{2}{98}}{1-0.98+\frac{2}{98}} = 25$

A.2.3. $\overline{DP} = \dfrac{1+r}{2r(1-p)+(1-r)} = \dfrac{1+0.98}{2 \times 0.981 - 0.98 + (1-0.98)}$

$= 33.44 \approx 33$

A.2.4. (i) $n \, ClCO - R - COCl + n \, NH_2 - R' - NH_2 \longrightarrow$
$\{OC - R - CO.NH - R'NH\}_n + 2_n \, HCl$

(ii) $n \, ClOC - R - COCl + n \, HO - R' - OH \longrightarrow$
$\{OC - R.CO - O - R' - O\}_n + 2_n \, HCl$

(iii) $-CH_2 - \underset{\underset{O}{\diagup}}{\overset{\diagdown}{CH}} - CH_2 + NH_2CH_2CH_2NH.CH_2CH_2NH_2 \longrightarrow$

$-CH_2 - \underset{OH}{CH} - CH_2NHCH_2CH_2NHCH_2CH_2NH -$

(iv) $n \, HOCH_2 \, CH_2OH + n \, OCN \, (CH_2)_6 \, NCO \longrightarrow$
$- OCH_2CH_2 \, O. \underset{\underset{O}{\|}}{C}. \, NH \, (CH_2)_6 \, NCO$

A.2.5. (i) Poly(amide), Proteins, (ii) Poly(urethane), (iii) Poly(ester).

A.2.6. (i) A linear polymer with the structure,

$\{OC - R - CO.O - R - O\}_n$, is obtained

(ii) The polymer will be cross linked and branched,

$$-O - \underset{\underset{O}{\underset{|}{|}}}{R'} - OOC - R - CO \{O - \underset{\underset{O}{\underset{|}{|}}}{R'} - O.CO - RCO -\}_n - O - \underset{\underset{O}{\underset{|}{|}}}{R'} -$$

(iii) A linear polymer will be produced,

$$-O-R-CO.O.-R-CO-O-(-OR-CO-)_n$$

A.2.7. $P_c = \dfrac{2}{f_{ave}}; f_{ave} = N_i f_i / N_i = \dfrac{12}{5}$

$P_c = \dfrac{2 \times 5}{12} = 0.533$

A.2.8 $f_{ave} = \dfrac{(1.5 \times 2) + (2 \times 0.98)}{1.5 + 0.98} = 2.39$

$P_c = \dfrac{2}{2.39} = 0.836$

(A.2.9)

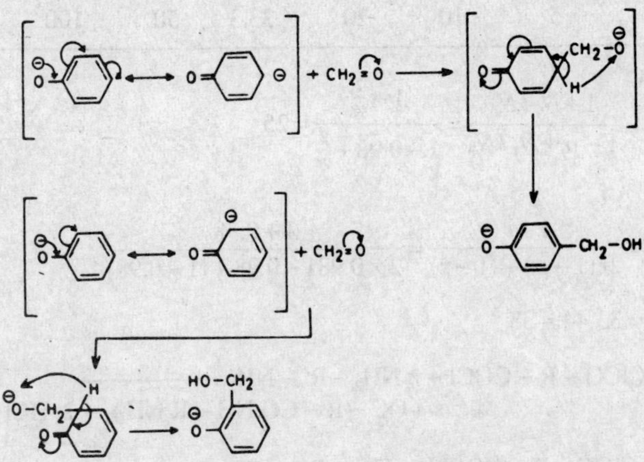

A.2.10 (i) $CH_3 - \underset{\underset{CH_3}{|}}{\overset{\overset{CH_3}{|}}{Si}} - Cl + HOH \longrightarrow CH_3 - \underset{\underset{CH_3}{|}}{\overset{\overset{CH_3}{|}}{Si}} - OH$

No polymer

(ii) $Cl - \underset{\underset{Cl}{|}}{\overset{\overset{CH_3}{|}}{Si}} - Cl + HOH \longrightarrow HO - \underset{\underset{OH}{|}}{\overset{\overset{CH_3}{|}}{Si}} - OH$

$\longrightarrow O - \underset{\underset{O}{|}}{\overset{\overset{CH_3}{|}}{Si}} - O - \left[\underset{\underset{O}{|}}{\overset{\overset{CH_3}{|}}{Si}} - O \right]$, a three dimensional Polymer.

(iii) $HO - \underset{\underset{CH_3}{|}}{\overset{\overset{CH_3}{|}}{Si}} - OH + HOH \longrightarrow \left[\underset{\underset{CH_3}{|}}{\overset{\overset{CH_3}{|}}{Si}} - O - \right]_n$

A linear polymer

CHAPTER 3

CHAIN (ADDITION) POLYMERIZATION

3.1 INTRODUCTION

Addition polymerization has the features of a chain reaction. The reactive centre may
be formed by reaction with a free radical, a cation or an anion. Polymerization occurs
through the propagation of the active species by the successive additions of a large
number of monomer molecules in a chain reaction occurring in a matter of a second
or so at the most, and usually in a much shorter time. By far the most common
example of chain polymerization is that of the polymerization of vinyl monomers.
The process can be depicted as:

Initiation

$$R^* + CH_2 = \underset{\underset{Y}{|}}{CH} \longrightarrow R - CH_2 - \underset{\underset{Y}{|}}{CH^*} \tag{3.1}$$

Propagation

$$R - CH_2 - \underset{\underset{Y}{|}}{CH^*} + n\, CH_2 = \underset{\underset{Y}{|}}{CH} \longrightarrow R \left(CH_2 - \underset{\underset{Y}{|}}{CH} \right)_n CH_2 - \underset{\underset{Y}{|}}{CH^*} \tag{3.2}$$

The growth of the polymer chain ceases when the reaction centre is destroyed by
one of a number of possible termination reactions. The basic individual steps in the
process of polymerization are thus:

1. Initiation.
2. Propagation, and
3. Termination.

The driving force for the reaction is provided by the relief in the steric strain by
the change of sp^2 bonding to sp^3 bonding (Fig. 3.1).

$$n\, CH_2 = \underset{\underset{Y}{|}}{CH} \longrightarrow \left(CH_2 - \underset{\underset{Y}{|}}{CH} \right)_n$$

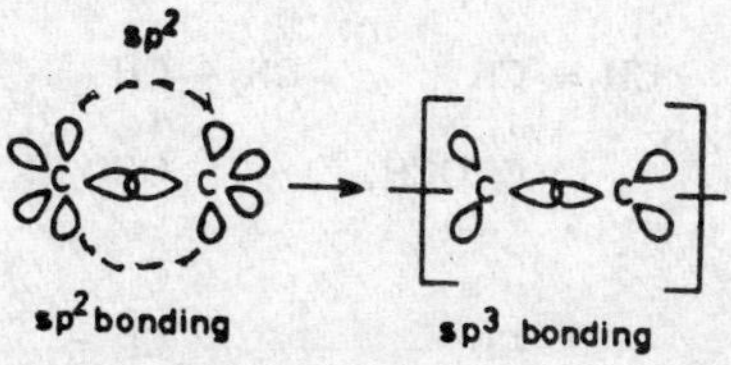

Fig. 3.1

Q.3.1. If 10 g of poly(ethylene) were completely burnt in the presence of excess air, how many moles of CO_2 will be produced?

3.2. ADDITION POLYMERIZATION

In addition polymerization the monomer is usually an ethylene derivative. Many of the industrially important monomers have the basic formula $CH_2 = C\genfrac{}{}{0pt}{}{X}{Y}$ and are usually known as vinyl monomers. In Table 3.1 are given the names of important addition polymers, their structures and important uses along with the structure of monomers from which they are produced.

Table 3.1 Important Addition Polymers

Polymer	Trade name	Monomer	Repeating Unit	Uses
Poly(ethy-lene)	Alkathene, Polythene	$CH_2 = CH_2$	$- CH_2 - CH_2 -$	Films, house-wares, electrical insulation
Poly(iso-butylene)	PIB, Vistanex	$CH_2 = \overset{\overset{CH_3}{\shortmid}}{\underset{\underset{CH_3}{\shortmid}}{C}}$	$- CH_2 - \overset{\overset{CH_3}{\shortmid}}{\underset{\underset{CH_3}{\shortmid}}{C}} -$	Cold flow rubbers, for making copolymers, in the preparation of butyl rubber
Poly(acry-lonitrile	Dynel	$CH_2 = CH - CN$	$- CH_2 - \underset{\underset{CN}{\shortmid}}{CH} -$	For making fibers, e.g., orlon
Poly(vinyl chloride	Breon, Darvic	$CH_2 = CH - Cl$	$- CH_2 - \underset{\underset{Cl}{\shortmid}}{CH} -$	Floor cover-ings, wire and cable insulation
Poly(sty-rene)	Distrene, Styron	$CH_2 = CH - C_6H_5$	$- CH_2 - \underset{\underset{C_6H_5}{\shortmid}}{CH} -$	Packaging, in-sulation, elec-trical insulat-ion, for making GRS rubber with butadiene
Poly(met-hyl metha-erylate)	Perspex, Plexiglas	$CH_2 = \overset{\overset{CH_3}{\shortmid}}{\underset{\underset{COOCH_3}{\shortmid}}{C}}$	$- CH_2 - \overset{\overset{CH_3}{\shortmid}}{\underset{\underset{COOCH_3}{\shortmid}}{C}} -$	Adhesives, au-tomobile parts (tail and signal light lenses etc.) display signs
Poly(vinyl acetate	Emultex, Vinnapas	$CH_2 = \underset{\underset{OCOCH_3}{\shortmid}}{CH}$	$- CH_2 - \underset{\underset{OCOCH_3}{\shortmid}}{CH} -$	Chewing gums, adhesives, tex-tile coatings, for making, polyvinyl alcohol
Poly(vinyli-dine chlo-ride)		$CH_2 = \overset{\overset{Cl}{\shortmid}}{\underset{\underset{Cl}{\shortmid}}{C}}$	$- CH_2 - \underset{\underset{Cl}{\shortmid}}{\overset{\overset{Cl}{\shortmid}}{CH}} -$	Co-polymers with vinyl chloride for flexibility and strength, car upholstry, brushes

Poly(tetra-fluroethy-lene)	Teflon, PTFE	$CF_2 = CF_2$	$-\overset{\displaystyle F}{\underset{\displaystyle F}{\overset{|}{\underset{|}{C}}}} - \overset{\displaystyle F}{\underset{\displaystyle F}{\overset{|}{\underset{|}{C}}}} -$		Chemically resistant films, moulded objects, electrical insulation etc.
Poly(isoprene)		$CH_2 = \underset{\displaystyle CH_3}{\overset{|}{C}} - CH = CH_2$	$- CH_2$ $\quad\quad$ CH_2 $C = CH$ H_3C		Natural rubber

Q.3.2 Identify the repeating unit in the following polymers and determine the structure of the monomer

(i) $- CH_2 - \underset{\displaystyle CH_3}{\overset{\displaystyle CH_3}{\overset{|}{\underset{|}{C}}}} - CH_2 - \underset{\displaystyle CH_3}{\overset{\displaystyle CH_3}{\overset{|}{\underset{|}{C}}}} - CH_2 - \underset{\displaystyle CH_3}{\overset{\displaystyle CH_3}{\overset{|}{\underset{|}{C}}}} -$

(ii) $- CH_2 - \underset{\displaystyle CN}{\overset{|}{C}}H - CH_2 - \underset{\displaystyle CN}{\overset{|}{C}}H - CH_2 - \underset{\displaystyle CN}{\overset{|}{C}}H -$

(iii) $- CH_2 - \underset{\displaystyle COOCH_3}{\overset{\displaystyle CH_3}{\overset{|}{\underset{|}{C}}}} - CH_2 - \underset{\displaystyle COOCH_3}{\overset{\displaystyle CH_3}{\overset{|}{\underset{|}{C}}}} - CH_2 - \underset{\displaystyle COOCH_3}{\overset{\displaystyle CH_3}{\overset{|}{\underset{|}{C}}}} -$

(iv) $\begin{array}{cccc} -CH_2 & CH_2 & CH_3 & CH_2 \\ \diagdown & \diagup & \diagup & \diagup \\ C = CH & C = & CH & \\ \diagup & \diagdown\diagup & \diagdown & \diagdown \\ H_3C & CH_2 & CH_2 & \end{array}$

Q.3.3. Write four units for the polymers obtained from the following monomers (assuming head-to-tail structure)

(i) $CH_2 = CH$

  $= O$ (N – Vinylpyrolidone)

(ii) $CH_2 = CH - CH = CH_2$ (Butadiene)

(iii) $CH_2 = \underset{\displaystyle COOCH_3}{\overset{|}{C}} - CN$ (α-Cyanomethylacrylate)

(iv) $CH_2 = CF_2$ (1, 1 – Difluoroethylene)

3.3. KINETICS AND MECHANISM OF FREE-RADICAL ADDITION POLYMERIZATION

As already stated, addition polymerization takes place by a chain-reaction, which consists of three main steps: initiation, propagation and termination.

Initiation

$$I \xrightarrow{k_d} 2R^{\cdot}$$

$$2R^{\cdot} + CH_2 = \underset{\underset{Y}{|}}{\overset{\overset{X}{|}}{C}} \longrightarrow 2\,R - CH_2 - \underset{\underset{Y}{|}}{\overset{\overset{X}{|}}{C^{\cdot}}}$$

Rate expression

$$R_i = \frac{d\,(\cdot RM\,\cdot)}{dt} -$$

$$= 2f\,k_d\,[I] \tag{3.3}$$

Here an initiator (I) such as benzoyl peroxide decomposes into two free radicals ($R^{\cdot}$). These radicals then attach themselves to the vinyl monomer and produce the new radical consisting of the initiator fragment and one monomer unit. If f is the fraction of free-radicals actually consumed in the reaction, the rate of initiation is given by Eq. 3.3.

Propagation

$$R - CH_2 - \underset{\underset{Y}{|}}{\overset{\overset{X}{|}}{C^{\cdot}}} + CH_2 = \underset{\underset{Y}{|}}{\overset{\overset{X}{|}}{C}} \longrightarrow R - CH_2 - \underset{\underset{Y}{|}}{\overset{\overset{X}{|}}{C}} - CH_2 - \underset{\underset{Y}{|}}{\overset{\overset{X}{|}}{C^{\cdot}}}$$

or more generally,

$$R\,M_n^{\cdot} + M \xrightarrow{k_p} R\,M_{n+1}^{\cdot} \tag{3.4}$$

Rate expression

$$R_p = k_p\,[M]\,[M\cdot]$$

The second step, viz., propagation is a very rapid one. No low molecular weight products are produced in the system if the reaction conditions are favourable to growth. This is illustrated in the above scheme and the rate expression is given by equation 3.4.

Termination

$$R - M_n^{\cdot} + R\,M_m^{\cdot} \xrightarrow{k_t} RM_{N+m}R \qquad\qquad \text{combination}$$

$$2R \left[CH_2\,\underset{\underset{Y}{|}}{\overset{\overset{X}{|}}{C}} \right]_{n-1} CH_2 - \underset{\underset{Y}{|}}{\overset{\overset{X}{|}}{C^{\cdot}}} \xrightarrow{k_t} \longrightarrow$$

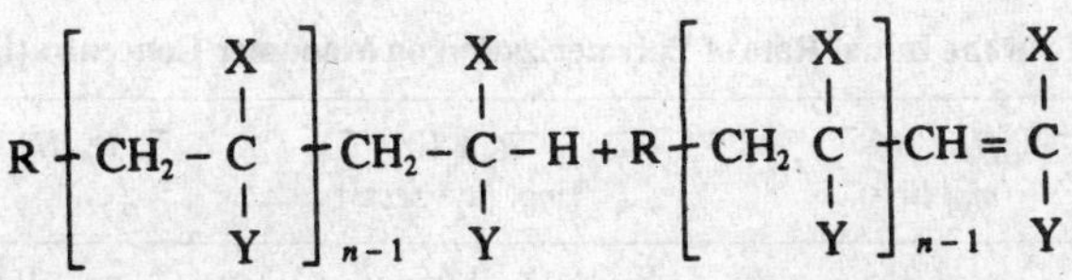

$$R \left[CH_2 - \underset{\underset{Y}{|}}{\overset{\overset{X}{|}}{C}} \right]_{n-1} CH_2 - \underset{\underset{Y}{|}}{\overset{\overset{X}{|}}{C}} - H + R \left[CH_2 \underset{\underset{Y}{|}}{\overset{\overset{X}{|}}{C}} \right]_{n-1} CH = \underset{\underset{Y}{|}}{\overset{\overset{X}{|}}{C}}$$

Disproportionation

Rate expression

$$R_t = 2 k_t [M^\bullet]^2 \tag{3.5}$$

Termination may occur by several processes. It may occur by combination of two growing radical chains. It may also occur by disproportionation which involves the transfer of a hydrogen atom from one growing chain to another. This is shown above and the corresponding rate expression is given by equation 3.5. In the above scheme R_i, R_p and R_t are the rates of initiation, propagation and termination and k_d, k_i, k_p and k_t are the rate constants for the decomposition of initiator, initiation, propagation and termination respectively. [I], [M] and [M$^\bullet$] are the concentrations of initiator, monomer and free radicals, respectively.

The energies of activation of the initiation, propagation and termination reactions are approximately 25-30 kcal per mole, 5-1 kcal per mole and 3-5 kcal per mole, respectively.

When the steady state sets in, the number of free radicals generated becomes equal to that of free radicals consumed, or,

$$R_i = R_t \quad \text{or} \quad 2f k_d [I] = 2k_t [M^\bullet]^2 \tag{3.6}$$

Hence,

$$[M^\bullet] = \left[\frac{f k_d I}{k_t} \right]^{1/2} \tag{3.7}$$

The quantity [M$^\bullet$] is difficult to measure. It can be assumed that the rate of propagation is the same as the rate of disappearance of the monomer. If the number of monomer units employed in the initiation step is small compared to the number used in the propagation step, the following equations can be written

$$R_p = \frac{d[M]}{dt} = k_p [M] [M^\bullet]$$

$$= k_p [M] \left[\frac{f k_d [I]}{k_t} \right]^{1/2}$$

$$= k_p \left[\frac{f k_d}{k_t} \right]^{1/2} [I]^{1/2} [M] \tag{3.8}$$

3.4. DEPENDENCE OF RATE ON THE INITIATOR AND MONOMER CONCENTRATIONS

Equation (3.8) predicts a square-root dependence of the initiator concentration on the initial rate of polymerization. Fig. 3.2 shows this graphically.

Data illustrating the relationship of the initial rate on the concentration of monomer at fixed concentration of the initiator are given in Table 3.2 for methyl methacrylate in benzene solution, using benzoyl peroxide as initiator at 50°C. If the efficiency of the utilization of primary free radicals is independent of the monomer concentration, the quantity $R_p / [I]^{1/2} [M]$ should be constant in the equation 3.8.

Table 3.2 Dependence of the Initial Rate of Polymerization on Monomer Concentration

$[M]$ (mol lit^{-1})	$[I] \times 10^3$ (mol lit^{-1})	$R_p \times 10^4$ (mol lit^{-1} sec^{-1})	$R_p /[I]^{1/2} [M]$
9.44	41.3	1.66	0.89
7.55	41.3	1.31	0.86
5.60	41.3	0.972	0.85
3.78	41.3	0.674	0.88
1.89	41.3	0.334	0.81
0.944	41.3	0.153	0.80

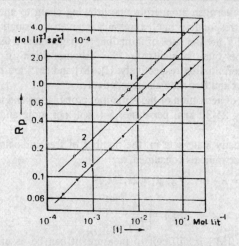

Fig. 3.2   Dependence of the steady state rate of polymerization on the initiator concentration (double logarithmic plot) [(1) Polymerization of methyl methacrylate with azobisisobutyronitrile at 50° (2) styrene with benzoyl peroxide at 60° and (3) Methyl methacrylate with benzoyl peroxide at 50°C.]

The results shown in Table 3.2 suggest a possible small decrease in f with dilution over the somewhat wide range covered in this case.

3.5. DEGREE OF POLYMERIZATION AND KINETIC CHAIN LENGTH

For addition polymers kinetic chain length (v) is proportional to the degree of polymerization ($\overline{DP}$). The proportionality constant is dependent on the mode of termination of the polymer chain. If the termination is by disproportionation, the number average degree of polymerization is equivalent to the kinetic chain length. On the other hand, if the termination is by combination, the number average degree of polymerization is equal to twice the kinetic chain length.

The kinetic chain length (v) of a radical chain polymerization is defined as the average number of monomer molecules consumed per free radical which initiates a polymer chain. The number average degree of polymerization, $\overline{P}$, is defined as the average number of monomer molecules contained in a polymer molecule (see Section 6.2).

$$\overline{P} = v \qquad \text{(for disproportionation)} \qquad (3.9)$$

$$\bar{P} = 2v \qquad \text{(for combination)} \tag{3.10}$$

For the steady state,

$$v = \frac{R_p}{R_i} = \frac{R_p}{R_t} \quad \text{(since } R_t = R_i) \tag{3.11}$$

Hence,

$$v = \frac{k_p \, [M] \, [M^{\cdot}]}{2k_t \, [M^{\cdot}] \, [M^{\cdot}]} = \frac{k_p \, [M]}{2k_t \, [M^{\cdot}]} \tag{3.12}$$

From Eq. 3.7 for initiated polymerization,

$$[M^{\cdot}] = \left[\frac{f \, kd \, [I]}{k_t} \right]^{1/2}$$

Hence,

$$v = \frac{k_p \, [M]}{2 \, (fk_d k_t)^{1/2} \, [I]^{1/2}} \tag{3.13}$$

3.6. FACTORS AFFECTING CHAIN-POLYMERIZATION

3.6.1. Monomer Structure

Ethylene, $CH_2 = CH_2$ is the only aliphatic hydrocarbon which polymerizes well under free-radical mode of initiation. Propylene and *iso*butylene are unreactive under these conditions. Electron density at the double bond may determine whether a particular monomer polymerizes by anionic, cationic or free-radical mechanism. (Table 3.3)

Table 3.3 Effect of Monomer Structure

Monomer structure	Electronic effect	Mode of initiation
$CH_2 = C$ X (inductive toward C)	Inductive or resonance	Cationic$^+$, e.g., butene, vinyl ethers, styrene
$CH_2 = C$ X (inductive away from C)	Inductive or resonance	Anionic$^-$, e.g., acrylonitrile, methyl methacrylate, styrene
$CH_2 = C$ X	–	Free-radical., e.g., acrylonitrile, methyl methacrylate, ethyl acrylate, vinyl acetate, vinyl chloride, styrene tetrafluoroethylene

Q.3.4.   Indicate by which mechanism the following monomers will polymerize – cationic, anionic or free-radical mechanism of polymerization?

(i) $CH_2 = \overset{\overset{\displaystyle CH_3}{|}}{\underset{\underset{\displaystyle CH_3}{|}}{C}}$ (ii) $CF_2 = CF_2$

(iii) $CH_2 = \overset{\overset{\displaystyle CN}{|}}{\underset{\underset{\displaystyle CN}{|}}{C}}$ (iv) $CH_2 = CHOCOCH_3$

3.6.2. Initiators

A number of polymerizations are initiated by heat and ultraviolet radiation. Usually, the initiator is a peroxide, hydroperoxide, an azo-nitrile or some closely related compound, which can readily be converted to free radical under the conditions suitable for the desired polymerization. Table 3.4 gives a list of the initiators and the temperatures at which they decompose to initiate the free-radical polymerization.

Table 3.4 Typical Initiators of Free-radical Polymerization

Name of initiator	Structure	Polymerization temp. range °C
1. Hydrogen peroxide	$H-O-O-H$	30-80
2. Potassium persulphate	$K^+O^- - \overset{\overset{\displaystyle O}{\|}}{\underset{\underset{\displaystyle O}{\|}}{S}} - O - \overset{\overset{\displaystyle O}{\|}}{\underset{\underset{\displaystyle O}{\|}}{S}} - O^- K^+$	30-80
3. Dibenzoyl peroxide	$C_6H_5 - \overset{}{\underset{\underset{\displaystyle O}{\|}}{C}} - O - O - \overset{}{\underset{\underset{\displaystyle O}{\|}}{C}} - C_6H_5$	40-100
4. Cumyl hydro-peroxide	$C_6H_5 - \overset{\overset{\displaystyle CH_3}{\|}}{\underset{\underset{\displaystyle CH_3}{\|}}{C}} - O - OH$	50-120
5. Di-tert. butyl peroxide	$CH_3 - \overset{\overset{\displaystyle CH_3}{\|}}{\underset{\underset{\displaystyle CH_3}{\|}}{C}} - O - O - \overset{\overset{\displaystyle CH_3}{\|}}{\underset{\underset{\displaystyle CH_3}{\|}}{C}} - CH_3$	80-150
6. Azobisbuty-roninitrile	$\overset{\displaystyle CH_3}{\underset{\displaystyle CH_3}{>}} \overset{\overset{\displaystyle CN}{\|}}{C} - N = N - \overset{\overset{\displaystyle CN}{\|}}{C} \overset{\displaystyle CH_3}{\underset{\displaystyle CH_3}{<}}$	70

Two typical initiators are *benzoyl peroxide* and *azobisisobutyronitrile* which decompose according to the scheme given below:

$$C_6H_5 - \underset{\underset{O}{\|}}{C} - O - O - \underset{\underset{O}{\|}}{C} - C_6H_5 \rightarrow 2\, C_6H_5 - \underset{\underset{O}{\|}}{C} - O^{\cdot} \rightarrow C_6H_5{}^{\cdot} + C_6H_5COO^{\cdot} + CO_2$$

(Benzoyl peroxide)   (Benzoyloxy radical)   (Phenyl radical)

(3.14)

$$\underset{\underset{CH_3}{|}}{\overset{\overset{CH_3}{|}}{\underset{\underset{(Azobisisobutyronitrile)\ (AIBN)}{}}{C}}} - N = N - \overset{\overset{CN}{|}}{\underset{\underset{CH_3}{}}{C}} \longrightarrow 2\, \overset{\overset{CN}{|}}{\underset{\underset{CH_3}{|}}{C}}{}^{\cdot} + N_2$$

(2-Cyano-2-propyl radical)

(3.15)

Benzoyl peroxide can undergo induced reactions to give secondary products whereas AIBN is free from such secondary reactions.

Besides the above mentioned initiations, heat and UV radiation, there is a third type known as *redox* initiation. Many oxidation-reduction reactions can produce radicals which can be used to induce polymerization. An advantage of the redox initiation is that the radicals can be produced at reasonable rates at very moderate temperatures (approx. 0–50°C). Examples of redox initiation are,

$$H_2O_2 + Fe^{2+} \longrightarrow \dot{O}H + OH^- + Fe^{3+}$$
(Hydrogen
peroxide)

(3.16)

$$ROOR + Fe^{2+} \longrightarrow R\dot{O} + RO^- + Fe^{3+}$$
(Dialkyl
peroxide)

(3.17)

$$ROOH + Fe^{2+} - R\dot{O} + OH^- + Fe^{3+}$$
(Alkyl
hydroperoxide)

(3.17a)

$$C_6H_5 - \underset{\underset{R}{|}}{N} - R + (C_6H_5 \underset{\underset{O}{\|}}{C} - O)_2 \longrightarrow \left[C_6H_5 - \underset{\underset{R}{|}}{N} - O - \underset{\underset{O}{\|}}{C} - C_6H_5 \right]^+$$

(Diakylaniline)   (Benzoyl peroxide)

$$C_6H_5COO^- \longrightarrow C_6H_5 - \underset{\underset{R}{|}}{N^+} - R + C_6H_5 - \dot{C}O_2 + C_6H_5CO^-{}_2$$

(3.18)

Q.3.5. Formulate the following redox reactions:

(i) $\bar{O} - \underset{\underset{O}{\|}}{\overset{\overset{O}{\|}}{S}} - O - O - \underset{\underset{O}{\|}}{\overset{\overset{O}{\|}}{S}} - \bar{O} + Fe^{2+} \longrightarrow$

(Persulphate ion)

(ii) $\bar{O} - \underset{\underset{O}{\|}}{\overset{\overset{O}{\|}}{S}} - O - O - \underset{\underset{O}{\|}}{\overset{\overset{O}{\|}}{S}} - \bar{O} + S_2O_3{}^{2-} \longrightarrow$

(iii) $R - CH_2OH + CO^{4+} \longrightarrow$

$$(iv)\ CH_3 - \overset{\overset{\displaystyle CH_3}{|}}{\underset{\underset{\displaystyle CH_3}{|}}{C}} - O - OH + Fe^{2+} \longrightarrow$$

3.6.3. Chain-Transfer

Chain-transfer may take place with the solvent, monomer, initiator, polymer or with a modifier.

$$
\begin{array}{ccccccc}
M_m{}^{\bullet} & + & XY & - & M_mX & + & Y^{\bullet} \qquad (3.19)\\
\text{(Growing} & & \text{(Transfer} & & \text{(Stabilised} & & \text{(Free-radical)}\\
\text{polymer} & & \text{reagent)} & & \text{polymer} & &\\
\text{chain)} & & & & \text{chain)} & &
\end{array}
$$

Transfer reactions

$$M_n^{\bullet} + M \longrightarrow M_n + M_1^{\bullet} \ \text{(monomer transfer } k_{tr}, M) \qquad (3.20)$$

$$M_n^{\bullet} + \underset{\text{Solvent}}{S} \longrightarrow M_n + \dot{S} \ \text{(Solvent transfer } k_{tr}, S) \qquad (3.21)$$

$$S + \underset{\text{Monomer}}{M} \longrightarrow M_1 + \dot{S}$$

$$M_n^{\bullet} + I \longrightarrow M_n + \dot{I} \ \text{(Initiator transfer } k_{tr}, I) \qquad (3.22)$$

where transfer constants are defined as,

$$C_M = \frac{k_{tr}, M}{k_p}, \ C_S = \frac{k_{tr}, S}{k_p} \ \text{and} \ C_I = \frac{k_{tr}, I}{k_p}$$

The degree of polymerization for small amounts of reaction is given by,

$$\bar{P} = \frac{\text{rate of chain} - \text{growth}}{\text{rate of termination by all processes}} \qquad (3.23)$$

Hence, $\dfrac{1}{\bar{P}} = \dfrac{\text{rate of termination by all process}}{\text{rate of chain} - \text{growth}}$

$$= \frac{k_{tr,M}[M][\dot{M}] + k_{tr,S}[S][\dot{M}] + k_{tr,I}[I][\dot{M}] + fk_d[I]}{R_p} \qquad (3.24)$$

Simplifying the equation, we get,

$$\frac{1}{\bar{P}} = C_M + C_S \ \frac{[S]}{[M]} + C_I\,(k_t/R_p^2 fk_d) \ \frac{R_p^2}{[M]^3} + (k_t/k_p^2) \ \frac{R_p}{[M]^2} \qquad (3.25)$$

3.6.3.1 *Transfer to solvent*

In cases where the chain-transfer to solvent becomes important, the second term on the right hand side of Eq. (3.25) makes the biggest contribution to the determination of the degree of polymerization. By an appropriate choice of the polymerization conditions, one can determine the value of C_s for various solvents. By using low concentrations of the initiator with negligible C_I values (e.g. AIBN), the third term in Eq. 3.25 can be made negligible. The last term on the right hand side of the above equation may be kept constant by keeping $R_p/[M]^2$ constant through the appropriate adjustment of the initiator concentration throughout the course of the reaction. Under these conditions Eq. 3.25 takes the shape,

$$\frac{1}{\bar{p}} = \frac{1}{\bar{p}_0} + C_s \frac{[S]}{[M]} \qquad (3.26)$$

where $1/p_0$ is the value of $1/p$ in the absence of chain transfer agent. $1/p_0$ is the sum of the first, third and fourth terms on the right hand side of Eq. 3.25. C_s is then determined as the slope of the linear plot of $\dfrac{1}{\bar{p}}$ vs $\dfrac{[S]}{[M]}$. Such plots are shown in Fig. (3.3) for several chain-transfer agents in the thermal polymerization of styrene. This method has been found to be of general utility in determining values of C_s. A few of these transfer constants for styrene, methyl methacrylate and vinyl acetate are given in Table 3.5.

Table 3.5 Chain-Transfer Constants of Radicals with Solvents at 80°C

Solvent	$C_s \times 10^5$		
	Styrene	*Methyl methacrylate*	*Vinylacetate*
Benzene	0.60	0.75	
Cyclohexane	0.66	1.00	
Toluene	3.10	5.20	920.00
Ethylbenzene	10.80	13.50	
i-Propylbenzene	13.00	19.00	
Carbon tetrachloride	1300.00	24.00	1000.00
n-Butylmercaptan	$22 \times 10^{5*}$	$0.67 \times 10^{5*}$	$48 \times 10^{5*}$

*at 60°.

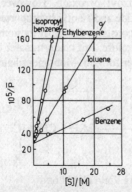

Fig. 3.3 Effect of several hydrocarbon solvents on the degree of polymerization of styrene at 100°C.

The presence of the weak benzylic hydrogen in toluene, ethylbenzene and dimethylbenzene leads to higher C_s values relative to benzene. The benzylic C–H bonds break easily because the resultant radical is resonance-stabilized (Eq. 3.27).

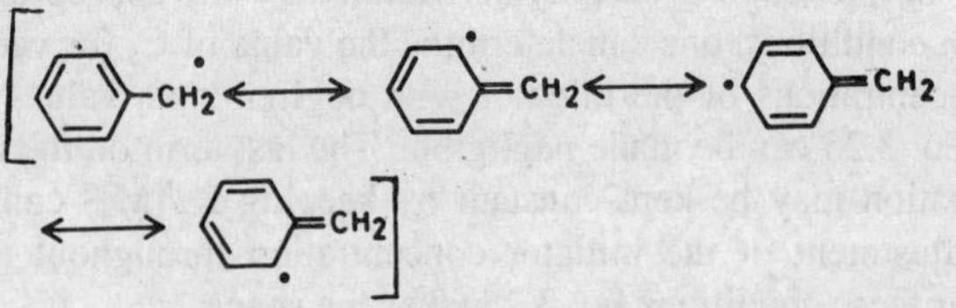

$$(3.27)$$

Q.3.6 (i) The chain transfer constant of styrene with the following hydrocarbons is greater than that with *t*-butylbenzene, *i-propylbenzene*, ethylbenzene and toluene; give a plausible explanation.

(ii) Carbon tetrachloride is a very strong transfer agent for styrene polymerization. Explain.

(iii) Disulphides like RSSH are very efficient transfer agents. Give reasons.

3.6.3.2 *Chain-transfer with Initiator and Monomer*

If chain-transfer to solvent is absent, the second term of Eq. (3.25) disappears.

$$\frac{1}{\bar{p}} = C_M + C_I \frac{k_t R_p^2}{k_p^2 k_d [M]^3} + \frac{k_t R_p}{k_p^2 [M]^2} \qquad (3.28)$$

This equation can be transformed into,

$$\frac{1}{\bar{p}} = C_M + C_I \frac{[I]}{[M]} + \frac{k_t R_p}{k_p^2 [M]^2} \qquad (3.29)$$

If $1/\bar{p}$ is plotted against R_p, the effect of chain-transfer to initiator can be observed. The initial portion of the plot is linear for several initiators, but at higher concentration of initiator and thus at higher values of R_p, the plot deviates from linearity as the contribution of transfer to initiator becomes considerable. It is observed that for azobisisobutyronitrile, there is no departure from linearity; a slight deviation is observed for benzoyl peroxide. However, considerable deviation is observed for cumyl hydroperoxide and *t*-butyl hydroperoxide initiated polymerization of styrene. The chain-transfer constants for most monomers are small, being mostly in the range of $10^{-4} - 10^{-5}$; therefore, self chain-transfer is not a problem limiting the molecular weight of the polymer formed. Monomer chain-transfer is low because the reaction,

$$M_n^\bullet + CH_2 = \underset{\underset{X}{|}}{\overset{\overset{H}{|}}{C}} \longrightarrow M_nH + CH_2 = \underset{\underset{X}{|}}{\overset{}{\dot{C}}} \qquad (3.30)$$

involves the breaking of a strong vinyl C–H bond.

3.6.3.3 *Chain-transfer to Modifiers*

Chain-transfer agents such as disulphides find use as modifiers in systems such as styrene-butadiene rubber, where dodecyl mercaptan is widely used to limit the molecular weight of the product formed to the required extent.

3.7. INHIBITION AND RETARDATION

If a foreign substance is added to a monomer and it leads to a reduction in the degree of polymerization $(\bar{P})$ without completely suppressing it, the added substance is known as *retarder*. An example is nitrobenzene. On the other hand, if it reacts with the chain-radical to yield non-radical products or radicals of such low reactivity that they are incapable of adding fresh monomer units, the compound added is called an *inhibitor*. After the consumption of an inhibitor, polymerization follows the normal course. An example of an inhibitor is benzoquinone. Fig. 3.4 illustrates the two modes.

Fig. 3.4   A comparison of the effects of 0.1 percent of benzoquinone (curve II), 0.5 per cent of nitrobenzene (curve III) and 0.2 per cent of nitrobenzene (curve IV) on the thermal polymerization of styrene at 100°C. Curve I represents the polymerization of pure styrene.

The reaction with an inhibitor can be represented as shown below:

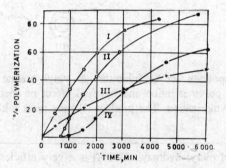

$$\text{(3.31)}$$

A shift of the hydrogen atom on radical I(a) converts it to the more stable hydroquinone radical I(b). Owing to resonance, the similar inhibiting radicals, Ib, II and III should be relatively stable and, therefore, not readily available to regenerate active chain radicals by adding monomer. They may be expected to disappear instead through reaction between pairs of these inactive radicals, in partial analogy with the normal bimolecular chain termination of the uninhibited polymerization.

The mechanism of radical termination by nitro groups may involve one of the following processes:

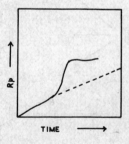

$$(3.32)$$

The nitro compounds act as inhibitors in vinyl acetate polymerization, as retarders in styrene polymerization and have no effect on methyl methacrylate or methyl acrylate polymerization. The product (I) may react with another radical or monomer molecule.

Q.3.7   (i) Diphenyl picryl hydrazyl (DPPH) is a very efficient radical scavenger. Explain.

(ii) Give equations to show the inhibitory action of benzoquinone.

(iii) Discuss the retarding action of nitro compounds.

3.8. VARIOUS POLYMERIZATION TECHNIQUES

The polymerization of vinyl monomers can be carried out by the following techniques in bulk, solution, suspension or emulsion. When a monomer is heated either in bulk or solution, at higher conversions, the gel effect is observed. This effect is also known as Tromorsdorf effect, after the name of its discoverer, and is shown in Fig. 3.5.

Fig. 3.5 Tromorsdorf or gel effect [. . . . . . Expected, – Actual]

In the radical chain polymerization of a vinyl monomer one would normally expect the reaction rate to fall with time since the monomer concentration decreases with conversion. However, the exact opposite behaviour is observed in many cases where the rate of polymerization increases with time as shown in Fig. 3.5. This is called gel effect. To cite an example, pure methyl methacrylate shows a dramatic auto-acceleration in the polymerization rate.

The gel effect is caused by a decrease in the termination rate constant with increasing conversion. As the polymerization proceeds the viscosity of the system increases the termination becomes increasingly slower. Although propagation is also hindered, this effect is much smaller since k_p values are smaller than k_t. Termination involves the reaction between two large growing radical chains, whereas propagation involves the reaction of a small monomer unit with a large radical. High viscosity, therefore, affects the former much more than the latter. Hence, the quantity $k_p/k_t^{1/2}$ increases and rate of polymerization increases with conversion (Eq. 3.8).

A second consequence of this effect is the increase in molecular weight with increasing conversion (Eq. 3.13).

When a monomer is polymerized in bulk, an initiator which can decompose to give free-radicals at a fairly good rate at the temperature of polymerization is used. There are advantages as well as disadvantages of this process. This method is now used for casting objects to be preserved such as an insect, a flower or any other thing. The process can be carried out at comparatively low temperature. Poly(methyl methacrylate) sheets and rods are made in this way. However, thermal control is difficult to achieve as also is any further isolation from the monomer.

In *solution* polymerization, on the other hand, improved thermal control is possible, but it is difficult to remove the solvent. Further, chain-transfer to solvent may limit the molecular weight of the polymer obtained. This limits the usefulness of the technique.

Emulsion and *suspension* polymerizations in aqueous media are free from the problems of heat transfer, chain transfer and viscosity. These two methods have assumed industrial importance. It is possible to get higher molecular weights with emulsion polymerization. In emulsion polymerization, the monomer is emulsified in water with an emulsifying agent which may be a soap or a detergent. Usually, a redox system is used. In a typical emulsion system a monomer(s), water, emulsifier (surface active reagent), a chain transfer agent and a water soluble initiator are usually used. This is illustrated by the recipe given in Table 3.6 for the preparation of styrene-butadiene rubber.

Table 3.6 Recipe for the Preparation for Styrene-Butadiene Rubber

Ingredient	Parts by weight
Butadiene	75.00
Styrene	25.00
Water	180.00
Soap	50.00
Dodecyl mercaptan	0.50
Potassium persulphate	0.30

3.9. IONIC ADDITION POLYMERIZATION

The polymerization of vinyl monomers can take place in the presence of a cation or an anion also.

3.9.1 Cationic Polymerization

Cationic polymerization differs from free-radical polymerization, specially in the initiation and termination steps. The initiators used are Lewis acids such as boron trifluoride, aluminium chloride, titanium tetrachloride, sulphuric acid and other strong protonic acids. All these, except possibly the strong protonic acids, need a co-catalyst to initiate polymerization. The co-catalyst supplies hydrogen, and the real initiator is believed to be a proton in every case. This means that one end group of the polymer is a hydrogen atom. The monomers which best polymerize by cationic initiation are olefins, substituted with electron releasing groups, such as isobutylene $\{CH_2 = C(CH_3)_2\}$, styrene ($C_6H_5CH = CH_2$), α-methylstyrene $\{C_6H_5C(CH_3) = CH_2\}$, vinyl alkyl ethers ($CH_2 = CHOR$) and dienes, which are to be considered as a separate group. Propylene does not yield high molecular weight polymer under cation initiated polymerization conditions. Polymerization proceeds most rapidly to give the high molecular weight polymers at very low temperatures. For example, i-butylene with boron trifluoride at $-100°C$ gives a high polymer in a fraction of a second. At room temperature a dimer is formed slowly.

Initiation

$$\underset{\text{(Catalyst)}}{TiCl_4} + \underset{\text{(Co-catalyst)}}{RH} \longrightarrow H^+ RTiCl_4^- \tag{3.33a}$$

$$H^+ RTiCl_4^- + \underset{\text{(i-Butylene)}}{(CH_3)_2C = CH_2} \longrightarrow \underset{\substack{\text{(Carbonium ion} \\ \text{or cation)}}}{(CH_3)_3C^+ + R\,TiCl_4^-} \tag{3.33b}$$

Propagation

$$(CH_3)_3C^+ + n\,(CH_3)_2C = CH_2 \longrightarrow (CH_3)_3 - C \{ CH_2 - \underset{\underset{CH_3}{|}}{\overset{\overset{CH_3}{|}}{C}} \}_n - CH_2 - \underset{\underset{CH_3}{|}}{\overset{\overset{CH_3}{|}}{C}} (+) \tag{3.34}$$

Termination

$$(CH_3)_3 C \{ CH_2 - \underset{\underset{CH_3}{|}}{\overset{\overset{CH_3}{|}}{C}} \}_{n-1} - CH_2 - \underset{\underset{CH_3}{|}}{\overset{\overset{CH_3}{|}}{C^+}} + RTiCl_4^- \longrightarrow$$

$$(CH_3)_3C \{ CH_2 - \underset{\underset{CH_3}{|}}{\overset{\overset{CH_3}{|}}{C}} \}_{n-1} CH_2 - \underset{\underset{CH_3}{|}}{\overset{\overset{CH_2}{||}}{C}} + H^+ R\,TiCl_4^- \tag{3.34a}$$

$$(CH_3)_3 C \{ CH_2 - \underset{\underset{CH_3}{|}}{\overset{\overset{CH_3}{|}}{C}} \}_{n-1} - CH_2 - \underset{\underset{CH_3}{|}}{\overset{\overset{CH_3}{|}}{C^+}} + RTiCl_4^- \longrightarrow$$

$$(CH_3)_3 C \{ CH_2 - \underset{\underset{CH_3}{|}}{\overset{\overset{CH_3}{|}}{C}} \}_{n-1} - CH_2 - \underset{\underset{CH_3}{|}}{\overset{\overset{CH_3}{|}}{C}} - R + TiCl_4 \tag{3.34b}$$

3.9.1.1 *Kinetics of Cationic Polymerization*

Most of the cationic polymerizations proceed so rapidly that it is difficult to establish the steady state. However, the following kinetic scheme seems to be valid. Let the catalyst be designated A and the co-catalyst by RH, then,

Initiation

$$A + RH \xrightarrow{\quad K \quad} H^+ + AR^-$$

$$H^+ + AR^- + M \xrightarrow{\quad k_i \quad} HM^+ + AR^- \tag{3.35}$$

Propagation

$$AR^- + HM^+ + (n-1)\,M \xrightarrow{\quad k_p \quad} HM_n{}^+ AR^- \tag{3.35a}$$

Termination

$$HM_n{}^+ AR^- \xrightarrow{\quad k_t \quad} M_n + H^+ AR^-$$

$$HM_n{}^+ AR^- + M \xrightarrow{\quad k_{tr} \quad} M_n + HM^+ AR^- \text{ (transfer)} \tag{3.35b}$$

Rate of initiation $R_i = K k_i\,[A]\,[RH]\,[M]$ \hfill (3.35c)

where [A] is the catalyst concentration. If the formation of $H^+ AR^-$ is the rate determining step, R_i can be independent of [M]. The kinetic scheme is then appropriately modified. Since termination is first order,

$$R_t = k_t\,[M^+] \tag{3.35d}$$

where $[M^+]$ is written as an abbreviation of $HM^+ AR^-$. Assuming steady state approximations, we may write $i\,R_i = R_t$ and,

$$[M^+] = \frac{Kk_i}{k_t}\,[A]\,[RH]\,[M] \tag{3.35e}$$

The overall rate of polymerization R_p is given by the expression,

$$R_p = k_p\,[M]\,[M^+] = K\,\frac{k_i k_p}{k_t}\,[A]\,[RH]\,[M] \tag{3.36}$$

If termination predominates over transfer, then,

$$\bar{p} = \frac{R_p}{R_t} = \frac{k_p\,[M]\,[M^+]}{k_t\,[M^+]} = \frac{k_p}{k_t}\,[M] \tag{3.36a}$$

Whereas transfer predominates, then

$$\bar{p} = \frac{R_p}{R_{tr}} = \frac{k_p\,[M]\,[M^+]}{k_{tr}\,[M]\,[M^+]} = \frac{k_p}{k_{tr}} \tag{3.36b}$$

An example illustrating cationic polymerization has already been given.

Q.3.8. Formulate the cationic polymerization of *i*-butylene initiated by BF_3 catalyst with water as the co-catalyst.

3.9.2 Anionic Polymerization

Monomers containing electron withdrawing groups (or electronegative groups) can polymerize by anionic mechanisms. It was found that sodium in liquid ammonia at $-75°$ was a particularly effective initiator for the polymerization of α-methyl acrylonitrile, acrylonitrile, methyl acrylate and methyl methacrylate were polymerized by this mechanism. Other effective initiators are Grignard reagents and triphenyl methyl sodium.

Initiation

$$Na + NH_3 \xrightarrow{} Na^+ NH_2^- \xrightarrow{K} Na^+ + NH_2^- \tag{3.37}$$

$$NH_2^- + M \xrightarrow{k_i} NH_2 M^- \tag{3.38}$$

Propagation

$$NH_2M^- + (n-1) M \xrightarrow{k_p} NH_2 M_n^- \tag{3.39}$$

Termination

$$NH_2 M_n^- + NH_3 \xrightarrow{k_{tr}} NH_2 M_n H + NH_2^- \tag{3.40}$$

According to usual kinetic analysis, rate and degree of polymerization are obtained as follows:

$$R_p = K \frac{k_i\, k_p}{k_{tr}} \frac{[NH_2^-]}{[NH_3]} [M^2] \tag{3.40a}$$

$$\bar{p} = \frac{k_p [M]}{k_{tr} [NH_3]} \tag{3.40b}$$

The following two examples show the scope of the anionic polymerization.

3.9.2.1 *Initiation by Alkyl Lithium Compounds*

Initiation

$$RLi + CH_2 = \underset{\underset{C_6 H_5}{|}}{CH} \longrightarrow R - CH_2^- - \underset{\underset{C_6 H_5}{|}}{CH^-} Li^+ \tag{3.41}$$

Propagation

$$R - CH_2 - \underset{\underset{C_6 H_5}{|}}{CH^-} Li^+ + (n-1) CH_2 = \underset{\underset{C_6 H_5}{|}}{CH} \longrightarrow$$

$$R \underset{\underset{C_6 H_5}{|}}{\left(CH_2 - CH \right)_n} \tag{3.42}$$

Polymerization proceeds in nomogeneous hydrocarbon solvents by nucleophilic addition to the double bond of the monomer.

3.9.2.2 *Initiation with Sodium and Naphthalene (living polymers)*

Initiation

$$Na + \underset{(Naphthalene)}{\bigcirc\!\!\bigcirc} \xrightarrow[\text{(Tetrahydro furan)}]{THF} \left[\bigcirc\!\!\bigcirc\right]^{-} Na^+ \qquad (3.43)$$

$$\left[\bigcirc\!\!\bigcirc\right]^{-} Na^+ + C_6H_5 CH = CH_2 \longrightarrow$$

$$\bigcirc\!\!\bigcirc + \underset{I}{\left[C_6H_5\overset{..}{C}H-\overset{.}{C}H_2 \longleftrightarrow C_6H_5\overset{..}{C}H-\overset{.}{C}H_2\right]} Na^+ \qquad (3.44)$$

Propagation

$$2\,I \longrightarrow Na^+[\,C_6H_5 - \overset{\overline{..}}{C}H - CH_2 - CH_2 \longrightarrow \overset{\overline{..}}{C}H - C_6H_5]\,Na^+ \quad (3.45)$$

$$Na^+ \left[\; \underset{C_6H_6}{:\overset{|}{\overline{C}}H - \overline{C}H_2 - CH_2 - \underset{C_6H_5}{\overset{|}{\overline{C}}H:} \;\right] Na^+ + (m+n)\,\underset{C_6H_5}{\overset{|}{CH_2 = CH}}$$

$$\longrightarrow Na^+ \left[\; :\underset{C_6H_5}{\overset{|}{\overline{C}}H} - CH_2 \!\left(\!\underset{C_6H_5}{\overset{|}{CH}} - CH_2\!\right)_{\!m}\! CH_2 - \right.$$

$$\left.\left(\underset{C_6H_5}{\overset{|}{CH_2 - CH}}\right)_{\!n} CH_2 - \underset{C_6H_5}{\overset{|}{\overline{C}H:}}\right] Na^+ \qquad (3.46)$$

These are also called living polymers.

Q.3.9. Consider the following catalysts and indicate the mechanism of polymerization which they can initiate
(i) Na + naphthalene, (ii) n-C_4H_9Li, (iii) H_2SO_4,
(iv) $TiCl_4$ + water, (v) benzoyl peroxide, and (vi) $(CH_3)_3COOH$

3.10. COORDINATION POLYMERIZATION *(Ziegler-Natta)*

There are a number of coordination catalysts which can polymerize olefins. The most common catalysts of this type are composed of aluminium trialkyls and titanium or vanadium chloride. These catalysts are mostly effective for the polymerization of 1-olefins. The following equations seem to best describe the reaction between titanium tetrachloride and a aluminium trialkyl.

$$TiCl_4 + AlR_3 \longrightarrow TiCl_3\,R + AlR_2\,Cl \qquad (3.47)$$

$$\text{TiCl}_3 \text{ R} \longrightarrow \text{TiCl}_3 + \text{R}^{\cdot} \tag{3.48}$$

$$\text{TiCl}_3 + \text{AlR}_3 \longrightarrow \text{TiCl}_2 \text{ R} + \text{AlR}_2 \text{ Cl} \tag{3.49}$$

$$\text{Cl}_2\text{TiR} + \text{X (monomer)} \longrightarrow \text{Cl}_2\text{Ti (monomer)}_x \text{ R} \tag{3.50}$$

Several important points have to be kept in mind in order to propose an acceptable mechanism:

(i) The polymerization reaction is stereo-selective as shown by the fact that isotactic polymers (see page 90) of 1-olefins are formed.

(ii) The reaction has the characteristics of anionic polymerization.

(iii) The reaction takes place on a solid surface.

Two types of structure have been put forward for the Zieglar-Natta initiator, a bimetallic (I) and a monometallic (II) one.

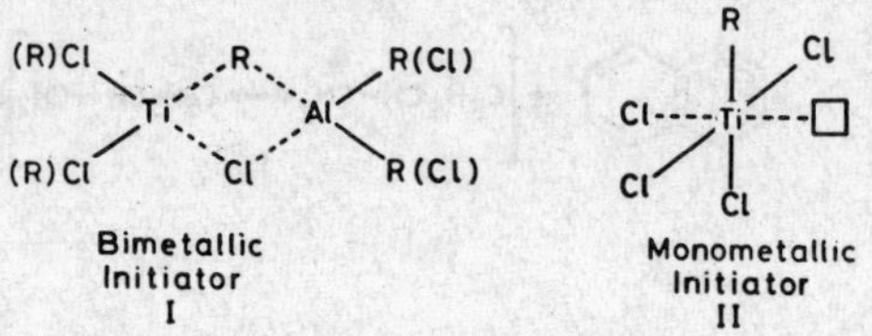

Structure (I) arises from the alkylation of titanium chloride by the organometallic compound. The symbol ▢ in (II) shows an unoccupied or vacant octahedral titanium site.

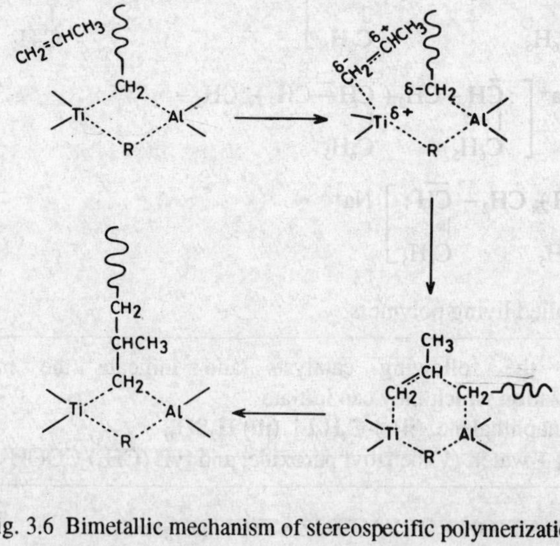

Fig. 3.6 Bimetallic mechanism of stereospecific polymerization.

Bimetallic mechanism (Fig. 3.6) involves a propagation step in which growth occurs at two metal centres. The monometallic mechanism (Fig. 3.7) is shown below.

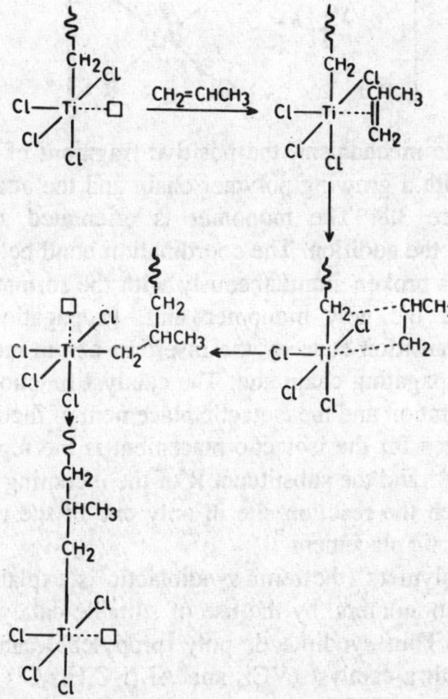

Fig. 3.7 Monometallic mechanism of stereospecific polymerization.

The mechanism of stereospecific polymerization by an initiator to yield an tactic polymer (for definition of the term tactic please see page 90) is shown in Fig. 3.8.

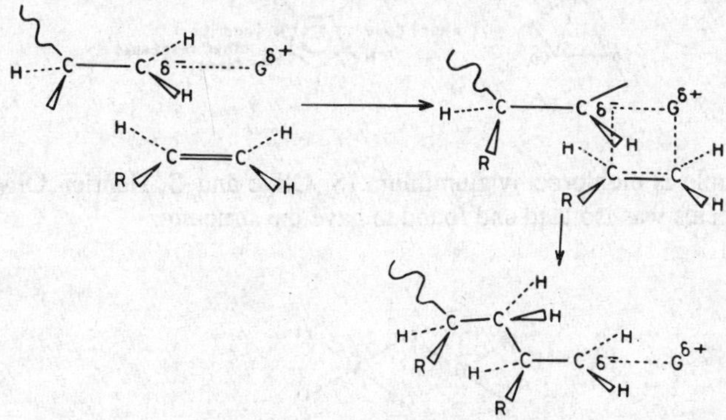

Fig. 3.8 Mechanism of stereospecific polymerization (isotactic).

The polymerization of a typical olefinic compound such as propylene to yield isotactic poly (propylene) may now be considered. It is known that when propylene is polymerized in heptane at 50° in the presence of a catalyst (a mixture of $TiCl_4$ and $Al\ (C_2H_5)_3$), the catalyst (I on page 52) can be considered to be composed of a positive and a negative fragment.

$$\left[\begin{array}{ccc} Cl & & R \quad\quad R \\ \diagdown & & \diagdown \;\; \diagup \\ & Ti^{\delta+} & Al^{\delta-} \\ \diagup & & \diagup \;\; \diagdown \\ Cl & & Cl \quad\quad R \end{array} \right]$$

In one plausible mechanism, the positive fragment of a catalyst (designated as $G^{\delta+}$) coordinates with a growing polymer chain and the attacking olefinic monomer, as shown in Figure 3.8. The monomer is orientated and held in position by coordination during the addition. The coordination bond between the catalyst and the propagating chain is broken simultaneously with the formation of bond between the polymer chain and the new monomer unit. Propagation proceeds in the four membered cyclic transition state by the insertion of the monomer unit between the catalyst and the propagating chain end. The catalyst functions as a template (mould) for the second orientation and the isotactic placement of incoming monomer units.

The driving force for the isotactic placement is the repulsive force between the catalyst fragment $G^{\delta+}$ and the substituent R of the incoming monomer. The monomer is forced to approach the reaction site in only one of the two possible orientations, which leads to isotactic placement.

Syndiotactic polymers (the term syndiotactic is explained on page 91) can be prepared for some monomers, by the use of suitable catalyst and under appropriate reaction conditions. Thus syndiotactic poly (propylene) can be obtained by reaction of the monomer with a catalyst (VCl_4 and Al (i-C_4H_9)$_2$Cl) in a mixture of solvents (anisole and toluene) at $-78°C$.

A good deal of information on the mechanism of Ziegler – Natta catalysts has been obtained with soluble catalysts in homogeneous systems. One such catalyst is cyclopentadienyltitanium dichloride ($C_{p2}TiCl_2$)

where Cp = (and its other resonance forms)

Another example is dichloroethylaluminium (S. Olive and G. Henrici- Olive, 1981). The active species was isolated and found to have the structure.

The polymerization of ethylene is supposed to take place at the vacant coordination site of the transition metal and inserted between the metal and the alkyl group (C_2H_5) or later the growing chain. The other ligands present have an important influence on the kinetics of reaction. The mechanism can be formulated as below:

3.11 SOME ADDITION POLYMERS

3.11.1 Poly(ethylene), $\{CH_2 - CH_2\}_n$

Low density poly(ethylene) (LDPE)

It can be prepared either by a batch process or a continuous process. Batch processes are not usually as useful as the continuous processes. In the continuous process, a better balance and control of the polymerization is achieved, and excessive branching is avoided. Reaction time is reduced to minutes when high pressures (500 atm. or more) and tabular reactors are employed. Only 15-30% of the monomer is converted to poly(ethylene) per pass and the unreacted monomer is recycled. The polymer is separated from the gas or/liquid phase. The process may involve multiple additions of monomer, initiator and chain-transfer agents along the tube length. In this high pressure process small amounts of oxygen (0.1 – 5%) are compressed with ethylene to at least 500 atm. A temperature of 200°C is maintained. Chain-transfer agents such as alcohols and ketones may be employed. The reactors employed are either tower or autoclave type. The polymer thus formed is extruded in the form of a ribbon and broken into chips.

High density poly(ethylene) (HDPE)

High density poly(ethylene) can be prepared by solution, slurry or gas phase processes. In the slurry process, a Ziegler-Natta type initiator is usually employed (aluminium trialkyl and titanium tetrachloride).

In a typical slurry process, ethylene at moderate pressure (1 – 20 atm) taken in a hydrocarbon solvent such as *n*-hexane or cyclohexane and at a moderate temperature (50 – 120°) is stirred in a reactor. The initiator is prepared separately as a slurry in the same solvent and then added to the reaction mixture. The reaction is carried out in an inert atmosphere, usually of nitrogen. As the polymerization proceeds, the viscosity of the system increases and polymer precipitates out. The initiator is deactivated by addition of water or alcohol. The polymer is filtered and the remaining diluent removed by passing steam. The remaining powder is dried by blowing hot air. Very little initiator is left in the polymer, which is used without further purification.

Properties

The properties of poly(ethylene) depend a great deal on its method of preparation. Table 3.7 lists the difference in physical properties exhibited by HDPE and LDPE.

Table 3.7 Comparison of Properties of HDPE and LDPE

Property	Low density poly(ethylene) (LDPE)	High density poly(ethylene) (HDPE)
Molecular weight	20,000 – 50,000	As low as 20,000 as high as 3,000,000
M_w/M_n	3-20	—
Melting point	111°C	127°C
Density	0.92	0.95-0.97
Per cent crystallinity	60%	90 – .95%
Methyl groups/ 1000 carbon atoms (branching)	21.5	1.5 – 3
Dielectric constant	2.3	2.3
Tensile strength at yield (p.s.i.)	1800 – 2000	3500 – 4500

Poly(ethylene) is highly resistant to chemicals and moisture. It possesses good physical, mechanical and dielectric properties. At room temperature, poly(ethylene) is insoluble in almost all solvents, but at temperatures above 70°C, it swells and dissolves in carbon tetrachloride, trichloroethylene and toluene. On cooling, it precipitates out from these solvents.

Applications

The low density poly(ethylene) is mainly used for producing extruded films and sheets, mainly for packaging and household uses. It is used for covering wires and submarine cables for insulation. It can be injection-moulded into toys, bottles and extruded into tubes useful for irrigation purposes. It can also be blow-moulded into laboratory ware and other objects.

3.11.2 Poly(styrene)

$$\{ CH_2 - CH \}_n$$
$$|$$
$$C_6H_5$$

The styrene monomer is obtained by the following process. Pure benzene and ethylene are reacted in the presence of anhydrous aluminium chloride to obtain ethylbenzene. This material is then dehydrogenated by passing through a furnace containing a catalyst consisting mainly of zinc oxide (80%), calcium oxide (5 – 7%) and aluminium oxide (10%). The temperature of reaction is about 600°C. The reactions that take place are:

$$C_6H_6 \quad + \quad C_2H_4 \quad \xrightarrow{\text{Anhydrous} \atop \text{AlCl}_3} \quad C_6H_5C_2H_5$$
(Benzene)   (Ethylene)   (Ethylbenzene)

$$C_6H_5C_2H_5 \xrightarrow[600°]{ZnO} C_6H_5CH = CH_2 + H_2$$

(Styrene)

Bulk polymerization

Styrene can be polymerized by bulk or suspension polymerization. In the bulk polymerization, poly(styrene) is made by pre-polymerization in mass to a thick consistency; it is fed to a vertical tower polymerizer and extruded at the bottom as a band, ground and packaged. The pre-polymerizer is a stirred aluminium alloy kettle with a heating and cooling coil inside. Styrene mixed with 0.2% glacial acetic acid is heated at 80°C for two days. Polymerization is carried out to 33 – 35% conversion, then the charge is bled off the bottom and fresh charge of styrene is added from the top. In one manufacturing process the viscous material is allowed to flow down the tower. The tower is usually six meters high with an inside diameter of 80 cm. It consists of six one-meter sections. The top two sections of the tower operate at 100 – 110°C, the central sections at 150°C and the lower two at 180°C. The lower sections, are equipped with electrical heating arrangement for additional heating, if desired. Two electrically heated discharge screws at the bottom extrude bands 30 mm wide and 2–3 mm thick. The bands travel under cooling rollers and to the grinder.

The function of glacial acetic acid is not clearly understood, but it reduces the final volatile content.

Suspension polymerization of styrene

A reaction kettle with arrangement for heating and cooling and vigorous stirring is used in this process. It is charged with 1.5 litres of water and 3 litres of styrene containing about 20 g of benzoyl peroxide. A suspension forming agent like gelatin or sodium poly(methacrylate) is added. The mixture is vigorously stirred with heating at about 80°C. Polymerization proceeds at this temperature and eventually the finely divided droplets which are suspended in the aqueous phase solidify into round beads of polystyrene. It is necessary that the agitation be effective until the beads attain a solid consistency, otherwise the droplets tend to fuse together. At the end of the reaction, the beads of poly(styrene) are filtered, washed with a large quantity of water and dried.

Crystalline poly(styrene) can be prepared by the polymerization of styrene over a titanium containing organometallic catalyst (titanium tetrachloride + lithium aluminium tetradecyl). The reaction proceeds slowly. The product formed has properties different from all other kinds of poly(styrene).

Properties

Poly(styrene) is a linear polymer and is amorphous. Isotactic polymers can also be prepared as described above. The latter are high melting but brittle and the processing conditions are more difficult for them. Hence, they are not of any commercial importance. Poly(styrene) is resistant to acids and alkalis, but is attacked by several solvents. It can be injection-moulded, but has poor heat resistance. It can be used for electrical insulation and has low dielectric loss factor at moderate frequencies. It has good optical properties, the refractive index being high (1.60).

Applications

One of the main defects of poly(styrene) is its poor mechanical properties. This defect can be removed by copolymerizing styrene with other monomers like butadiene (SBR rubber) or acrylonitrile (Buna N rubber) or blending it with other polymers. The optical properties can be improved by the addition of ultraviolet light absorbers. Poly(styrene) can be easily nitrated and sulphonated. This makes it possible to use it in ion- exchange resins. Polymer foams of styrene are used for heat insulation.

3.11.3 Poly(vinyl chloride) $+CH_2 - CH+_n$
$$\begin{array}{c} | \\ Cl \end{array}$$

The monomer vinyl chloride ($CH_2 = CH.Cl$) can be produced by the reaction of dry hydrochloric acid and pure dry acetylene in the presence of mercuric chloride (on charcoal) catalyst.

$$CH \equiv CH + HCl \longrightarrow CH_2 = CHCl$$
(Acetylene)   (Vinyl chloride)

It can be produced by dehydrohalogenation of ethylene dichloride.

$$CH_2 = CH_2 + Cl_2 \longrightarrow \quad ClCH_2 CH_2 Cl$$
(Ethylene)   (1, 2-Dichloroethane)

$$\begin{array}{ccc} CH_2 - & CH_2 & \xrightarrow{\ -\ HCl\ } CH_2 = CH \\ | & | & | \\ Cl & Cl & Cl \end{array}$$
(1,2 – Dichloroethane)   (Vinyl chloride)

Polymerization

The polymer is usually produced by suspension polymerization. In one formulation 100 parts by weight of monomer (acetylene free), 100 parts of water (carefully purified), 4 parts of soap and 0.4 parts of 40% hydrogen peroxide are used. The soap is dissolved in water and added to the autoclave, followed by the addition of monomer and catalyst. No other modifying reagents are added, but 1–2% urea could be added to the polymer after manufacture. Polymerization is done in closed rotating autoclaves. Reaction takes four hours to start at, temperatures between 40 and 50°. In the initial stages of the reaction, cooling is needed. At the end of the reaction period (usually 24 hours), the pressure drops and 10% of the vinyl chloride monomer blows out as waste when the autoclave is opened; yield around 90%. The product is spray dried without washing in spray driers. The polymer is obtained as a fine powder.

Properties

Poly(vinyl chloride) is a tough and horny material and does not lend itself to handling as a plastic material. When, however, it is mixed with a plasticizer such as tricresyl phosphate, dibutyl phthalate or dioctyl phthalate, it can be extruded or calendered (see Chapter 10). The polymer has been shown to have a head-to-tail structure:

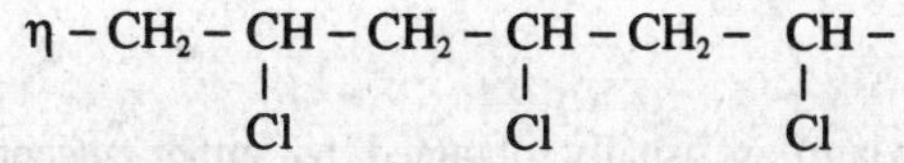

$$\eta - CH_2 - CH - CH_2 - CH - CH_2 - CH -$$
$$\qquad\qquad | \qquad\qquad | \qquad\qquad |$$
$$\qquad\qquad Cl \qquad\quad Cl \qquad\quad Cl$$

Poly (vinyl chloride) is atactic and has very little crystallinity. It is susceptible to dehydrohalogenation (see Chapter 11) in the presence of base, light or heat. Some of the physical properties of the plasticized and unplasticized polymer are given in Table 3.8.

Table 3.8 Physical Properties of Poly(vinyl chloride)

Property	Unplasticized PVC	Plasticized PVC
Density	1.4	1.3
Dielectric constant (60 C.p.s.)	3.2	6.9
Tensile strength	9000	2600
% Elongation	5–25	340

The molecular weight of PVC ranges between 30,000 and 80,000.

Applications

PVC has found use as a covering for wires and cables. This purpose is usually achieved by extrusion. Sheets for rain-coats and other purposes are obtained by calendering. Copolymers of vinyl chloride with vinyl acetate and vinylidine chloride find use in upholstery and fibers.

3.11.4 Poly (methyl methacrylate)
$$\left\{ CH_2 - \underset{\underset{CH_3}{|}}{\overset{\overset{COOCH_3}{|}}{C}} \right\}_n$$

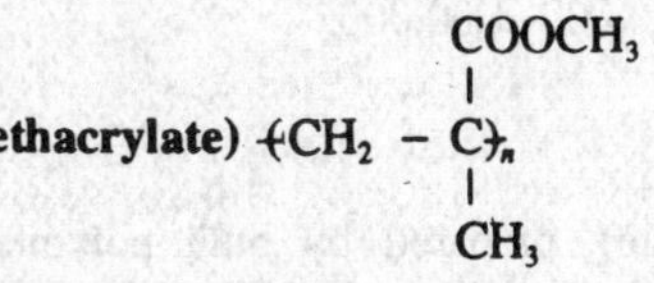

The monomer, methyl methacrylate is obtained from acetone cyanhydrin by reacting it with sulphuric acid and methyl alcohol.

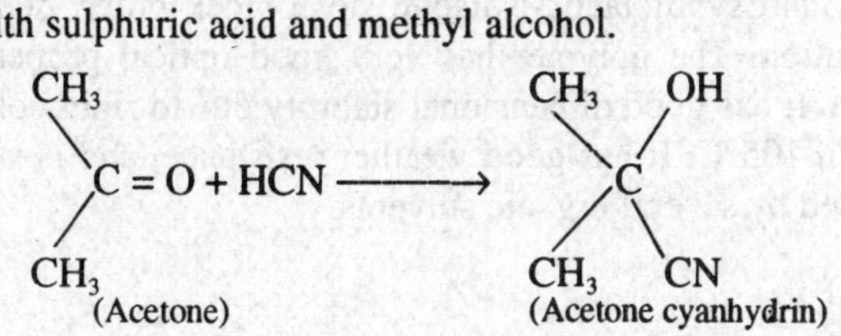

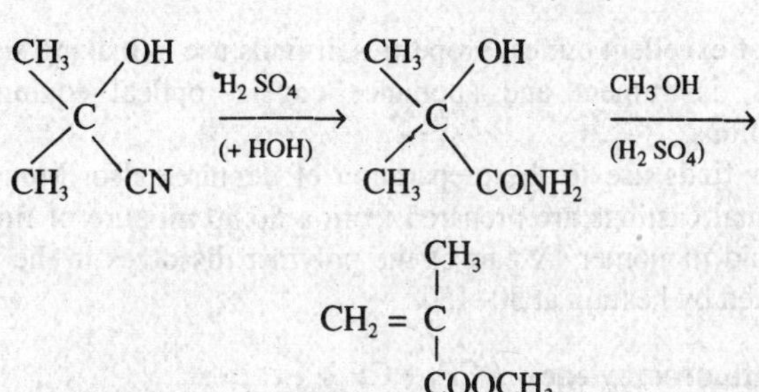

$$\underset{CH_3}{\overset{CH_3}{C}} = O + HCN \longrightarrow \underset{\underset{CH_3 \quad CN}{}}{\overset{CH_3 \quad OH}{C}}$$
$$\text{(Acetone)} \qquad\qquad \text{(Acetone cyanhydrin)}$$

$$\underset{CH_3 \quad CN}{\overset{CH_3 \quad OH}{C}} \xrightarrow[\text{(+ HOH)}]{H_2SO_4} \underset{CH_3 \quad CONH_2}{\overset{CH_3 \quad OH}{C}} \xrightarrow[\text{(H}_2SO_4)]{CH_3 OH}$$

$$CH_2 = \underset{\underset{COOCH_3}{|}}{\overset{\overset{CH_3}{|}}{C}}$$
$$\text{(Methyl methacrylate)}$$

It is distilled to get a 99.9% pure product, removing dimethyl ether and methyl oxyisobutyrate present as impurities.

Polymerization

Poly(methyl methacrylate) is usually obtained by either suspension or bulk polymerization. On account of its transparency and good optical properties, sheets, rods and tubes are prepared by a casting process.

The glass casting cell is prepared as follows: Two heat resistant glass plates are carefully cleaned, washed with very dilute nitric acid, and then with distilled water. The plates are further washed with 1% stannous chloride solution and finally with distilled water. The plates are drained and then dried in an oven at 40°. The gasket for the cell is made from a flexible tubing of say 3/8th inch diameter. If rubber tubing is used, it must be covered by wrapping it with cellophane. The cell is made by placing the gasket along the edge of one dry glass plate, folding it back to leave a small gap, then placing the second glass plate on top and keeping the whole assembly together by means of four spring clips.

Benzoyl peroxide (5%, w/v) is dissolved in pure and dry methyl methacrylate monomer and the solution heated carefully on a water bath or in an air-oven at 80-85° for 15–30 minutes. This stage is critical and bubble formation should be avoided. The syrup, which should be just pourable, is then carefully added to the cell through the small gap left open, care being taken to avoid entrapping air bubbles. Finally, the gap is closed. The filled cell is then placed in an oven in an upright position and allowed to stand at a temperature of 40° for 24 hours. After cooling, the cell is dismantled and the sheet trimmed, annealed and covered with a paper adhesive to avoid scratching.

Poly(methyl methacrylate), for injection or extrusion moulding, is obtained by suspension polymerization. Typical ingredients of the recipe include monomer, water, benzoyl peroxide and a suspending agent. The polymer has a more uniform molecular weight, usually in the range 50,000–1,00,000.

Properties

Poly(methyl methacrylate), prepared by bulk polymerization, is amorphous (atactic) and has high molecular weight (upto 16^6). It can be obtained in the form of isotactic, syndiotactic, and syndiotactic-isotactic stereoblock order, depending upon the method of preparation. The polymer has very good optical properties, but has poor scratch resistance. It has good dimensional stability due to rigid polymer chains. Its glass temperature in 105°C. It has good weather resistance, and is stable to acids and alkalis. It is attacked by several organic solvents.

Applications

On account of excellent optical properties, it finds use in making signs, skylights, aircraft canopies, instrument and appliance covers, optical equipment, lighting fixtures and partitions.

The polymer finds use in the preparation of dentures also. Monomer-polymer mouldings in dental castings are prepared from a 50:50 mixture of finely powdered polymer and liquid monomer. A part of the polymer dissolves in the monomer and the slush is finished by heating at 80–150°.

3.11.5 Poly(tetrafluoroethylene), $\{CF_2 - CF_2\}$

The monomer is prepared by the following series of reactions:

$$CHCl_3 + 2HF \longrightarrow CHClF_2 + 2HCl$$
(Chloroform) \qquad (Chlorodifluoromethane)

$$2CHClF_2 \xrightarrow[\text{furnace}]{\text{pyrolysis}} CF_2 = CF_2 + 2HCl$$

(Tetrafluoro- ethylene)

It is a non-toxic gas, (b.p. $-76°C$).

Tetrafluoroethylene polymerizes to a wax-like polymer at elevated pressures in the presence of free radical initiators like persulphate or hydrogen peroxide in aqueous medium. Redox initiators have also been used.

Properties

Poly(tetrafluoroethylene) or teflon is a highly crystalline material (93–98%). It is a linear polymer with very little branching. The density of teflon is 2.30 g/cm^3 and its crystalline melting point is 327°C. According to indirect methods, its number average molecular weight is of the order of several millions. As C-F bond is very difficult to break, according to all available evidences, the structure of the polymer is:

$$- CF_2 - CF_2 - CF_2 - CF_2 - CF_2 -$$

Teflon is unaffected by corrosive chemicals and solvents. However, it is attacked by alkali metals, either molten or dissolved in liquid ammonia. It has cold flow, which enables it to be shaped in a two step process:

(i) The monomer is pressed cold into the desired shape, and
(ii) then it is sintered to a temperature of about 380°C (above the crystalline melting point).

Teflon, however, decomposes at elevated temperatures.

Applications

Teflon is very tough, has excellent electrical properties, is heat-resistant and has low friction coefficient. Ball bearings made of teflon are used where lubrication is not possible. It is used for insulation of motors, transformers, generators, coils and capacitors. It finds wide use in appliances where resistance to high temperatures, toughness and resistance to chemicals, oils and solvents is required.

3.11.6 Poly(vinyl acetate) and poly(vinyl alcohol)

3.11.6.1 *Poly(vinyl acetate)*, $+CH_2 - CH\cdot)_n$
$$\underset{OCOCH_3}{\mid}$$

The vinyl acetate monomer, is prepared by the vapour phase addition of acetic acid to acetylene in the presence of a catalyst.

$$HC \equiv CH + CH_3 COOH \longrightarrow CH_2 = CHOCOCH_3$$
(Acetylene) \qquad (Acetic acid) \qquad (Vinyl acetate)

Purified acetylene is blown through acetic acid, maintained at 60°C, to get vapour containing 85 volumes of acetylene to 15 volumes of acetic acid. The reaction is exothermic and the temperature has to be controlled. The vapours are heated to 170°C with super-heated steam and passed over a catalyst consisting of charcoal impregnated with zinc acetate and traces of mercury salts. The temperature is gradually increased from 180°C to 210°C. The product thus obtained is fractionated in columns to separate it from the low boiling products and acetic acid. The yield of vinyl acetate based on acetylene is 92 – 95%.

Polymerization

The bulk polymerization of vinyl acetate is difficult to control and leads to extensive branching both on the α-carbon atom of the acetic group and the tertiary carbon atom of the backbone chain as shown below.

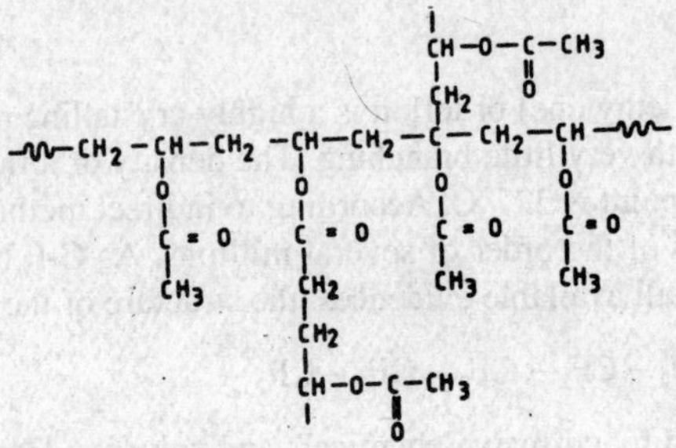

Branching in poly (vinyl acetate)

The preferred methods of polymerization are solution and emulsion polymerization.

Solution polymerization

Vinyl acetate is purified by distillation and dried by passing it through a silica gel column just before use. A reaction kettle is equipped with a stirring device, a reflux condenser and air inlet tube for passing nitrogen. The kettle is charged with 2 litres of vinyl acetate and 3 litres of dry benzene. The mixture is heated under reflux and a stream of nitrogen is passed through it. The polymerization is started by the addition of 2 g of azobisisobutyronitrile and allowed to proceed for about 2 hr under reflux. The unreacted monomer and the solvent are distilled off, the polymer broken, filtered and finally dried.

Properties and uses

Poly(vinyl acetate) is soft and tacky and cannot be easily moulded. As is to be expected, it is atactic and has head-to-tail structure. It has a density of 1.19 g/cm^3. Its glass transition temperature is 40°C and, thus, it can be easily deformed even at room temperature. It is combined with other polymers to make lacquers, adhesives and material for the manufacture of gramophone records. The co-polymers of vinyl acetate with vinyl chloride are of commercial importance.

3.11.6.2 *Poly(vinyl alcohol)*, $+CH_2 - CH+_n$
$$\underset{OH}{|}$$

Preparation

Poly(vinyl alcohol) is prepared by the alcoholysis of poly(vinyl acetate).

$$+CH_2 - CH+_n \longrightarrow + nCH_3 OH \longrightarrow +CH_2 - CH+_n + nCH_3 COOCH_3$$
$$\underset{\underset{\text{Poly(vinyl acetate)}}{OCOCH_3}}{|} \qquad\qquad\qquad \underset{\underset{\text{Poly(vinyl alcohol)}}{OH}}{|}$$

Commercial grade poly(vinyl acetate) (PVA) is dissolved in dry methanol by refluxing it. In the presence of added potassium hydroxide, *trans*-esterification takes place and poly(vinyl alcohol) precipitates from the solution. It is filtered and washed with several small portions of methanol and air-dried.

Properties

Poly(vinyl alcohol) is atactic and has head-to-tail structure. It is crystalline and has similar crystal lattice as ethylene, with the repeat distance of 2.57°A. The presence of crystal lattices in an atactic molecule is perhaps due to the small size of the hydroxyl groups, which can fit into the lattice. Poly(vinyl alcohol) does not melt to a thermoplastic, but loses water through condensation of two adjacent hydroxyl groups when heated above 150°C.

Applications

(i) PVOH is water soluble and finds use as a thickening agent in the emulsion polymerization of several monomers. It is also used as a packaging film where water solubility is important.

(ii) A second use of PVOH depends on making it insoluble. This is done by wet spinning of PVOH in a concentrated solution of sodium sulphate containing formaldehyde and sulphuric acid. Some aldehydic groups are thus formed and a few cross-links are introduced. This modified PVOH has higher water absorption than other fibers and is used for replacing cotton in clothing.

(iii) A third use of PVOH is the manufacture of safety glass. A sheet of poly(vinyl butyral), 0.02 – 0.05 cm thick, is placed between two sheets of glass. This is sealed by pressure and further consolidated by autoclaving the laminated sheets by the application of heat and pressure.

Poly(vinyl butyral) is made by condensing poly(vinyl alcohol) with butyraldehyde in the presence of sulphuric acid as a catalyst.

$$+CH_2 - \underset{\underset{OH}{|}}{CH} - CH_2 - \underset{\underset{OH}{|}}{CH}+_n \; + \; nC_3H_7CHO \longrightarrow$$

Poly (vinyl alcohol) (Butyraldehyde)

$$+CH_2 - \underset{\underset{O}{|}}{CH} - CH_2 - \underset{\underset{O}{|}}{CH}+_n \; + \; nH_2O$$
$$\underset{\underset{C_3H_7}{CH}}{\diagdown\;\diagup}$$

Poly (vinyl butyral)

3.11.7 Poly (acrylonitrile), $+CH_2 - CH +_n$
$$\underset{CN}{|}$$

The monomer acrylonitrile $(CH_2 = CHCN)$ can be prepared by two methods:

(1) Addition of hydrocyanic acid to acetylene in the presence of a catalyst, usually cuprous chloride.

$$HC \equiv CH + HCN \xrightarrow[H_2O]{Cu_2Cl_2} \underset{\text{(Acrylonitrile)}}{CH_2 = CHCN}$$

(2) Catalytic dehydration of ethylene cyanohydrin.

$$\underset{\underset{\text{(Ethylene oxide)}}{O}}{CH_2 - CH_2} + HCN \longrightarrow \underset{\text{(Ethylene \quad cyanohydrin)}}{HOCH_2 - CH_2 \, CN} \xrightarrow[\text{dehydration}]{\text{catalytic}} \underset{\text{(Acrylonitrile)}}{CH_2 = CHCN}$$

The dehydration can be done in either liquid or vapour phase.

Polymerization

Acrylonitrile is soluble in water to the extent of 7.5% at room temperature. It can be polymerized in aqueous solution in the presence of redox initiators (see page 41) when it precipitates out as an insoluble powder. However, it is necessary to exclude oxygen from the system as it is an inhibitor of acrylonitrile polymerization and leads to long induction periods.

Properties and reactions

Poly (acrylonitrile) is insoluble in most of the common organic solvents. It is soluble in dimethyl formamide and tetramethylene sulphone. Both dry and wet spinning are practised with these solvents {for poly(acrylonitrile) fibers see Section 9.6.1}. On heating poly (acrylonitrile) above 20°C it becomes red and finally at 350°C very dark.

The action of heat on poly (acrylonitrile) is a complex reaction (see Scheme 11.6). A special rubber known as Buna-N (GR-N or perbunan) is obtained by copolymerization of 25 parts acrylonitrile and 75 parts butadiene. This rubber is resistant to oils and solvents and has better abrasion resistance than natural rubber (see Section 8.3.4).

3.11.8 Poly (propylene), $+CH_2 - CH +_n$
$$\underset{CH_3}{|}$$

The monomer is obtained from the cracking of petroleum and as a by-product in oil refineries.

Polymerization

Isotactic poly(propylene) can be obtained by the polymerization of propylene in the presence of Ziegler-Natta catalyst (aluminium trialkyl and titanium tetrachloride) in the same manner as described for the preparation of high density poly (ethylene) (see Section 3.11.1).

Properties

Poly (propylene) can be obtained in isotactic, syndiotactic and atactic forms. Isotactic poly(propylene) is highly crystalline and has a crystalline m.p. of 176°C and density of 0.92. The mechanical properties of poly (propylene) depend upon the degree of crystallinity, molecular weight and molecular weight distribution. The stereochemistry of poly(propylene) depend upon the degree of crystallinity, molecular weight and molecular weight distribution. The stereochemistry of poly (propylene) has been discussed in detail elsewhere (see Section 5.5.2). Sections of poly(propylene) (atactic and isotactic chains) have also been illustrated in Fig. 5.4.

High crystallinity imparts a high tensile strength, stiffness and hardness to isotactic poly (propylene). The high melting point enables it to be sterilized. It has excellent electrical properties, inertness to chemicals and moisture resistance. However, at low temperatures, its impact strength is impaired.

Applications

It can be injection-moulded into automotive parts. Another important use is in the manufacture of filaments (ropes etc.) and staple for carpenting. Poly(propylene) compares favourably with nylon in its fiber properties.

3.12 ION EXCHANGE RESINS

If poly(styrene) is cross-linked by a suitable method and then sulphonated, it would merely swell in water but will not dissolve. Let us represent such a resin by,

$$R - \overset{\overset{\displaystyle O}{\|}}{\underset{\underset{\displaystyle O}{\|}}{S}} - O^- H^+$$

If we pack this resin in a column and pass a solution of caustic soda through it, the H^+ ions will be gradually replaced by Na^+ ions, until all the negatively charged positions in RSO_2^- are completely neutralized. If we now add to the top of the column a solution of Ba $(OH)_2$, the Ba^{2+} ions will tend to displace the Na^+ ions. NaOH will be liberated at the bottom of the column. This method can, therefore, be used for the removal of ions from solution. The oldest and the most established use of these cation-exchange resins is in the softening of water by the removal of Ca^{2+} ions. The presence of the latter ions leads to the precipitation of calcium soap from ordinary soap. Here a cationic resin, containing sodium is used. When the hard water enters the column, the Ca^{2+} ions displace the Na^+ ions. When the column is completely saturated with calcium, these ions can be displaced by passing a strong solution of sodium chloride until the column regains its original composition.

Anionic exchange resins can similarly be formed by substituting the cross-linked poly(styrene) resin with basic groups such as NH_2. These resins find a variety of uses including removal of copper ions from factory effluents, and purification of blood so that it does not clot.

A polymer of styrene and divinylbenzene (approx. 5% of DVB) in the form of beads is sulphonated to the extent of 1–1.77 sulphite groups per ring to get a cation exchange resin (Dowex). This is used for water softening. The structure of Dowex is shown in Fig. 3.9.

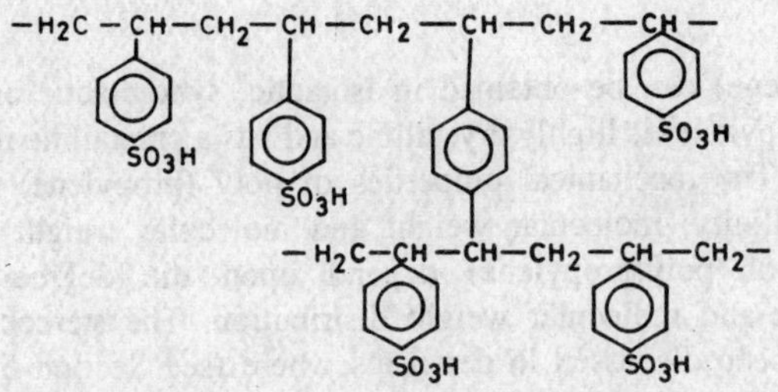

Fig. 3.9 A cation exchange resin.

To obtain an anion exchange resin, the beads of cross-linked styrene are chloro-methylated with chloromethyl methyl ether. These are then treated with a tertiary amine to get a quaternary ammonium salt.

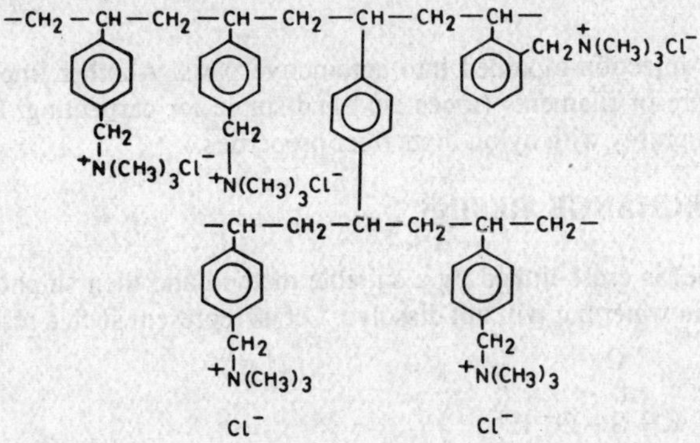

Fig. 3.10 Anion exchange resin.

Other ion exchange resins based on phenol-formaldehyde and melamine-formaldehyde resins are also known and used.

3.13 CONDENSATION POLYMERIZATION VERSUS ADDITION POLYMERIZATION

Condensation polymerization, also called a step reaction, has the features of a small reaction. Here two monomer units, which must be at least bi-functional, react together to give a dimer. A dimer then reacts with another monomer molecule or a dimer to produce a trimer or a tetramer. Thus, a high molecular weight compound is gradually built up. Monomer disappears much faster in condensation polymerization as dimers, trimers and tetramers are formed at the same time. The molecular weight increases in the course of the reaction and high molecular weight polymer is obtained towards the end of the reaction. Long reaction time are, therefore, necessary for high percentage conversions and high molecular weights.

On the other hand, addition polymerization, initiated by a free radical, cation or an anion is essentially a chain type reaction. The reaction proceeds rapidly and high molecular weight polymers are produced from the very beginning of the reaction. The reaction consists of three discrete steps, viz., initiation, propagation and termination. The concentration of the monomer decreases throughout the course of the reaction. At any particular instant, the reaction mixture consists of the high polymer, the growing chains and the unchanged monomer.

ANSWERS TO QUESTIONS

A.3.1 $+CH_2 - CH_2 +_n + 3nO_2 \longrightarrow 2nCO_2 + 2nH_2O$ •

10 g of poly(ethylene) is $10/28 = 0.357$ mole repeating unit of poly(ethylene) which gives twice the number of moles of CO_2 or 0.714 mole

A.3.2. (i) Repeating unit:
$$-CH_2 - \overset{\displaystyle CH_3}{\underset{\displaystyle CH_3}{\overset{|}{\underset{|}{C}}}} -$$

Monomer:
$$CH_2 = \overset{\displaystyle CH_3}{\underset{\displaystyle CH_3}{\overset{|}{\underset{|}{C}}}}$$

(ii) Repeating unit:
$$-CH_2 - \overset{\displaystyle }{\underset{\displaystyle CN}{\overset{}{\underset{|}{CH}}}} -$$

Monomer:
$$CH_2 = \overset{\displaystyle }{\underset{\displaystyle CN}{\overset{}{\underset{|}{CH}}}}$$

(iii) Repeating unit:
$$-CH_2 - \overset{\displaystyle CH_3}{\underset{\displaystyle COOCH_3}{\overset{|}{\underset{|}{C}}}} -$$

Monomer:
$$CH_2 = \overset{\displaystyle CH_3}{\underset{\displaystyle COOCH_3}{\overset{|}{\underset{|}{C}}}}$$

(iv) Repeating unit:
$$-\overset{\cdot}{C}H_2 - \overset{\displaystyle }{\underset{\displaystyle CH_3}{\overset{}{\underset{|}{C}}}} = CH - CH_2 -$$

Monomer:
$$CH_2 = \overset{\displaystyle }{\underset{\displaystyle CH_3}{\overset{}{\underset{|}{C}}}} - CH = CH_2$$

A.3.3. (i)

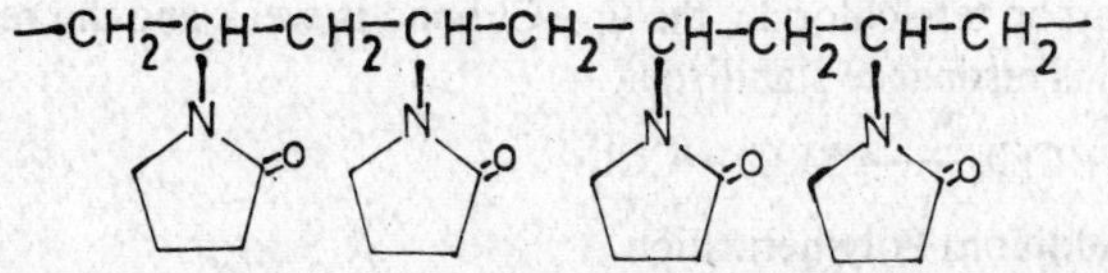

(ii) (a) $- CH_2 - CH = CHCH_2 - CH_2 - CH = CHCH_2 - CH_2 - CH =$
(Poly(1, 4-butadiene)
$$CH-CH_2 - CH_2CH = CHCH_2 -$$

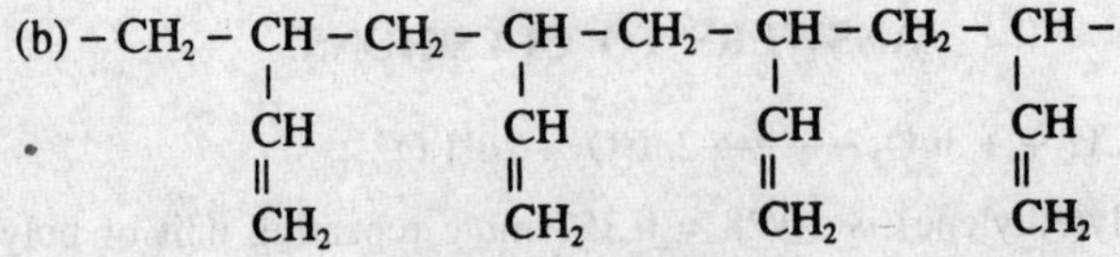

(b) $- CH_2 - CH - CH_2 - CH - CH_2 - CH - CH_2 - CH -$
$\qquad\qquad |\qquad\qquad |\qquad\qquad |\qquad\qquad |$
$\qquad\qquad CH\qquad\quad CH\qquad\quad CH\qquad\quad CH$
$\qquad\qquad \parallel\qquad\qquad \parallel\qquad\qquad \parallel\qquad\qquad \parallel$
$\qquad\qquad CH_2\qquad\quad CH_2\qquad\quad CH_2\qquad\quad CH_2$

(Poly(1, 2-butadiene))

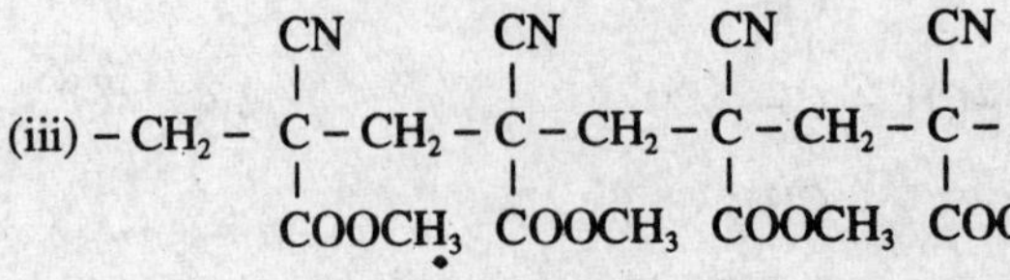

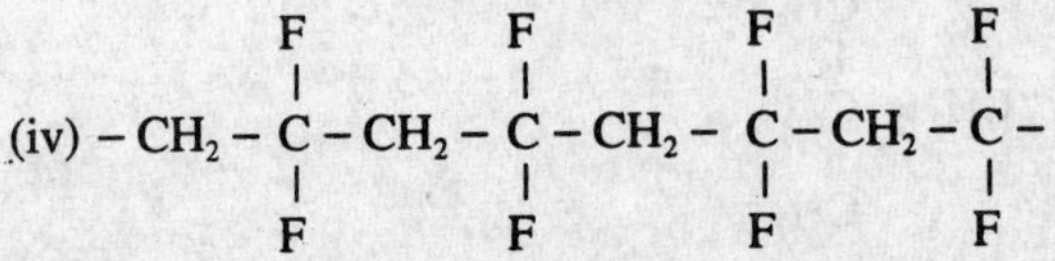

$\qquad\quad CN\qquad\quad CN\qquad\quad CN\qquad\quad CN$
$\qquad\quad |\qquad\qquad |\qquad\qquad |\qquad\qquad |$
(iii) $- CH_2 - C - CH_2 - C - CH_2 - C - CH_2 - C -$
$\qquad\qquad |\qquad\qquad |\qquad\qquad |\qquad\qquad |$
$\qquad\quad COOCH_3\ COOCH_3\ COOCH_3\ COOCH_3$

$\qquad\quad F\qquad\quad F\qquad\quad F\qquad\quad F$
$\qquad\quad |\qquad\qquad |\qquad\qquad |\qquad\qquad |$
(iv) $- CH_2 - C - CH_2 - C - CH_2 - C - CH_2 - C -$
$\qquad\qquad |\qquad\qquad |\qquad\qquad |\qquad\qquad |$
$\qquad\quad F\qquad\quad F\qquad\quad F\qquad\quad F$

A.3.4. (i) Cationic, (ii) Free radical, (iii) Anionic, (iv) Free radical.

A.3.5. (i) $\overline{S}O_3 - O - O - \overline{S}O_3 + Fe^{2+} \longrightarrow \dot{S}O_4^- + SO_4^{2-} + Fe^{3+}$

(ii) $\overline{S}O_3 - O - O - \overline{S}O_3 + S_2 O_3^{2-} \longrightarrow SO_4^{2-} + \overline{S}O\dot{\,}_4^- + S_2O_3^-$

(iii) $R - CH_2 - OH + Ce^{4+} \longrightarrow R - \dot{C}H - OH + Ce^{3+} + H^+$

(iv) $(CH_3)_3 COOH + Ce^{4+} \longrightarrow (CH_3)_3\dot{C}O + OH^- + Ce^{3+} + H^+$

A.3.6 (i) In toluene, ethylbenzene and isopropylbenzene, weak benzylic hydrogen $(C - H)$ is present, which leads to resonance stabilization of the resultant radical. This is absent in benzene and *t*-butylbenzene.

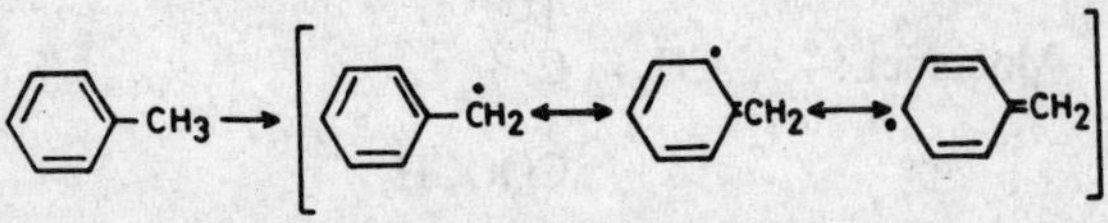

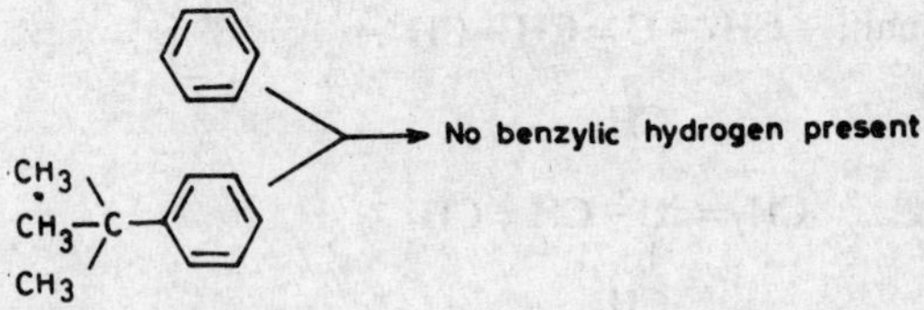

(ii) In carbon tetrachloride, the $C - Cl$ bond is weak and the resultant radical CCl_3 is resonance stabilized.

$M_n^{\cdot} + CCl_4 \longrightarrow M_nCl + CCl_3^{\cdot}$

Chain (Addition) Polymerization

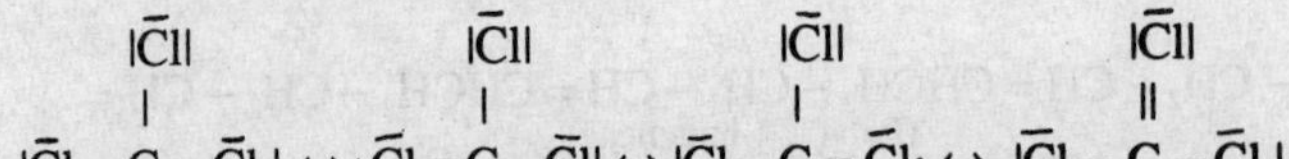

(iii) In disulphide also the C–S bond is weak and can easily break,

$$M_n{}^\cdot + RSSH \longrightarrow M_n\,SH + RS^\cdot$$

A.3.7. (i) Diphenyl picryl hydrazyl (DPPH) is a very stable free radical and can react with another free radical instantaneously. Hence, it is also called a radical scavenger.

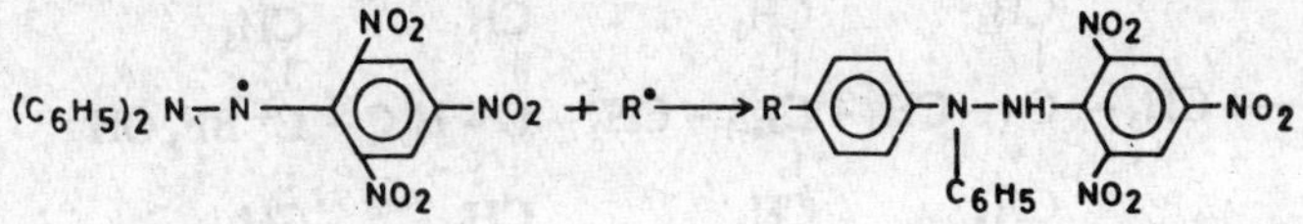

(ii) See under Section 3.7.

(iii) See under Section 3.7.

A.3.8. *Initiation*

$$\underset{\text{(Catalyst)}}{BF_3} + \underset{\text{(Co–catalyst)}}{HOH} \longrightarrow H^+\,BF_3\,OH^-$$

$$BF_3OH^-\,H^+ + CH_2 = \underset{\underset{CH_3}{|}}{\overset{\overset{CH_3}{|}}{C}} \longrightarrow CH_3 - \underset{\underset{CH_3}{|}}{\overset{\overset{CH_3}{|}}{C^+}} + BF_3\,OH^-$$

Propagation

$$CH_3 - \underset{\underset{CH_3}{|}}{\overset{\overset{CH_3}{|}}{C^+}}\,BF_3\,OH^- + n\,CH_2 = C\,(CH_3)_2 \longrightarrow$$

$$CH_3 - \underset{\underset{CH_3}{|}}{\overset{\overset{CH_3}{|}}{C}} \left(CH_2 - \underset{\underset{CH_3}{|}}{\overset{\overset{CH_3}{|}}{C}} \right)_{n}\,CH_2 - \underset{\underset{CH_3}{|}}{\overset{\overset{CH_3}{|}}{C^+}}\,(BF_3\,OH^-)$$

Termination

(i) *Kinetic termination*

$$CH_3 - \underset{\underset{CH_3}{|}}{\overset{\overset{CH_3}{|}}{C}} \left(CH_2 - \underset{\underset{CH_3}{|}}{\overset{\overset{CH_3}{|}}{C}} \right)_{n} - CH_2 - \underset{\underset{CH_3}{|}}{\overset{\overset{CH_3}{|}}{C^+}}\,BF_3\,OH^- \longrightarrow$$

$$CH_3 - \underset{\underset{CH_3}{|}}{\overset{\overset{CH_3}{|}}{C}} \left(CH_2 - \underset{\underset{CH_3}{|}}{\overset{\overset{CH_3}{|}}{C}} \right)_{n-1} - CH_2 - \underset{\underset{CH_3}{|}}{\overset{\overset{CH_2}{||}}{C}} + H^+\,BF_3\,OH^-$$

(ii) *Monomer termination*

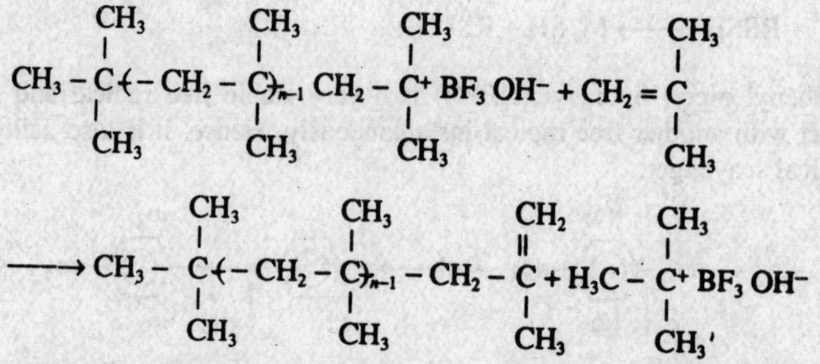

$$CH_3 - \underset{\underset{CH_3}{|}}{\overset{\overset{CH_3}{|}}{C}} \{ - CH_2 - \underset{\underset{CH_3}{|}}{\overset{\overset{CH_3}{|}}{C}} \}_{n-1} CH_2 - \underset{\underset{CH_3}{|}}{\overset{\overset{CH_3}{|}}{C}}{}^+ BF_3 OH^- + CH_2 = \underset{\underset{CH_3}{|}}{\overset{\overset{CH_3}{|}}{C}}$$

$$\longrightarrow CH_3 - \underset{\underset{CH_3}{|}}{\overset{\overset{CH_3}{|}}{C}} \{ - CH_2 - \underset{\underset{CH_3}{|}}{\overset{\overset{CH_3}{|}}{C}} \}_{n-1} - CH_2 - \underset{\underset{CH_3}{|}}{\overset{\overset{CH_2}{\|}}{C}} + H_3C - \underset{\underset{CH_3'}{|}}{\overset{\overset{CH_3}{|}}{C}}{}^+ BF_3 OH^-$$

A.3.9 (i) Anionic, (ii) Anionic, (iii) Cationic, (iv) Coordination, (v) Free radical, (vi) Free radical.

CHAPTER 4

COPOLYMERIZATION

4.1 INTRODUCTION

In homogeneous chain polymerization of only one monomer, one gets a product containing one type of repeating unit only. It is possible to prepare copolymers containing two or more different monomer units by chain-polymerization processes. There are various types of co-polymers as indicated in Chapter one (Section 1.5).

4.2 CO-POLYMERIZATION COMPOSITION EQUATION

The composition of the co-polymer in most instances is found to be different from that of the co-monomer feed from which it is produced. In other words, different monomers have differing tendencies towards undergoing co-polymerization. Monomers M_1 and M_2 can each add either to a propagating chain ending in M_1 or M_2.

$$M_1{}^* + M_1 \xrightarrow{\ k_{11}\ } M_1{}^* \tag{4.1}$$

$$M_1{}^* + M_2 \xrightarrow{\ k_{12}\ } M_2{}^* \tag{4.2}$$

$$M_2{}^* + M_2 \xrightarrow{\ k_{22}\ } M_2{}^* \tag{4.3}$$

$$M_2{}^* + M_1 \xrightarrow{\ k_{21}\ } M_1{}^* \tag{4.4}$$

where k_{11} is the rate constant for the addition of an $M_1{}^*$ radical to M_1, k_{12} is the rate constant for the addition of an $M_1{}^*$ radical to M_2, k_{21} is the rate constant for the addition of an $M_2{}^*$ radical of M_1, and k_{22} is the rate constant for the addition of an $M_2{}^*$ radical to M_2.

The rate of disappearance of monomer M_1, neglecting any loss in the initiation process, is given by Eq. 4.5.

$$\frac{dM_1}{dt} = k_{11}\,[M_1{}^*]\,[M_1] + k_{21}\,[M_2{}^*]\,[M_1] \tag{4.5}$$

$$\frac{dM_2}{dt} = k_{12}\,[M_1{}^*]\,[M_2] + k_{22}\,[M_2{}^*]\,[M_2] \tag{4.6}$$

Dividing Eq. (4.5) by Eq. (4.6), we get

$$\frac{dM_1}{dM_2} = \frac{k_{11}[M_1^*][M_1] + k_{12}[M_2^*][M_1]}{k_{12}[M_1^*][M_2] + k_{22}[M_2^*][M_2]} \tag{4.7}$$

Assuming steady state approximation, we get

$$k_{21}[M_2^*][M_1] = k_{12}[M_1^*][M_2]. \tag{4.8}$$

and defining reactivity ratios as

$$r_1 = \frac{k_{11}}{k_{12}} \quad \text{and} \quad r_2 = \frac{k_{22}}{k_{21}} \tag{4.9}$$

we get,

$$\frac{dM_1}{dM_2} = \frac{[M_1]}{[M_2]} \left[\frac{k_{11}[M_1^*] + k_{21}[M_2^*]}{k_{12}[M_1^*] + k_{22}[M_2^*]} \right] \tag{4.10}$$

Multiplying throughout by $\dfrac{1}{k_{21}} \times \dfrac{[M_2]}{[M_2^*]}$, we get

$$\frac{dM_1}{dM_2} = \frac{[M_1]}{[M_2]} \left[\frac{\dfrac{k_{11}}{k_{21}} \dfrac{k_{11}([M_1^*][M_2])}{[M_2^*]} + [M_2]}{\dfrac{k_{12}}{k_{21}} \dfrac{([M_1^*][M_2])}{[M_2^*]} + \dfrac{k_{22}}{k_{21}}[M_2]} \right]$$

Since $\quad [M_1] = \dfrac{k_{12}}{k_{21}} \dfrac{[M_1^*][M_2]}{[M_2^*]}$

Hence, $\quad \dfrac{dM_1}{dM_2} = \dfrac{[M_1]}{[M_2]} \left[\dfrac{r_1[M_1] + [M_2]}{r_2[M_2] + [M_1]} \right] \tag{4.11}$

Equation (4.11) is called the co-polymer composition equation. It is not necessary to assume steady state approximation as Eq. (4.11) has been obtained by a statistical method also. This equation can be expressed in terms of mole fraction also in place of concentration. If f_1 and f_2 are the mole fractions of the monomers M_1 and M_2 in the feed, and F_1 and F_2 are the mole fractions in the copolymer formed,

$$f_1 = 1 - f_2 = \frac{[M_1]}{[M_1] + [M_2]} \tag{4.12}$$

and

$$F_1 = 1 - F_2 = \frac{d[M_1]}{d[M_1] + d[M_2]}. \tag{4.13}$$

and

$$F_1 = \frac{r_1 f_2^2 + f_1 f_2}{r_1 f_1^2 + 2f_1 f_2 + r_2 f_2^2} \tag{4.14}$$

Hence,

$$\frac{f_1(1 - 2F_1)}{F_1(1 - f_1)} = r_2 + r_1 \frac{f_1^2(F_1 - 1)}{F_1(1 - f_2)^2} \tag{4.15}$$

Here r_2 is the intercept and r_1 is the slope of the plot between the left hand side of Eq. 4.15 and $f_1^2 (F_1 - 1)/F_1 (1 - f_2)^2$.

Q.4.1. Using the co-polymer composition equation, calculate the composition of the co-polymer that would be formed at low conversion for an equimolar mixture of monomers. The following data are provided.

	M_1	M_2	r_1	r_2
1.	Styrene	Maleic anhydride	0.05	0.0
2.	Styrene	Vinylidine chloride	2.0	0.1
3.	Styrene	Vinyl methyl ketone	0.29	0.35
4.	Acrylonitrile	Styrene	0.05	0.40
5.	Acrylonitrile	Vinyl acetate	5.0	0.10
6.	Acrylonitrile	Vinyl chloride	3.3	0.04

4.3. APPLICATIONS OF CO-POLYMER COMPOSITION EQUATION

The terms r_1 and r_2 are called monomer reactivity ratios. If r_1 is greater than unity, it means that the tendency of M_1^* is to add preferentially to M_1 instead of M_2. If r_1 has a value of less than unity, it means that M_1^* adds to M_2 preferentially. If $r_1 = 0$ it shows that M_1^* is incapable of undergoing homopolymerization independently and prefers to alternate with another monomer. If $r_1 = r_2 = 1$, the composition of the co-polymer is preferably random. If $r_1 r_2 = 1$, it is considered to be an ideal system. If $r_1 = r_2 = 0$, an alternating co-polymer whose composition will be 1 : 1 is formed. Most monomer pairs polymerize when $0 < r_1 r_2 < 1$.

The co-polymerization composition equation has been verified in innumerable cases. It is equally applicable to anionic, cationic and free-radical polymerizations although the values of r_1 and r_2 can be drastically different, depending on the mode of initiation. Fig. 4.1 illustrate this.

Fig. 4.2 shows the variation in co-polymer composition as a function of the co-monomer feed composition for different values of r_1. The term ideal co-polymerization was introduced to show the analogy between the curves in Fig. 4.2 and those for vapour-liquid equilibria in ideal liquid mixtures. The copolymer is richer in M_1 when $r_1 > 1$ and is poorer in M_1 when $r_1 < 1$. The term ideal copolymerization does not mean a desirable type of co-polymerization. A very important practical consequence of ideal co-polymerizations is that it becomes progressively more difficult to produce co-polymers containing appreciable amounts of both the monomers as the difference between the reactivities of the two monomers increases.

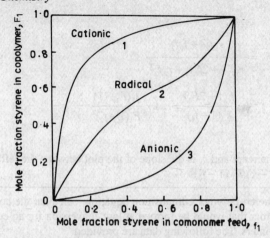

Fig. 4.1 Dependence of the instantaneous co-polymer composition F_1 on the co-monomer feed composition f_1 for styrene-methyl methacrylate in cationic (curve 1) radical (curve 2), and anionic (curve 3) copolymerizations initiated by $SnCl_4$, benzoyl peroxide, adn Na/liquid NH_3, respectively.

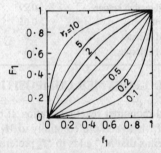

Fig. 4.2. Plot of mole fraction F_1 of co-monomer 1 in the composition as a fraction of the mole fraction f_1 of composition 1 in the feed for co-polymerization in which $r_1r_2 = 1$ showing compositional variations for several indicated values of r_1 (co-monomer 1 refers to a uniform co-polymer having (A) same reactivity ratio, (B) low conversion, or (C) constant co-monomer composition).

The perfectly random case is illustrated by the co-polymerization of ethylene and vinyl acetate (curve I). The ideal behaviour in co-polymerization (i.e., when $r_1r_2 = 1$) has already been discussed above. When the product $r_1r_2 = 1$ approximately (1.17 in curve II), the co-polymerization is almost ideal. Curve III in Fig. 4.3 is indicative of what happens with the alternating system. Curve IV illustrates another aspect of radical co-polymerization. The point P where the curve IV crosses the perfectly random case line is referred to as azeotropic. This term is borrowed from azeotropic distillation. If, however, it is desired to determine r_1 and r_2 experimentally, the concentration of M_1 and M_2 in several polymerizations, which are taken to low conversions, the composition of the polymers formed is determined. Eq. (4.15) gives a method of determining r_1 and r_2.

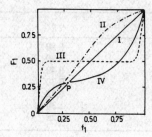

Fig. 4.3 Variations of F_1 with f_1 for co-polymerizations which are: (I) completely random, (II) almost ideal (i.e. $r_1r_2 = 1.17$) (III) regular alternating and (IV) intermediate between alternating and random (i.e., $0 < r_1r_2 < 1$).

Q.4.2. Using Eq. (4.14), calculate F_1 for different values of f_1. The following data are provided: M_1 = Styrene; M_2 = Butadiene; $r_1 = 0.78$; $r_2 = 1.39$ and $r_1r_2 = 1.08$. Construct the curve and show that the curve shows an ideal behaviour.

Q.4.3. Using Eq. (4.14) construct a curve between F_1 and f_1 from the following data: M_1 = Styrene; M_2 = Maleic anhydride; $r_1 = 0.05$; $r_2 = 0.0$ and $r_1r_2 = 0$. Show that the curve depicts an alternating behaviour in co-polymerization.

Q.4.4. Distinguish between the random, alternating and ideal behaviour in co-polymerization.

4.4. MONOMER AND RADICAL REACTIVITY ($Q - e$ scheme)

T. Alfrey Jr. and C.C. Price have put forward a relationship for the rate constant of reaction between a radical-monomer M_1 and another monomer (M_2) in the form,

$$k_{12} = P_1Q_2 \exp(-e_1e_2) \qquad (4.16)$$

where P_1 is a measure of resonance stabilization of monomer radical $M_1^{\cdot}$, Q_2 is similarly a measure of resonance stabilization of monomer M_2, e_1 is a measure of the polar properties of radical $M_1^{\cdot}$, and e_2 is a measure of the polar properties of monomer M_2.

If it is assumed that the same value of e applies to both the monomer and its radical, it is possible to write expressions for k_{11}, k_{22} and k_{21}. These can then be used to derive expressions for the reactivity ratios r_1 and r_2,

$$r_1 = \frac{Q_1}{Q_2} \exp[-e_1(e_1 - e_2)] \qquad (4.17)$$

$$r_2 = \frac{Q_2}{Q_1} \exp[-e_1(e_2 - e_1)] \qquad (4.18)$$

The $Q-e$ scheme as given above is at best an empirical approach to placing monomer reactivity on a quantitative basis.

The following values (Table 4.1) of Q and e have been calculated for a new monomers assuming $Q = 1$ and $e = -0.80$ for styrene.

Table 4.1 Q and e Values of Some Monomers

Monomer	e	Q
Styrene (reference)	−0.80	1
Acrylonitrile	1.20	0.60
Methyl Methacrylate	0.40	0.70
Vinyl chloride	0.20	0.044
Vinyl acetate	−0.22	0.026
Acrylamide	−1.30	1.18
Maleic anhydride	2.25	0.23

It is possible to draw a number of useful generalizations from the Q and e data provided in Table 4.1. In the cases where values are very different, co-polymerization between monomers proceeds very poorly. Co-polymerization between monomers having similar Q values, specially high, is more likely. Ideal co-polymerization is more likely between monomers having similar Q and e values. Alternating polymerization occurs readily between monomers having same Q values but e values of opposite signs.

4.5. BLOCK AND GRAFT CO-POLYMERS

The terms *block* and *graft* co-polymers have already been explained and illustrated in Chapter one (Section 1.5.1).

4.5.1 Synthesis of Graft Co-polymers

4.5.1.1 *By Chain-Transfer*

The following reaction sequence describes the chain-transfer which occurs when styrene is polymerized in presence of dissolved poly (acrylate).

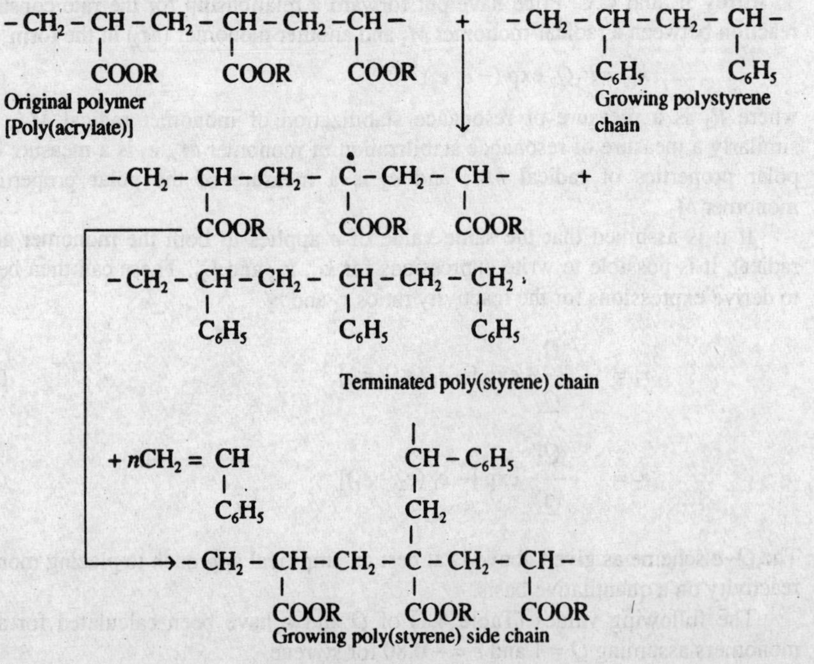

Scheme 4.1

4.5.1.2 *Incorporation of Peroxide Group*

In this method peroxide or azo-group is created in the chain. For example, poly(styrene) can be partially converted into poly(para-isopropyl styrene) by reaction with isopropyl chloride and anhydrous $AlCl_3$. The resultant polymer can be easily converted, with oxygen, to polymeric hydroperoxide, similar to the formation of cumene hydroperoxide. The resulting hydroperoxide decomposes (in the presence of Fe^{2+} as reductant of the redox pair) into radicals which can initiate the polymerization of a monomer, for example methyl methacrylate, to produce a graft polymer.

4.5.1.3 *By Redox Initiation*

Poly(vinyl alcohol), when reacted with Ce^{4+}, gives rise to a free radical on which a homopolymer can be grafted.

$$-CH_2-\underset{\underset{OH}{|}}{CH}- + Ce^{4+} \longrightarrow -CH_2-\underset{\underset{OH}{|}}{\overset{\bullet}{C}}-$$

$$+ Ce^{3+} + H^+$$

Other hydroxyl containing polymers such as cellulose can be grafted with, say, acrylonitrile.

4.5.1.4 *By Anionic Initiation*

A polymer such as poly(1, 4-butadiene) can give rise to an active anionic radical by reaction with a catalyst like *t*-butyl lithium.

$$-CH_2-CH=CH-CH_2- \xrightarrow{\text{\textit{t}-Bu Li}} -\overset{-}{C}H-CH=CH-CH_2$$

4.5.1.5 *By Ethoxylation*

If one reacts ethylene oxide under high pressure and temperature with a polymer containing –NH groups in the chain, such as finely divided polyamide, graft polymerization takes place as expected (Scheme 4.2).

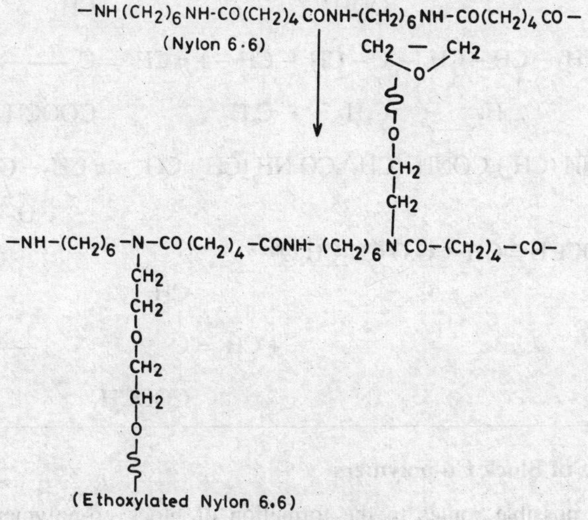

$$-NH(CH_2)_6\,NH\text{-}CO(CH_2)_4\,CONH\text{-}(CH_2)_6\,NH\text{-}CO(CH_2)_4\,CO-$$

(Nylon 6·6)

$$-NH-(CH_2)_6-\underset{\underset{\vdots}{|}}{N}-CO(CH_2)_4-CONH-(CH_2)_6\,\underset{\vdots}{N}\,CO-(CH_2)_4-CO-$$

(Ethoxylated Nylon 6.6)

Scheme 4.2

4.5.1.6 *Graft Co-polymers through Co-polymerization*

In this method, one dissolves an unsaturated polymer in the monomer to be grafted on and allows the monomer to polymerize. The double bond of the polymer is then incorporated into the growing chain of the polymerizing monomer. An important example is the graft polymerization of an unsaturated poly(ester) casting resin (Scheme 4.3).

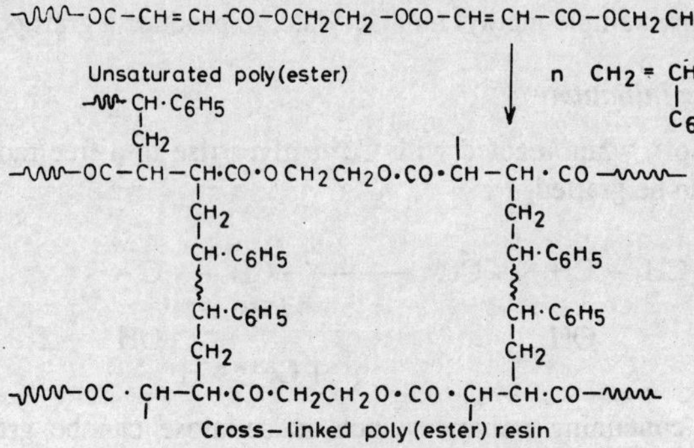

$$-\!\!\sqrt{\!\sqrt{}}\!-OC-CH=CH-CO-OCH_2CH_2-OCO-CH=CH-CO-OCH_2CH_2O-\!\!\sqrt{\!\sqrt{}}\!-$$

Unsaturated poly(ester)

$-\!\!\sqrt{\!\!\sqrt{}}\!\!-CH\cdot C_6H_5$
$\qquad\quad|$
$\qquad CH_2$

$\qquad\qquad\qquad\qquad\qquad\qquad\qquad n\ CH_2=CH$
$\qquad\qquad\qquad\qquad\qquad\qquad\qquad\qquad\qquad\quad |$
$\qquad\qquad\qquad\qquad\qquad\qquad\qquad\qquad\qquad\ C_6H_5$

$-\!\!\sqrt{\!\sqrt{}}\!-OC-CH-CH\cdot CO\cdot O\,CH_2CH_2O\cdot CO\cdot CH-CH\cdot CO-\!\!\sqrt{\!\sqrt{}}\!-$
$\qquad\qquad\qquad\qquad\ |\qquad\qquad\qquad\qquad\qquad\qquad\quad |$
$\qquad\qquad\qquad\quad CH_2\qquad\qquad\qquad\qquad\qquad\qquad CH_2$
$\qquad\qquad\qquad\quad CH\cdot C_6H_5\qquad\qquad\qquad\qquad\ CH\cdot C_6H_5$
$\qquad\qquad\qquad\quad CH\cdot C_6H_5\qquad\qquad\qquad\qquad\ CH\cdot C_6H_5$
$\qquad\qquad\qquad\quad CH_2\qquad\qquad\qquad\qquad\qquad\qquad CH_2$
$-\!\!\sqrt{\!\sqrt{}}\!-OC\cdot CH-CH\cdot CO\cdot CH_2CH_2O\cdot CO\cdot CH-CH\cdot CO-\!\!\sqrt{\!\sqrt{}}\!-$

Cross–linked poly(ester) resin

Scheme 4.3
The above is an example of cross-linked graft co-polymer

4.5.1.7 *Grafting by Irradiation*

Radiations such as X-rays, γ-rays, etc., are capable of producing radicals in organic molecules. If one irradiates the solution of a polymer in a monomer, one obtains the same type of reaction that occurs in chain-transfer, i.e., the polymer chain of the primary polymer radicals is formed. This then becomes the point of initiation for the side-chain.

Q.4.5. Formulate the following reactions:

(i)

$$
-CH_2-CH-CH_2-\overset{\displaystyle OOH}{\underset{\displaystyle C_6H_5}{\overset{|}{\underset{|}{C}}}}-CH_2-CH-+n\,CH_2=\overset{\displaystyle CH_3}{\underset{\displaystyle COOCH_3}{\overset{|}{\underset{|}{C}}}}\longrightarrow
$$

with C_6H_5 on first and third carbon.

(ii) $-NH(CH_2)_5CONH.(CH_2)_5CO\,NH(CH_2)_5CO^- + n\,CH_2-CH_2 \overset{\diagdown\ \diagup}{\underset{O}{\quad}} \longrightarrow$

(iii) $-OCCH=CH-CO.O\,CH_2CH_2O^-$

$$+\,CH_2=\overset{\displaystyle CH_3}{\underset{\displaystyle COOCH_3}{\overset{|}{\underset{|}{C}}}}\qquad\longrightarrow$$

4.5.2 Synthesis of Block Co-polymers

There are two possible routes to the formation of block co-polymers, which are similar to those used in the formation of graft polymers.

1. One can polymerize a second monomer onto an existing polymer chain. An example is given in Scheme 4.4.

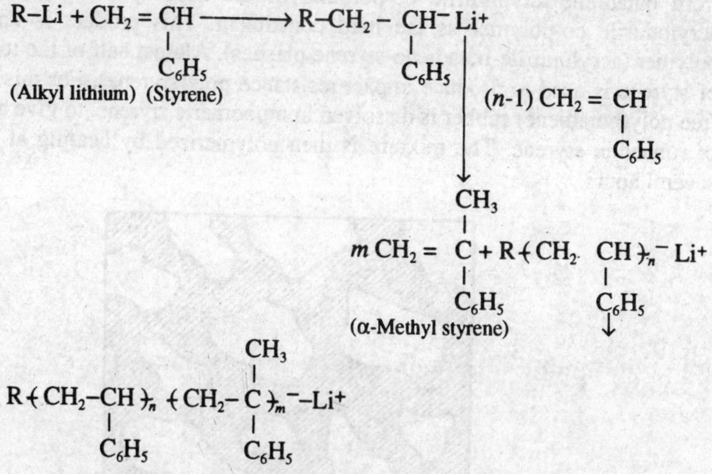

$$R–Li + CH_2 = CH \longrightarrow R–CH_2 – CH^- Li^+$$

$$\underset{\text{(Alkyl lithium)}}{\underset{|}{C_6H_5}} \quad \underset{\text{(Styrene)}}{} \qquad \underset{|}{C_6H_5}$$

$(n-1) CH_2 = CH$

$|$

C_6H_5

$$m\,CH_2 = \underset{|}{\overset{CH_3}{\underset{C_6H_5}{C}}} + R\{CH_2\ CH\}_n^- Li^+$$

$$\underset{(\alpha\text{-Methyl styrene})}{} \qquad \underset{|}{C_6H_5}$$

$$R\{CH_2–CH\}_n \{CH_2–\overset{CH_3}{\underset{|}{\underset{C_6H_5}{C}}}\}_m^{\,-}–Li^+$$

$$\underset{|}{C_6H_5}$$

Scheme 4.4

2. If poly(styrene) containing carboxylic groups is reacted with the ethylene glycol-adipic acid condensation product, a block copolymers is obtained (Scheme 4.5).

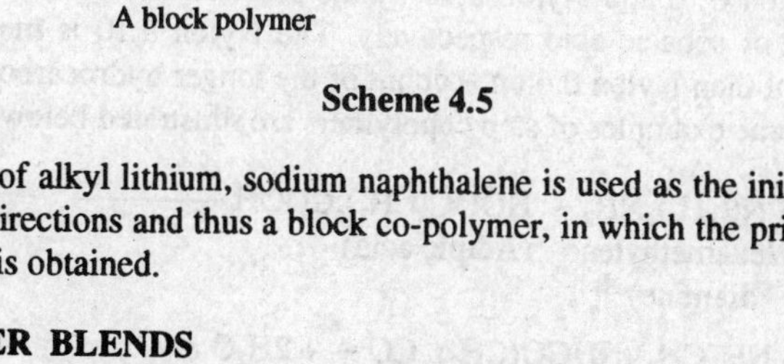

$$HOCH_2. CH_2–\{–O.CO (CH_2)_4 CO.O (CH_2)_2 \}_n OH+$$

$$HOOC \xrightarrow{\text{poly (styrene)}} COOH \longrightarrow$$

$$–O–\underset{\overset{\|}{O}}{C} \xrightarrow{\hspace{2cm}} \underset{\overset{\|}{O}}{C}–O–$$

A block polymer

Scheme 4.5

If instead of alkyl lithium, sodium naphthalene is used as the initiator, the chains grow in both directions and thus a block co-polymer, in which the primary polymer is in the middle, is obtained.

4.6 POLYMER BLENDS

In order to modify the properties of a polymer, two or more polymers are mixed together or blended. This method of processing is assuming increasing importance industrially. Example of this is the blending of poly(styrene) which is brittle, with natural rubber or with butadiene-acrylonitrile copolymer, to get products of superior impact strength. Whereas a polymer or co-polymer is homogeneous, polymer blends are heterogeneous, with the dispersed phase of one polymer in the continuous phase of the other polymer, as shown in Fig. 4.4

For the commercial preparation of polymer blends, the polymers are either ground together or melt-processed. In this way polymer blends of poly(vinyl chloride) − poly(methyl methacrylatate) (50:50 ratio) and poly(vinyl chloride)-

poly(vinyl acetate) (50:50 ratio) are obtained. This ratio can be varied according to the properties desired in the polymer blend. Polymer blends are similarly prepared from butadiene-acrylonitrile co-polymer as the rubbery component and styrene-acrylonitrile co-polymer as the hard constituent. This product is known as ABS polymer (acrylonitrile-butadiene-styrene plastics). Almost half of the total production of styrene is used to produce impact-resistance poly(styrene). For this purpose a gel free poly(butadiene) rubber is dissolved in monomeric styrene, to give a 10% solution of rubber in styrene. The mixture is then polymerized by heating at 80–140°C for several hours.

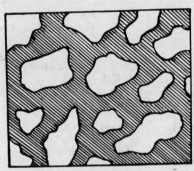

Fig. 4.4. Schematic representation of a polymer blend, showing dispersed phase of a polymer in the continuous phase of another polymer.

4.7. STEP CO-POLYMERIZATION

Step or condensation polymerization can be utilized in the synthesis of such products as poly(amides) and poly(esters). The properties of these polymers can differ according to the reactants used in the condensation reaction. Thus, for example, one can produce Nylon 6, 6 and Nylon 6,10 by the reaction of hexamethylene diamine with adipic acid or sebacic acid respectively. The Nylon 6,10 is more flexible and moisture-resistant than Nylon 6,6 on account of the longer hydrocarbon chain present in the former. Some examples of step copolymers are illustrated below.

$$H_2N(CH_2)_6NH_2 + HOOC(CH_2)_4COOH \longrightarrow$$
(Hexamethylene   (Adipic acid)
   diamine)

$$- NH(CH_2)_6NHCO(CH_2)_4\,CO - + 2H_2O$$
(Nylon 6,6)

$$H_2N\,(CH_2)_6NH_2 + HOOC\,(CH_2)_8COOH \longrightarrow$$
(Hexamethylene   (Sebacic acid)
   diamine)

$$- NH(CH_2)_6NHCO(CH_2)_6\,CO \longrightarrow$$
$$+ 2H_2O$$
(Nylon 6,10)

Aromatic polyamides can also be prepared similarly.

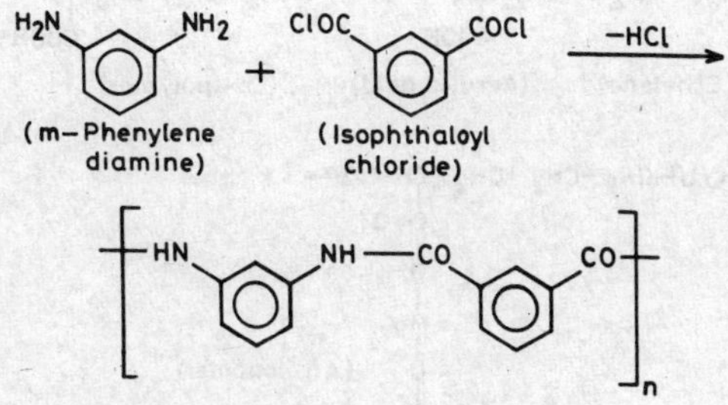

Poly(*m*-phenylene isophthalamide) is highly crystalline and has a fiber strength exceeding that of steel. A related polymer, poly(p-phenylene terephthalamide), is now being commercially produced.

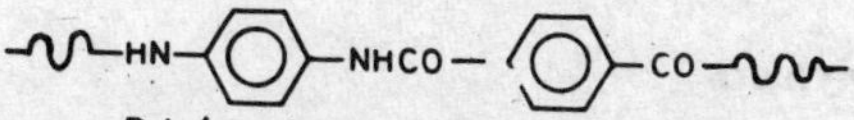

Poly (p—phenylene terephthalamide)

4.8. SOME IMPORTANT CO-POLYMERS

The term co-polymer used here denotes co-polymers having two or more monomers incorporated in the same polymer chain. Most of these polymers are random co-polymers, distinct from block co-polymers, whose synthesis has already been discussed (see Section 4.5.2). The structures of different types of polymers have been illustrated in Section 1.5.

4.8.1 Ethylene Co-polymers

(i) *Ethylene-propylene co-polymers*—These are prepared in the presence of a Ziegler type catalyst. A highly dispersed catalyst obtained from vanadium oxytrichloride and aluminium trihexyl has been found to be most suitable for this co-polymerization. The reactivity ratios are 7.08 and 0.086 for ethylene and propylene, respectively. Ethylene-propylene co-polymers containing about 30% propylene find use as elastomers. Terpolymerization of a small amount of a diene such as isoprene with ethylene and propylene yields an elastomer with occasional unsaturation. This enables its vulcanization with sulphur to introduce sulphur cross-links. Both of these co-polymers find use as general purpose rubbers.

(ii) *Ethylene-vinyl acetate co-polymers*—Both ethylene and vinyl acetate monomers singly are polymerized with difficulty to homopolymers. However, in the presence of peroxides and in emulsion, they undergo co-polymerization readily. Co-polymers of ethylene with small amounts of vinyl acetate or ethyl acrylate are used for packaging material due to their increased flexibility at room temperature. Co-polymers containing upto 40% vinyl acetate are rubbery and tough.

(iii) *Ionomer*—A co-polymer of ethylene with acrylic or methacrylic acid, on treatment with magnesium acetate, gives a product known as an *ionomer*.

$$nH_2C = CH_2 + nCH_2 = \underset{\underset{COOH}{|}}{CH} \longrightarrow \text{-w-} CH_2 - CH_2 - CH_2 - \underset{\underset{COOH}{|}}{CH} \text{-w-} \xrightarrow{Mg^{++}}$$

(Ethylene)   (Acrylic acid)   (Co-polymer)

$$\text{-w-} CH_2 - CH_2 - CH_2 - \underset{\underset{\underset{\underset{\underset{\underset{\underset{-\text{w-}CH_2-CH_2-CH_2-CH\text{-w-}}{|}}{\underset{C=O}{|}}}{O}}{Mg}}{|}}{\overset{|}{C=O}}}{CH} \text{-w-}$$

(An Ionomer)

Due to the labile metal-carboxylate bond, the ionomers are thermoplastic at higher temperatures. At ambient temperatures, however, they are thermosetting (for definitions of the terms thermoplastic and thermosetting see Section 10.1).

4.8.2 Styrene Co-polymers

Poly(styrene) is brittle as its chains are stiff on account of phenyl-phenyl interactions. Consequently, it has poor impact strength. It has a low glass transition temperature of 85°C and thus its use at higher temperatures is limited. It is attacked by many solvents. In order to obviate these defects, two important co-polymers have been developed.

(i) *Styrene-butadiene co-polymer (SBR, GR-S, Buna-S)*—A co-polymer containing 25 parts of styrene and 75 parts of 1, 3-butadiene, known as styrene-butadiene rubber, is obtained by the emulsion polymerization technique. A recipe for its preparation is given in Table 3.6 (page 47). It is used as a synthetic rubber (see Section 8.3.3). It can be vulcanized in the same manner as natural rubber. A typical recipe for vulcanization of SBR rubber consists of 100 parts of rubber, 50 parts of carbon black, 2 parts of sulphur, 1.2 parts of captax accelerator (see page 132), 5 parts of zinc oxide and 1.5 parts of stearic acid. By the use of carbon black filler, good tensile strength and other desirable physical properties are obtained. The main use of this co-polymer is in the tyre industry, where it is used for tread rather than the carcass of the tyres. It has good abrasion resistance. Styrene-butadiene co-polymers with higher percentage of styrene (40 – 75%) are used for latex paints.

(ii) *Styrene-acrylonitrile co-polymer*—This co-polymer prepared from styrene and acrylonitrile (10–40%) in the presence of a free-radical initiator, has higher tensile and impact strength. The resistance to solvents is also increased. These improved properties are perhaps due to increased inter-atomic forces of the $-C \equiv N$ group.

(iii) *ABS Co-polymer*—This has already been described under polymer blends (Section 4.6).

4.8.3 Vinyl Chloride Co-polymers

The polymer obtained from vinyl chloride (PVC) is a horny and tough material due to rigid poly(vinyl chloride) chains occasioned by bulky chlorine-chlorine interactions. Before it can be used, the homopolymer PVC chains have to be made flexible. This is done either by plasticization or by co-polymerization with monomers such as vinyl acetate or vinyldine chloride. The plasticizers used are dibutyl or dioctyl phthalate or tricresyl phosphate. They act by lubricating the chains of the polymer, but suffer from the fact that they can vaporize and migrate. Co-polymers do not suffer from this drawback.

(i) *Vinyl chloride-vinyl acetate co-polymers*—The use of the polymer containing 25% vinyl acetate has been described as a fiber under the trade name, *vinyon*. Co-polymers containing about 5% vinyl acetate are used for the preparation of gramophone records, extruded pipes and tank lining sheets. Co-polymers containing upto 40% of vinyl acetate are used for wire insulation, sheets, shower curtains, etc.

(ii) *Vinyl chloride-Vinylidine chloride co-polymers*—These co-polymers under the trade name *Saran* contain a high percentage of vinylidine chloride monomer. They are used as fibers, packaging material and in coatings. They are resistant to fats, oils, water and solvents.

4.8.4. Acrylonitrile Co-polymers

Acrylonitrile-butadiene co-polymers are marketed under the trade names perbunan and GR–N. A description of GR–N synthetic rubber is given under Section 8.4.4 (page 135).

ANSWERS TO QUESTIONS

4.1 $[M_1] = [M_2]$

Using equation (4.11) we get,

S.No.	:	1	2	3	4	5	6
$\dfrac{d[M_1]}{d[M_2]}$	:	1.05	2.12	0.956	0.75	5.45	4.13

4.2 $r_1 = 0.78$; $r_2 = 1.39$

f_1	:	0	0.20	0.40	0.60	0.80	1.00
F_1	:	0	0.154	0.33	0.53	0.15	1.00

See Fig. 4.5

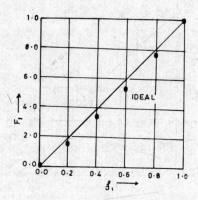

Fig. 4.5

4.3. $r_1 = 0.05$, $r_2 = 0.0$

f_1 :	0.0	0.10	0.20	0.40	0.60	0.80	0.90	1.0
F_1 :	0.0	0.501	0.503	0.508	0.518	0.545	0.592	1.0

See Fig. 4.6

4.4 When $r_1 r_2 = 1$, ideal co-polymerization occurs. Here the two types of propagating species show the same preference for adding the one or the other type of monomer.

When $r_1 r_2 = 0$, each of the propagating species adds preferentially to the other type of monomer. This is called alternating type of co-polymerization. When $r_1 = r_2 = 1$, the composition of the co-polymer is completely random.

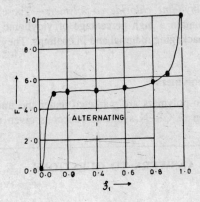

Fig. 4.6

4.5

(ii) $-NH(CH_2)_5CO.NH(CH_2)_5CO.NH(CH_2)_5CONH - + CH_2 - CH_2 \longrightarrow$

$$-NH(CH_2)_5CO-N-(CH_2)_5CONH(CH_2)_5CO \cdot N-$$

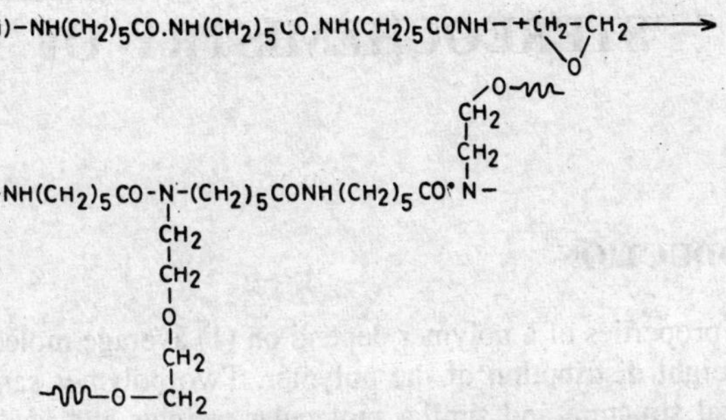

(iii) $-C-CH=CH-C-O\ CH_2\ CH_2O - + n\ CH_2 = C \longrightarrow$
with CH_3, $COOCH_3$

$$-C-CH-CH - C-OCH_2CH_2O -$$
$(CH_2-C)_n$ CH_3 $COOCH_3$ $-CH-CH-C-$ $COOCH_2CH_2O-$

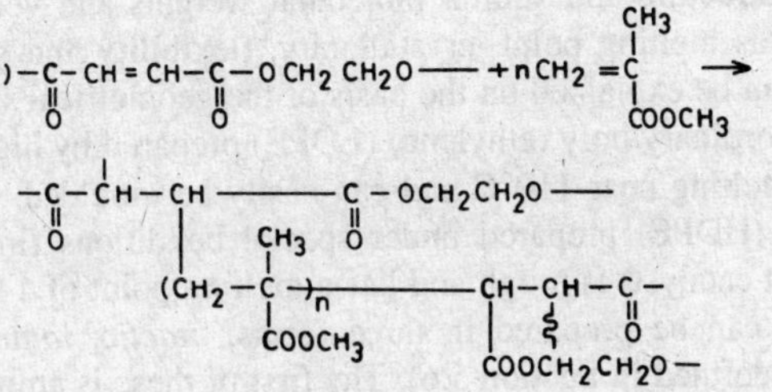

CHAPTER 5

STEREOCHEMISTRY OF POLYMERS

5.1. INTRODUCTION

Many of the properties of a polymer depend on (1) average molecular weight and (2) molecular weight distribution of the polymer. Two polymer samples may have the same chemical structure and similar molecular weights and yet may differ in many properties such as melting point, crystallanity, flexibility and solubility in different solvents. This can be explained on the basis of the geometrical structure of a polymer molecule. Thus ordinary poly (ethylene) (LDPE) prepared by high pressure technique is amorphous, melting near 110°C and can easily be moulded. However, crystalline poly (ethylene) (HDPE) prepared under special conditions (in the presence of K. Zigler - G. Natta catalyst) is tough and has a melting point of 144 - 150°C. Similarly (polypropylene) can be prepared in three forms, *atactic*, *iotactic* and *syndiotactic* (these terms are defined in Section 5.6). The first of these is amorphous and the other two are crystalline. The factors on which these properties are dependent are explained in subsequent pages.

5.2. CONFIGURATION AND CONFORMATION

When two carbon atoms are joined together by a single bond, there is free-rotation of one carbon with respect to the other. For example, in the molecule of ethane, if one carbon atom is kept stationary and the other rotated, say by one degree, this would constitute a second *conformation* of ethane. Thus, there can be 360 conformations of ethane. In actual practice only those conformations are considered which are energetically favourable.

If, however, the two carbon atoms are joined by a double bond as in the case of maleic and fumaric acids, only these two forms are possible and they cannot be easily transformed into each other. These two forms are spoken of as the two *configurations*.

5.3 FISCHER, SAWHORSE AND NEWMAN PROJECTIONS

Glyceraldehyde has the structure: $OHCH_2CH_2$ (OH).CHO, (1), and its systematic name would be 2, 3-dihydroxypropanal. Two forms of glyceraldehyde can be written as (2) and (3):

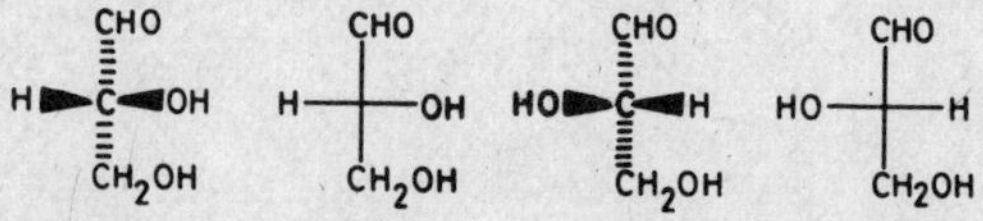

In the above forms, broken lines indicate below the plane of the paper and thick lines those above the plane of the paper. In Fischer projection, 2(a) and 3(a) these lines are transformed into full lines.

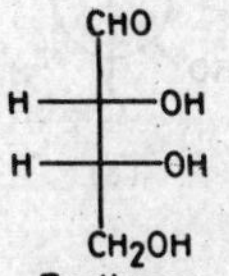

CHO
H————OH
H————OH
CH$_2$OH
Erythrose
(Fischer projection)
(4)

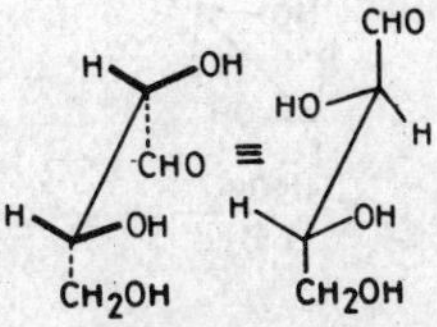

CHO
H————OH
HO————H
CH$_2$OH
Threose
(Fischer projection)
(6)

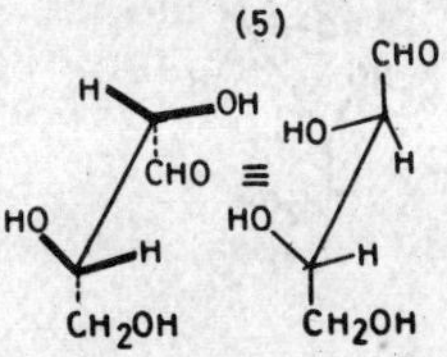

Erythrose
(Sawhorse projection)
(5)

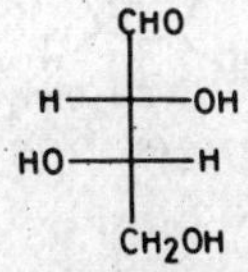

Threose
(Sawhorse projection)
(7)

There is another type of notation for projecting structures called the sawhorse projection. Sawhorse projections are obtained by the transformation of Fischer projection as illustrated for the erythrose and threose molecules [Figs. (5) and (7)]. In the conversion of Fischer projection (4) of erythrose to a sawhorse projection (5), the disposition of groups in the Fischer projection is maintained. The thick lines lie in a plane above and the broken lines below the plane of the paper. Since the carbon chain in acylic compounds assumes a zigzag conformation, the erythrose structure can be represented by the equivalent structure with full lines as shown in Fig. (5), in which the rear half of the structure has been rotated with respect to the front.

There is yet another method of representation called Newman projection. Like the sawhorse projection, Newman projection shows the special relationship between the ligands attached to the two adjacent atoms. Sawhorse projection (5) can be transformed to Newman projection (8). Since there is free rotation around the single bond joining the two reference atoms, (8) is one of the many ways in which the Newman conformation of erythrose can be written. Similar considerations apply to threose structure (Fig. 9).

In the Newman structure, one is looking down the C–C bond and the plane of projection of the other six bonds. The six bonds are arranged like the spokes of a wheel. In the Newman projection of erythrose (8) and threose (9), the ligands attached to the nearer carbon atom are shown by full lines and those attached to the farther carbon atom are represented by lines joined to the outer circumference of the circle.

5.4 CONFORMATIONS OF *n*-BUTANE

In the end on view of the models for *n*-butane (Scheme 5.1), the various conformations obtained, in performing a 360° rotation about the central C–C bond, the following conformations are worth considering.

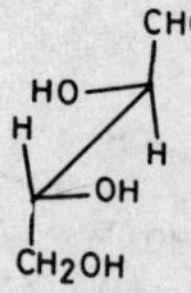

CHO
HO——
H
——OH
CH₂OH
(5)

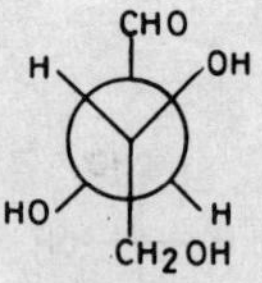

CHO
H   OH
HO   H
CH₂OH
(8)

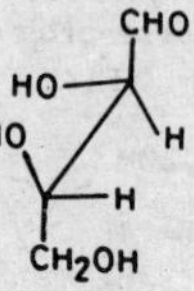

CHO
HO——
HO——H
——H
CH₂OH
(7)

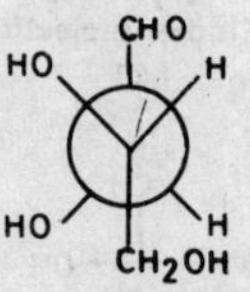

CHO
HO   H
HO   H
CH₂OH
(9)

In conformation (13), the methyl groups are distributed as far away from one another as possible, and the conformation is known as fully *staggered*. (11) and (15) are two equivalent conformations derived by a rotation of 60° and 300° about the central bond. These are known as *skew* conformations. In these conformations, the methyl groups are closer than allowed by Van der Waal's radii, and so are somewhat of higher energy than the staggered conformation (13). (12) and (14) are known as *eclipsed* conformations in which H and methyl are opposed. (10) (or equivalent 16) is known as *fully eclipsed* conformation. In this conformation, the two methyl groups are opposed.

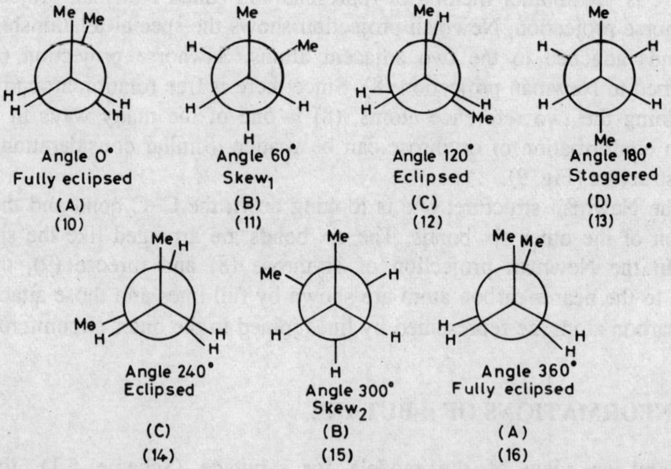

Me Me

H   H
H   H
Angle 0°
Fully eclipsed
(A)
(10)

Me
H   Me
H   H
H
Angle 60°
Skew₁
(B)
(11)

Me H
H   H
H   Me
Angle 120°
Eclipsed
(C)
(12)

Me
H   H
H   H
Me
Angle 180°
Staggered
(D)
(13)

Me H
Me   H
H   H
Angle 240°
Eclipsed
(C)
(14)

Me
Me   H
H   H
H
Angle 300°
Skew₂
(B)
(15)

Me Me
H   H
H   H
H
Angle 360°
Fully eclipsed
(A)
(16)

Scheme 1

5.4.1 Energy Differences Between the Different Conformations

The energy of the various conformations of *n*-butane molecule can be plotted against the torsional angle and it has been shown that it follows a curve (Fig. 5.1). The fully eclipsed conformation (10) has the maximum energy, when the torsional angle between the two methyl groups = 0. This is followed by two eclipsed conformations (12) and (14) where the torsional angle between the two methyl groups is 120° and 200° respectively. The fully staggered conformation (13) (torsional angle = 180°) has the minimum energy, followed by the two skew conformations (11) and (15) (torsional angle = 60° and 300° respectively). The difference between the two conformations with eclipsed groups results from *steric interaction* between non-bonded groups (Van der Waal's forces). These are considered to be insignificant between the two hydrogen atoms, small between hydrogen and methyl groups and larger for two methyl groups. The energy distribution shown (Fig. 5.1) is consistent with the view that torsional strain is greater than steric strain.

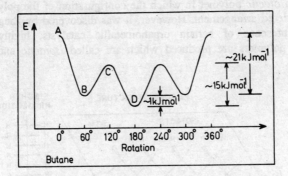

Fig. 5.1. Potential Energy diagram for *n*-butane.

Q.5.1 Draw the Fischer, Sawhorse and Newman projections for the two optical forms of tartaric acid and the *meso* form.

Q.5.2 Draw the Newman projections for the various important conformations of 1, 2-dichloroethane.

5.5. SHAPE OF POLYMER MOLECULES

The molecular weights of linear polymers may, in most cases, extend upto 1,000,000. In an extended form, the diameter of such a molecule may be between 5×10^{-8} cm and 10×10^{-8} cm and length of the order of 10^{-4} cm. There are three possibilities for the arrangement of the molecule: (i) a random coiled arrangement (Fig. 5.2a), (ii) like the length of a string, an extended rigid, rod like structure (Fig. 5.2b) and (iii) a regular helical coil like structure (Fig. 5.2c).

The type of arrangement which a polymer molecule may take up depends upon its conformation, configuration and stickiness. All these factors will be discussed while considering the structures of different polymers.

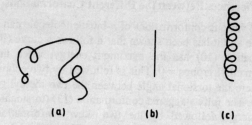

(a)   **(b)**   **(c)**

Fig. 5.2. Shapes of polymer molecules.

5.6. TACTICITY

If one takes monosubstituted ethylene and polymerizes it under normal conditions, one obtains an *atactic* polymer in which the configuration of the polymer does not follow any ordered arrangement. However, it was discovered by Ziegler and Natta that in the presence of certain organometallic catalysts, highly ordered or *stereoregular* polymers are produced which are called *isotactic* and *syndiotactic* polymers.

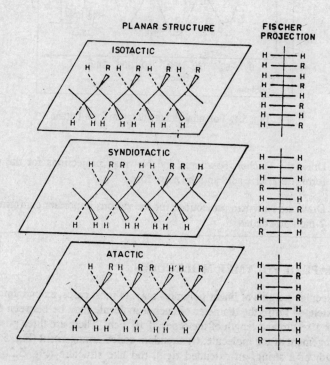

Fig. 5.3. Isotactic, syndiotactic and atactic polymers.

If the substituent R is on the same side of the carbon backbone in the Fischer projection, the polymer is called isotactic and if it is arranged alternately, the polymer is called *syndiotactic* (Fig. 5.3). Stereoregular polymers have a high degree of crystallinity, and well defined density and melting point.

Some specific examples of polymer molecules of different geometrical arrangements are considered below. (After C.C. Price)

5.6.1. Poly(ethylene)

We have considered the stability of the various conformations of *n*-butane. On energetic grounds, the various forms follow the following order of energies (Sec. 5.3):

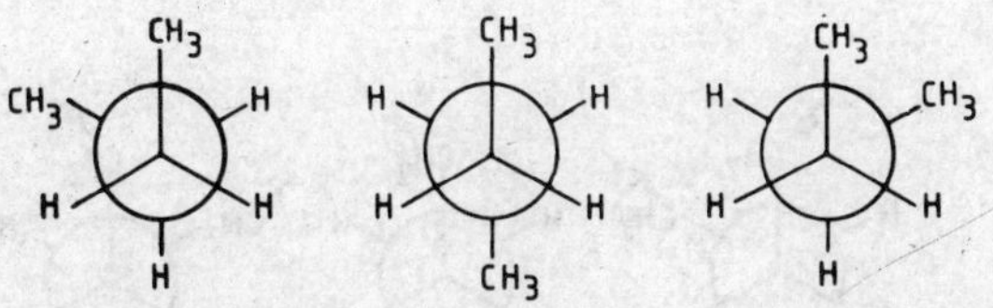

$$\underset{(19)}{\text{Skew}_1} \quad < \quad \underset{(18)}{\text{Trans}} \quad > \quad \underset{(17)}{\text{Skew}_2}$$

The various conformations of poly(ethylene) may be envisaged by replacing the CH_3 groups by CH_2. The enhanced stability of the *trans* conformation leads to the preferred linear-extended (and, therefore, easily crystallisable) poly(ethylene) arrangement.

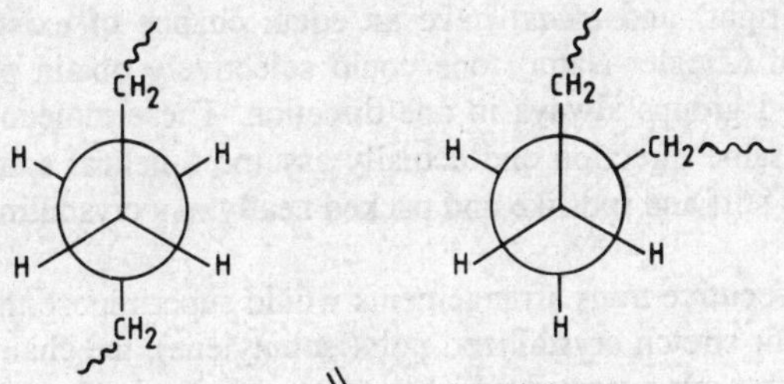

$$\underset{(20)}{\text{Trans}} \quad \ggg \quad \underset{(21)}{\text{Skew}}$$

This *trans* arrangement is thus energetically favoured over other rotational isomers such as the skew. These extended rod like molecules pack together in neatly ordered crystalline arrays which are stable and resist dissolution or distortion at temperatures below the m.p. of 140°. This is revealed by X–ray studies.

5.6.2 Poly(propylene)

Poly(propylene) prepared under normal conditions is a rubbery material.

$$n \; CH_2 = \underset{\underset{H}{|}}{\overset{\overset{CH_3}{|}}{C}} \rightarrow \left(CH_2 - \underset{\underset{H}{|}}{\overset{\overset{CH_3}{|}}{C}} \right)_n$$

(Propylene)   [Poly(propylene)]

However, when prepared in the presence of Ziegler-Natta catalyst, a highly, crystalline material, melting at 170° is obtained which is a typical plastic. The three stable conformations (22, 23 and 24) of poly(propylene) are shown below:

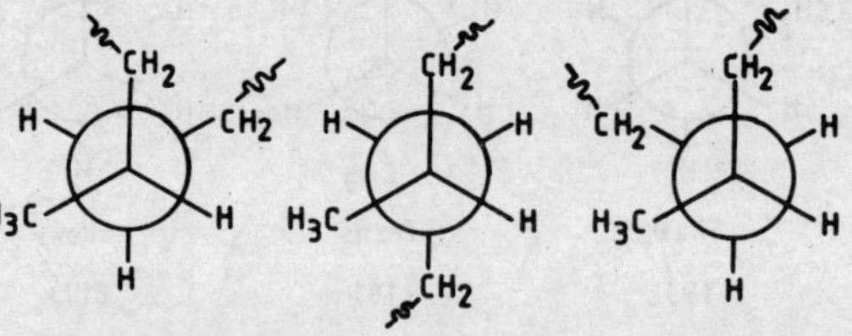

Skew (right) 〜 Trans   》   (Skew (left)

(22)   (23)   (24)

The skew (left) form of poly(propylene) (24), in which CH_2 attached to the back carbon atom is placed between two large groups of the front carbon atom (CH_3 and CH_2), is comparatively unstable as compared to forms (22) and (23) and, hence, these forms [skew (right) and *trans*] have an equal chance of existence. With the new catalyst system (Ziegler-Natta), one could selectively obtain poly(propylene) units with the methyl groups always in one direction. These molecules then tend to coil always in the same direction and actually assume a helical spiral arrangement. The regular helix is stiff and rod-like and packed neatly in a crystalline pattern.

Since consecutive trans arrangements would superimpose the 1, 3 methyl (found unfavourable for stretch crystallized poly(isobutylene), the chain prefers an alternate trans-skew (right) arrangement which leads to a helical arrangement involving three propylene units, with a repeat distance of about 7 Å.

It may be noted that for atactic poly (propylene), the configuration of the methyl groups on the units would be randomly left or right and would lead to randomly-coiled rather than helical chain. The atactic, randomly-coiled poly(propylene) is in fact found to be amorphous and rubbery; the isotactic helically-coiled poly(propylene) is a high melting (170°C) crystalline material readily formed into fibers.

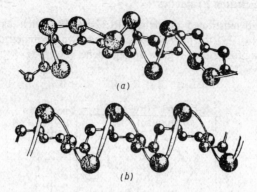

Fig. 5.4. Section of a poly(propylene) chain: (a) atactic, (b) isotactic.

5.6.3. 1,1-Disubstituted Ethylene [Poly(isobutylene), Poly(methyl methacrylate) and Poly(vinylidine chloride)]

The three conformations of poly(isobutylene) are as follows (25, 26 and (27):

Skew (right) ⇌ Trans ⇌ Skew (left)

(27)   (26)   (25)

In the case of 1,1-disubstituted ehtylene (CH_2==CRR') stereo-isomerism does not exist if R and R′ groups are the same as in isobutylene and vinylidine chloride. In poly(isobutylene) the methylene groups on the back carbon atom do not see any difference between the two methyl groups in the front (*trans* form, 26) or between the CH_3 and CH_2 groups on the front carbon atom (skew left 25 and skew right 27). The chain thus assumes a randomly coiled conformation. This condition is necessary for a rubbery polymer. On stretching, an X-ray diffraction pattern is obtained which is like that of a crystalline polymer.

When R and R′ are different, as in the case of methyl methacrylate (CH_3 and $COOCH_3$), the polymer behaves in a manner similar to poly(propylene). The methyl groups can be located on the same side of the carbon chain or alternately on the two sides of the chain, giving rise to isotactic and syndiostatic arrangements respectively.

If the methyl groups are not arranged in an orderly sequence, atactic polymer is formed. The second substituent present has no effect on the situation, as the steric effect of the first substituent automatically fixes the second.

5.6.4. 1,2,-Disubstituted Ethylenes

In the case of disubstituted ethylenes, RCH=CHR', such as pentene-2 where R=CH₃ and R'=C₂H₅, the polymer obtained has two asymmetrical centres and can have the following four stereo-regular arrangements (Fig. 5.5).

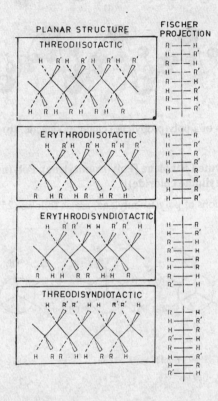

Fig. 5.5. Steroregular structures of 1,2-disubstituted polymers $+ RCH - CHR' +_n$

Q.5.3 Distinguish between isotactic, syndrotactic and atactic polymers.

Q.5.4 Distinguish between the terms stereoregular, stereospecific and stereoselective.

5.6.5. Poly(butadienes)

The polymerization of 1,3-butadiene leads to both optical and geometrical isomerisms. Thus, butadiene can yield *cis*-1,4- *trans*-1,4-poly(butadiene) as well as isotactic, syndiotactic and atactic 1,2-poly(butadienes).

Table 5.1. Various Stereoisomers of Poly(butadiene)

1. $$\left[\begin{array}{c} CH_2 \qquad\qquad CH_2 \\ CH\!=\!\!CH \end{array}\right]_n \qquad \textit{cis}\text{-1, 4-poly(butadiene)}$$

2. $$\left[\begin{array}{c} CH_2 \\ CH\!=\!\!CH \\ \qquad\qquad CH_2 \end{array}\right]_n \qquad \textit{trans}\text{-1, 4-poly(butadiene)}$$

3. $$\left[\begin{array}{cccc} -CH_2- & CH- & CH_2- & CH- \\ & | & & | \\ & CH & & CH \\ & \| & & \| \\ & CH_2 & & CH_2 \end{array}\right]_n \qquad \text{Isotactic poly-(1,2-butadiene)}$$

4. $$\left[\begin{array}{cccc} & & & CH_2 \\ & & & \| \\ & & & CH \\ & & & | \\ CH_2- & CH- & CH_2- & CH- \\ & | \\ & CH \\ & \| \\ & CH_2 \end{array}\right]_n \qquad \text{Syndiotactic poly-(1,2-butadiene)}$$

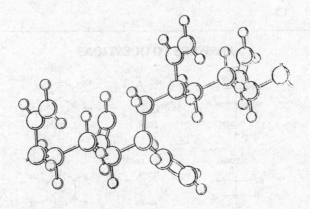

Fig. 5.6 Isotactic poly-(1, 2-butadiene).

All these can be synthesized by the use of proper catalyst and reaction conditions. The structures of isotactic and syndiotactic forms of poly(1, 2-butadiene) are shown in Figs. 5.6 and 5.7 respectively.*

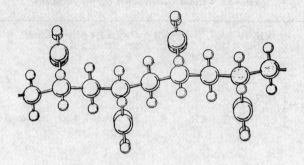

Fig. 5.7. Syndiotactic poly(1, 2-butadiene).

Q.5.5 Which of the following monomers will give stereoregular polymers. Draw the structures.

(i) $CH_2 == CH$
$\quad\quad\quad\quad\quad\;\; |$
$\quad\quad\quad\quad\quad\; CH_3$

(ii) $CH_2 == C$
$\quad\quad\quad\quad\quad\quad |$
$\quad\quad\quad\quad\quad\; CH_3$
   with top CH_3 :
$CH_2 == C$ with CH_3 above and CH_3 below

(iii) $CH_2 == C$
with CH_3 above and $COOCH_3$ below

(iv) $CH_2 == CH - CH == CH_2$

(v) $CH_2 == C - CH == CH_2$
$\quad\quad\quad\quad\quad\quad |$
$\quad\quad\quad\quad\quad\; CH_3$

(vi) $CH_2 == C - CH == CH_2$
$\quad\quad\quad\quad\quad\; |$
$\quad\quad\quad\quad\; Cl$

ANSWERS TO QUESTIONS

5.1

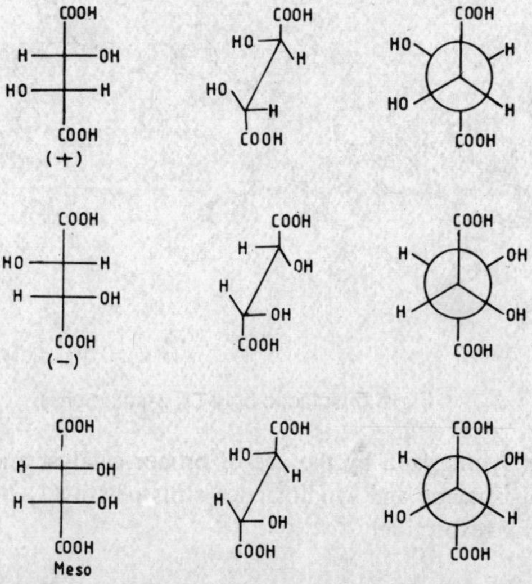

(Structures for 5.1 showing Fischer projections and Newman projections:)

Row 1:
- COOH / H—OH / HO—H / COOH (+)
- COOH / HO, H / HO, H / COOH
- COOH / HO—H, HO—H / COOH (Newman)

Row 2:
- COOH / HO—H / H—OH / COOH (−)
- COOH / H, OH / H, OH / COOH
- COOH / H—OH, H—OH / COOH (Newman)

Row 3:
- COOH / H—OH / H—OH / COOH Meso
- COOH / HO, H / H, OH / COOH
- COOH / H—OH, HO—H / COOH (Newman)

5.2

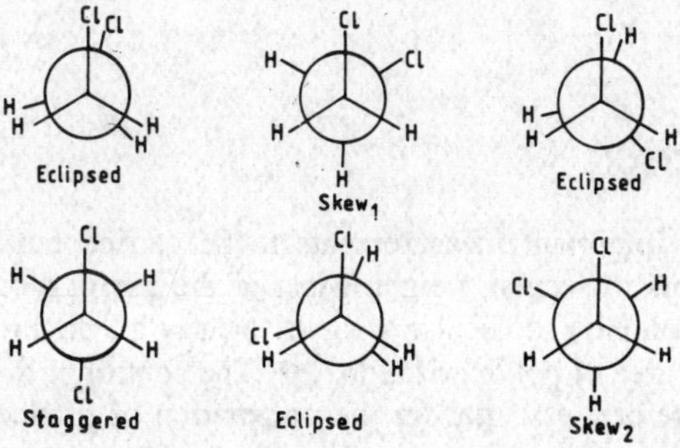

Eclipsed   Skew₁   Eclipsed

Staggered   Eclipsed   Skew₂

5.3 If a monosubstituted ethylene with a substituent R is polymerized under different conditions, one can obtain polymers of different configurations. In an ordered arrangement of atoms in a polymer, if all the R groups are on the same side of the carbon backbone, it is called an *isotactic* polymer. If they are arranged alternately on the two sides of the carbon backbone, it is called *syndiotactic* polymer. If the group R does not follow any ordered sequence, it is an *atactic* polymer.

5.4 Polymerizations which yield ordered structures (isotactic or syndiotactic) are called stereospecific polymerizations.

These ordered structures are called *stereoregular* polymers.

Stereoselective polymerization is one in which one type of ordered structure is preferentially formed in contrast to the other.

5.5 (i) Poly(propylene) can occur with isotactic and syndiotactic structures. See structures (a) and (b) of Fig. 5.3.

(ii) Poly(isobutylene) cannot show stereoregularity as there are no asymmetric centres in the molecule.

(iii) Poly (methyl methacrylate) exhibits a behaviour similar to a monosubstituted ethylene like poly (propylene). The methyl group can be located on the same side or alternatively on one or the other side of the c-c chain. Thus isotactic or syndiotactic structures can result. The second substituent has no effect as the steric effect of the first substituent automatically fixes the second.

(iv) Poly (1,2-butadiene) can occur in isotactic and syndiotactic configurations. (see table 5.1)

(v) Poly (1,4-isoprene) can occur in *cis* and *trans* forms; 1,2-poly (isoprene) can occur as isotactic and syndiotactic form; 3,4-poly (isoprene) can occur in isotactic and syndiotactic forms with R = $-(CH_3) = CH_2$

(vi) It behaves similar to isoprene, the substituent at position 2- is chlorine in place of CH_3.

CHAPTER 6

MOLECULAR WEIGHT DETERMINATION

6.1 INTRODUCTION

One of the most important measurements to be carried out on a polymer is the determination of its molecular weight. Most of the useful mechanical properties of polymers depend on their molecular weights. Nearly all the properties of a polymer change with its degree of polymerization, $\overline{DP}$. The control of the molecular weight of a polymer is, therefore, essential for the preparation of products with different end uses.

6.2 MOLECULAR WEIGHT AVERAGES

In contrast to the low molecular weight compounds, a polymer is usually a complex mixture of molecules of different molecular weights. The polymers are thus polydisperse and heterogeneous in composition. Therefore, the molecular weight of a polymer is actually an average of the molecular weights of constituent molecules. Different averages are obtained depending on the method of measurement of the molecular weight.

The number average molecular weight $\overline{M}_n$, is obtained by the measurement of the colligative properties of a polymer by osmometry or end group analysis and is defined as,

$$\overline{M}_n = \frac{\sum N_i M_i}{\sum N_i} \tag{6.1}$$

where N_i is the number of molecules of molecular weight M_i.

The weight average $(\overline{M}_w)$ molecular weight is obtained from light scattering measurements and is defined as,

$$\overline{M}_w = \frac{\sum N_i M_i^2}{\sum N_i M_i} \tag{6.2}$$

To explain these molecular weight averages, one can take an example. Suppose there are 50 polymer molecules of molecular weight 10^2, 200 polymer molecules of molecular weight 10^3, and 100 molecules of molecular weight 10^4, then,

$$\overline{M}_n = \frac{50 \times 10^2 + 200 \times 10^3 + 100 \times 10^4}{50 + 200 + 100}$$

$$= 3443 \approx 3440 \quad \text{approx.}$$

$$\overline{M}_w = \frac{50 \times 10^4 + 200 \times 10^6 + 100 \times 10^8}{50 \times 10^4 + 200 \times 10^3 + 100 \times 10^4}$$

$$= 8465 \approx 8470 \text{ approx.}$$

It is evident from this example that the number average and weight average molecular weights are not the same except in monodisperse systems where $\overline{M}_n = \overline{M}_w$; $\overline{M}_n$ is very sensitive to the molecules of lower molecular weight whereas $\overline{M}_w$ is sensitive to the presence of molecules of higher molecular weight.

The viscosity average molecular weight, $\overline{M}_v$ is obtained by viscosity measurements and is defined as,

$$\overline{M}_v = \left[\frac{\sum N_i M_i^{(1+\alpha)}}{\sum N_i M_i} \right]^{1/\alpha} \qquad (6.3)$$

where α is the exponent in the Mark-Howink-Sakurada equation: $[\eta] = KM^\alpha$ (see Section 6.5 and Eq. 6.12).

α can vary between 0.5 and 1.0. The following inequality occurs:

$$\overline{M}_n < \overline{M}_v \leq \overline{M}_w$$

The z-average molecular weight is defined as,

$$\overline{M}_z = \frac{\sum N_i M_i^3}{\sum N_i M_i^2} \qquad (6.3a)$$

Q.6.1.   (i) If two polymers of molecular weights 10,000 and 100,000 are mixed together in equal parts by weight, determine the number average and weight average molecular weights. (ii) If the above polymers are mixed so that equal number of molecules are added, determine $\overline{M}_n$ and $\overline{M}_w$.

Q.6.2.   If you mix 1000 g of a polymer of molecular weight 1000 g/mole and 1000 g of a polymer of molecular weight 10^6 g/mole, what is the ratio $\overline{M}_w/\overline{M}_n$?

6.3. MOLECULAR WEIGHT DISTRIBUTION (ADDITION POLYMERS)

When sufficiently sharp fractions have been isolated from a polymer and the molecular weights of the fractions have been determined, the distribution of the fractions of the various molecular weights may be represented in a number of ways. This is illustrated by taking a sample of poly(styrene). R. Signer and H. Gross took a 100 g sample of poly(styrene) and separated the various fractions with the aid of ultra-centrifuge and determined their molecular weights. The original molecular weight of the sample of poly(styrene) was 80,000. The molecular weights of different fractions are listed in Table 6.1.

Table 6.1. Molecular Weight Distribution Data on Poly(styrene)

Fraction number	Wt. of polymer in (g)	Percent of polymer used on total	Cumulative percent of polymer	Mol wt. x 10^{-3}
1	0.2	0.2	0.2	25 – 35
2	1.7	1.7	1.9	35 – 45
3	3.6	3.6	5.5	45 – 55
4	8.4	8.4	13.9	55 – 65
5	20.0	20.0	33.9	65 – 75
6	23.8	23.8	57.7	75 – 85
7	20.2	20.2	77.9	85 – 95
8	10.4	10.4	88.3	95 – 105
9	6.0	6.0	94.3	105 – 115
10	3.3	3.3	97.6	115 – 125
11	1.7	1.7	99.3	125 – 135
12	0.5	0.5	99.8	135 – 145
13	0.2	0.2	100.0	145 – 155

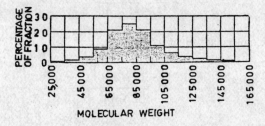

Fig. 6.1. Population curve of a poly(styrene) sample.

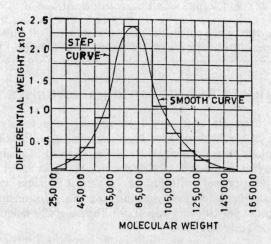

Fig. 6.2. Differential weight distribution curve of a poly(styrene) sample.

The data in Table 6.1 is represented in Fig. 6.1 as a population curve, in Fig. 6.2 as a differential weight distribution curve, in Fig. 6.3 as differential number distribution curve and in Fig. 6.4 as integral weight distribution curve. The data of Table 6.1 has been plotted as population curve in Fig. 6.1 by plotting the percentage weights of the fractions against their molecular weights. This is not of much practical value. The integral distribution curve (Fig. 6.4) is obtained by plotting the cumulative weight percentages of the polymer fractions against their molecular weights. Such a plot results in a step curve through which a continuous curve is obtained. The differential weight distribution curve shown in Fig. 6.2 is obtained by graphic differentiation of the continuous integral weight distribution curve (Fig. 6.4) by dividing the change in the weight fraction by ΔMW, the change in molecular weight.

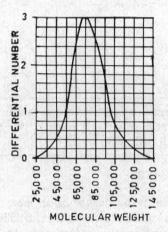

Fig. 6.3. Differential number distribution function of a poly(styrene) sample.

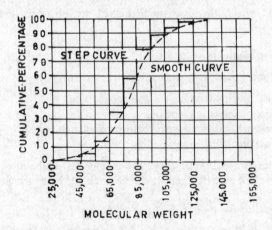

Fig. 6.4. Integral weight distribution curve of a poly(styrene) sample.

Q.6.3.   From the following data, construct the integral and differential distribution curves for 2-vinyl pyridine.

Sample No.	Wt. of polymer (g)	$\overline{DP}$
1.	3.57	466
2.	1.17	840
3.	3.43	1000
4.	3.74	1130
5.	4.93	1270
6.	8.88	1480
7.	2.82	1480
8.	4.72	1640
9.	1.89	1650
10.	2.78	1810
11.	4.80	2420
12.	4.64	2880

Q.6.4.   If a polymerization reaction is carried out so that initiation takes place at a constant rate, the chain grows at a constant rate and there is no chain termination, what will be the ratio $\overline{M}_w/\overline{M}_n$?

6.4. FRACTION OF POLYMERS

Separation of a polymer sample into fractions which are more homogeneous is known as polymer fractionation. In suitable solvents the solubility of a homogeneous polymer fraction decreases with increasing molecular weight. Thus, the addition of a non-solvent to a solution of heterogeneous polymer precipitates out fractions in the order of decreasing molecular weights. Fractionation procedures are of two types:

(i) precipitation in which actual fractions are isolated for further study, and
(ii) analytical in which the distribution curves are obtained without isolating fractions.

6.4.1. Precipitation Methods

Following are the main precipitation methods:

(A) fractional precipitation by non-solvent addition,
(B) triangular precipitation method,
(C) fractional precipitation by evaporation of solvent, and
(D) fractional precipitation by cooling.

These methods are self-explanatory. The method (B) is the most elaborate and can lead to more or less pure fractions. It will be discussed here in some detail.

6.4.1.1 *Fractional Precipitation by Non-solvent Addition*

In this procedure, usually a solution of 0.5% or lesser concentration of the high polymer to be fractionated is prepared in a suitable solvent. Precipitation of the polymer molecules in solution is then carried out by the slow addition of a precipitant until an initial turbidity is reached at a controlled temperature.

The mixture is then warmed slightly after the initial turbidity is observed to aid the attainment of thermal equilibrium conditions. It is then cooled slowly to the original precipitation temperature. The precipitate is then recovered usually by centrifuging to effect a clearer separation. The average molecular weight as well as the weight percent is determined for the fraction.

After separating the first fraction from the polymer molecules still remaining in the solution, a further increment of precipitant is now added as before to the remaining solution, and the process repeated to obtain the second fraction, and so forth, for as many fractions as are desired or required.

The solubility of a chain polymer in suitable solvents or a mixture of solvents falls off with increasing molecular weight. Hence, the longer the chain length of the polymer, the smaller is the volume of the precipitant required to cause precipitation.

6.4.1.2 *Triangular Precipitation Method*

In this method sufficient precipitant is added to polymer solution (*P*) at the first addition to bring about the precipitation of the polymer to get two phases: gel (a) (precipitated phase) and solution (b) (supernatant layer). After separation, the gel (a) is brought into solution to get solution (a). The two solutions (a) and (b) are treated similarly as the original solution (*P*) was treated to get four approximately equal fractions. The solution (a) gives gel (c) and the solution (d), and the solution (b) gives gel (e) ad solution (f). Solution (d) and gel (e) are combined to get the solution d/e, and gel (c) is again dissolved. Each of the three solutions thus obtained are subjected to further fractionation in a way shown in Fig. 6.5. This procedure is continued until a number of fractions are obtained. The larger the number of fractions, the more accurate is the fractionation.

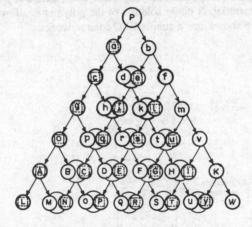

Fig. 6.5. The triangular fraction scheme.

6.4.1.3 *Fractional Precipitation by Evaporation of the Solvent*

An alternative method for precipitation of a polymer fraction from a solution is the preferential removal of solvent by evaporation. Certain advantages over the non-solvent addition procedure are:

(i) Local complete precipitation is avoided.

(ii) The volume of the system decreases throughout the fractionation.

(iii) Fraction size can be controlled from the turbidity developed.

6.4.1.4 *Fractional Precipitation by Cooling*

The solubility of a high polymer in a solvent is temperature-dependent. As the temperature is lowered, for example, the thermodynamic properties of the solvent change, making it a poorer solvent. Eventually, a temperature will be reached at which precipitation of the least soluble components will occur because at that stage the solvent and polymer are no longer miscible in all proportions. A stepwise precipitation and recovery of respectively precipitated fractions may then be performed and the data handled as in fractional precipitation procedure.

6.4.2. Analytical Methods

These include methods like turbidimetric titrations/ultra centrifugation, gradient elution analysis and gel permeation chromatography. The last mentioned method will be described in detail.

6.4.2.1 *Gel Permeation Chromatography*

This is one of the latest techniques employed and is accurate. It is used with advantage for fractionating polymer samples and also for determining molecular weight distribution without collecting the different fractions. The fractionation is done in a chromatographic column packed with rigid `gel' beads. These gel beads are made of porous highly cross-linked poly(styrene). These gels and porous glass beads form the packing material. A dilute solution of the polymer is allowed to flow down the column. Fig. 6.6 shows how a simple GPC column works.

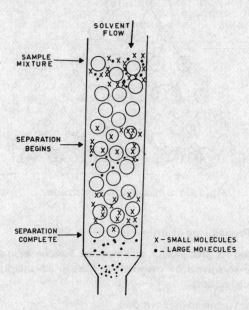

Fig. 6.6. The working of a simple GPC column.

6.4.2.1.1 *Principle of the Method*

When a dilute solution of the polymer to be fractionated is allowed to stream down the GPC column, followed by pure solvent, the desired number of fractions can be collected and characterized. In this process one obtains the fraction of the polymer with highest molecular weight first and that with the lowest molecular weight at the last. This is because the 'gel' beads used in the column are full of capillaries. These capillaries or pores are of different diameters. Of these capillaries, some are accessible only to very small polymer molecules present in the solution, some others are accessible to polymer molecules of medium size or molecular weight. A few are accessible to macromolecules of very high molecular weights. If $V_{(\bar{P}=100)}$ is the total volume of the pores which are just accessible only to polymer molecules of degree of polymerization ($\bar{P} = 100$) and $V_{(\bar{P}=10)}$ is the total volume of pores accessible to molecules of $\bar{P} = 10$, then it becomes evident that the larger the $V_{\bar{P}}$ the smaller is $\bar{P}$, that is,

$$V_{\bar{P}=10} \quad > \quad V_{\bar{P}=100} \quad > \quad V_{\bar{P}=1000} \quad > \quad V_{\bar{P}=100,000}$$

The highly cross-linked poly(styrene) gel of the above micrograph (Fig. 6.7) was prepared by polymerizing 30% styrene and 10% divinylbenzene in the presence of a 60% diluent (mixture of 20% diethylbenzene and 80% of isoamyl alcohol).

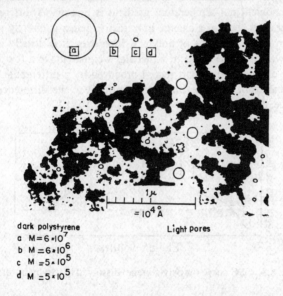

dark polystyrene   Light pores
a $M = 6 \cdot 10^7$
b $M = 6 \cdot 10^6$
c $M = 5 \cdot 10^5$
d $M = 5 \cdot 10^5$

Fig. 6.7. Electron micrograph of a highly cross-linked poly(styrene) gel.

Thus, macromolecules with degree of polymerization 10 can enter all the capillaries in the gel, whereas polymers with $\bar{P} = 10,000$ can enter into very few capillaries (those with diameter large enough to permit entering of macromolecules with $\bar{P} = 10,000$ or more). It, therefore, follows that when a dilute solution of a polymer sample is allowed to flow through a column, the probability of a molecule entering the capillaries (available volume) is the larger and, hence, the residence time is the greater for macromolecules with lower degree of polymerization than for those with higher molecular weight (greater $\bar{P}$). Hence, macromolecules with higher $\bar{P}$ are eluted first and those of lowest $\bar{P}$ at the last.

This method can be used for a wide variety of polymers and solvents. When poly(styrene) gels are used, nonpolar polymers can be analysed, using solvents such as tetrahydrofuran or toluene. The results of carefully performed GPC fractionations for molecular weight distribution agree closely with those obtained by other methods such as solvent gradient elution method. The accuracy of the method is further established by the fit obtained between the experimentally obtained data and the theoretically calculated distribution curve on the basis of polymerization kinetics.

6.4.2.1.2. Determination of Molecular Weight Distribution

In many cases, fractionation is carried out to determine the molecular weight distribution. This is conveniently done by collecting fractions, drying them, weighing to determine the amount of polymer and measuring its molecular weight by some independent method. A major advantage of GPC is that these steps can be eliminated by use of a detector that determines the concentration of the polymer in the effluent since neither a solvent nor temperature gradient is present. A differential refractometer is usually used because a change in refractive index is directly proportional to concentration in dilute solutions of polymer. Moreover, it is usually independent of the molecular weight in the range of molecular weights above a few hundreds. Fig. 6.8 shows an example of the chart record produced by a differential refractometer. The displacement from the base line is proportional to the difference in refractive index and, hence, the concentration.

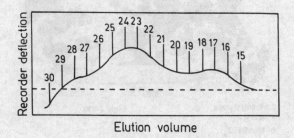

Fig. 6.8. GPC curve for poly(styrene) dissolved in tetrahydrofuran.

The vertical lines measure the elution volume. A siphon at the end of the column is dumped when 5 cc of the effluent has been collected. This dumping action sends an impulse to the recorder to automatically produce the spike on the curve. Thus, if the concentration is recorded directly, then the molecular weight remains to be determined. This may be done by reference to a calibration curve (Fig. 6.9). If the calibration is done in terms of molecular size parameter, for example $[\eta]\,M$, it can be applied to a wide variety of both linear and branched polymers.

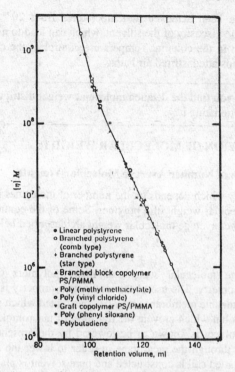

Fig. 6.9. Calibration curve for gel permeation chromatography based on hydrodynamic volume expressed by the product $[\eta]\,M$.

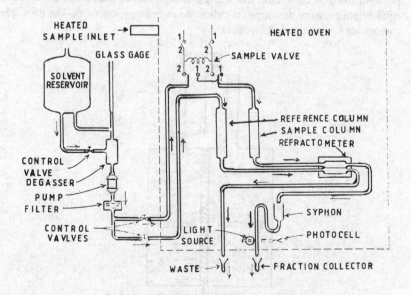

Fig. 6.10. Apparatus for gel permeation chromatography.

A suitable apparatus for gel permeation chromatography is shown in Fig. 6.10.

Separations have been achieved at temperatures from 20° to 150°C; higher temperatures lower the viscosity of the solvent, which can lead to more rapid analysis and higher resolution in the column. Temperature control of the columns is usually achieved in the thermostatted, stirred air baths.

Q.6.5.   How would you find the detailed molecular weight distribution of a polymer without fractionating it?

6.5. DETERMINATION OF MOLECULAR WEIGHTS

6.5.1. Determination of Number Average Molecular Weights

A colligative property which depends on the number of molecules is measured to get number average molecular weight of a polymer. Some of the common methods used for determining number average molecular weight are described below.

6.5.1.1. *Osmometry*

The number average molecular weight of many polymers can be measured conveniently by osmometry. The use of this colligative property is based on the fact that certain semi-permeable membranes may be constructed which permit penetration by solvent molecules but which prevent the transport of macromolecules. The Gibb's free energy of the solvent is known to be lowered by the presence of solute. Pure solvent, thus, passes through the membrane in order to lower the free energy of the system. If a thermostatted cell is constructed and pure solvent is placed on one side of the membrane while a solution of the polymer in the same solvent is placed on the other side of the membrane, a pressure gradient will develop. The process will continue until an equilibrium is reached in which the free energy change due to the pressure rise just equals the free energy change due to dilution of the solution. The equilibrium pressure developed is called the osmotic pressure (π). The principle of an osmometer is illustrated in Fig. 6.11.

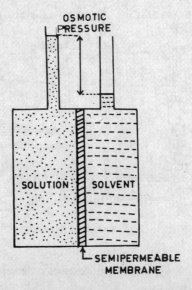

Fig. 6.11. The basic principle of an osmometer.

Application of elementary thermodynamics to the osmotic equilibrium gives the following relationship.

$$\frac{\pi}{c} = RT\left(\frac{1}{\overline{M}_n} + A_2c + A_3c^2 + \ldots\right)$$

(6.4)

where c is the concentration of the solution in g/cc, R is the gas constant and T is the absolute temperature. It is assumed in this derivation that solutions are dilute and temperature is constant. It is noted that Eq. (6.4) is a power series in the concentration. In order to obtain the number average molecular weight ($\overline{M}_n$) an extrapolation procedure yielding π/c at the limit of infinite dilution is required. As c approaches zero, all intermolecular interactions vanish and the theory becomes exact. Eq. (6.4) shows that a graphical treatment of the osmometric data produces a line having zero concentration intercepts inversely proportional to $\overline{M}_n$. Often $(\pi/c)^{1/2}$ versus c plot is more nearly linear and provides exactly the same extrapolated point.

6.5.1.1.1. Osmometers

The membranes usually used in a membrane osmometer are collodion (11-13.5% nitrogen containing nitrocellulose), regenerated cellulose, gell cellophane, poly(vinyl alcohol), poly(urethane) etc. The choice of a membrane is critical for the accurate determination of molecular weight by this method. The presence of low molecular weight polymers in the unfractionated polymer vitiates the results. These have to be removed by fractionation or extraction. A simple calculation will show that 0.1% (Wt.) of an impurity having molecular weight of 100 g/mole greatly distorts the osmotic result of a polymer having M.W. = 10^5 g/mole (an $\overline{M}_n$ of slightly greater than 50,000 g/mole is the result).

In the *static* equilibrium procedure, the osmotic pressure is determined by measuring the difference in heights of the two capillaries (connected to the cells) due to the diffusion of the solvent across the semi-permeable membrane. The main disadvantage of these osmometers is the length of time required to attain equilibrium.

In the *dynamic* method, the counter pressure required to prevent the diffusion of the solvent through the membrane is measured. In high speed membrane osmometers, an optical system in the solvent chamber detects flow through the membrane and automatically adjusts pressure by means of an electro-mechanical system to prevent any net flow. Since practically no flow of the solvent is needed to establish osmotic pressure, the whole process takes only minutes and causes no dilution of the solution on the other side of the membrane. This not only shortens the overall time per determination but also gives better results for materials which can diffuse through the membrane.

6.5.1.2 *End Group Analysis*

The number average molecular weight of linear polymers can be determined by estimating the number of end groups by chemical analysis. For the successful application of the end group method, the number and nature of end groups per polymer molecule should be reliably known. This method is applicable to linear condensation polymers and also to addition polymers. For the method to be reliable, the molecular weight of the polymer should not exceed 25,000.

6.5.1.2.1. *Condensation Polymers*

Condensation polymers usually contain functional groups which can lend themselves to chemical analysis. Thus, poly(esters) and poly(amides) contain carboxy groups, which can be estimated by titration with a base in alcoholic or phenolic solutions. The amino groups in poly(amides) can be estimated by titration against an acid under similar conditions. For the determination of hydroxyl groups in a polymer, it is reacted with a titrable reagent.

A recent method involves derivatization and use of NMR spectroscopy. For example, a poly(carbonate) derived from Bisphenol-A was capped at the ends by a trimethyl silyl ether. By observing the ratio of the peak intensity of methyl groups bonded to silicon to that of the aromatic protons of Bisphenol-A, it was possible to get a very good estimate of $\overline{M}_n$, the number average molecular weight.

6.5.1.2.2. *Addition Polymers*

As already stated the end-group method can be applied to addition polymers also when the polymerization is initiated by initiators containing identifiable groups or radioactive labelled groups. The initiator fragments get attached to one or both the ends of the chain depending on the mode of termination. If the mechanism of termination involves disproportionation, only one fragment gets attached. On the other hand, if it is by coupling, two such fragments usually get attached.

6.5.1.3. *Vapour Phase Osmometry*

This method has been found to be useful in determining the number average molecular weight of those polymers whose molecular weight is too low for measurement in a membrane osmometer. In an instrument called `vapour pressure osmometer', use is made of the small temperature difference resulting from different rates of solvent evaporation from and condensation onto droplets of pure solvent and polymer solution maintained in an atmosphere of solvent vapour. The principle of the method is that the temperature difference is proportional to the vapour pressure lowering of the polymer solution at equilibrium point and, thus, to the number average molecular weight. As there are heat losses, the measurement is conducted at several concentrations and results extrapolated to zero concentration. To get accurate results, the instrument is calibrated with low molecular weight standards. Further, such variables as drop size and line of measurement have to be carefully standardised. The method is useful in measuring $\overline{M}_n$ upto 40,000.

In a suitable instrument there is a vapour phase chamber. Droplets of solvent and polymer solution are placed with the aid of hypodermic syringes on the beads of two thermistors used as temperature sensing elements. These are maintained in equilibrium with an atmosphere of solvent vapour. The lower activity of the solvent in the solution droplet leads to an excess of solvent condensation over evaporation there, compared to the droplet of pure solvent. The excess heat of evaporization thus leads to a rise in temperature.

Ebulliometry and cryoscopy have also been used to determine number average molecular weights by using accurate temperature sensing devices.

Q. 6.6.   Explain why colligative property measurements give number average molecular weight.

6.5.2. Determination of the Weight Average Molecular Weight

6.5.2.1. *Light Scattering Measurements*

If a beam of light is passed through a colloidal solution, it is possible to see the light beam from the sides. This is the well known Tyndall effect which results from the scattering of a part of the beam of light by the colloidal particles in all directions. Light scattering measurements can be used to determine the weight average molecular weights of polymers.

6.5.2.1.1. *Light Scattering from Particles with Diameter less than λ/20*

In the case of macromolecules, the colloidal particles are highly solvated and the gel particles consist of about 90 to 99% of the solvent. The refractive index of such colloidal particles is not very different from that of the solvent.

According to P. Debye, the following power function applies in such cases,

$$\Delta\tau = \frac{32\,\pi^3}{3\,\lambda^4} \frac{RT\,c}{N_0} \left(\frac{n\,dn}{dc}\right)(d\pi/dc) \tag{6.5}$$

where π is the turbidity, $\Delta\pi$ represents the excess turbidity of the solution over that of the pure solvent, λ is the wave length and n is the refractive index. In the absence of absorption, π is related to the primary beam intensity before and after it passes through the scattering medium. If the incident intensity I_0 is reduced to I in a length l of the sample,

$$\frac{I}{I_0} = e^{-\tau l} \tag{6.6}$$

The turbidity, which is the total scattering integral over all angles, is often referred by the Rayleigh ratio $\bar{R}\,(\theta)$ [$\bar{R}\,(\theta) = R\,(\theta) - R\,(\theta)_{solvent}$] which relates the scattered intensity at angle θ to the incident beam intensity. For particles small compared to λ, the Debye equation is obtained as,

$$\frac{Kc}{R(\theta)} = \frac{Hc^\cdot}{\tau} = \frac{1}{M} + 2A_2c + \dots$$

$$K = \frac{2\pi^2 n^2}{N_0\lambda^4}\left(\frac{dn}{dc}\right)^2 \text{ and } H = \frac{32\pi^3 N^2}{3N_0\lambda^4}\left(\frac{dn}{dc}\right)^2 \tag{6.7}$$

where N_0 is the Avogadro number.

Eq. (6.7) forms the basis of determination of polymer molecular weight by light scattering. Beyond the measurement of τ or $\bar{R}\theta$, only the refractive index (n) and the specific refractive index increment (dn/dc) require experimental determination. The latter quantity is constant for a given polymer, solvent and temperature and is measured with an interferometer or differential refractometer.

Eq. (6.7) is correct only for vertically polarized incident light and for optically isotropic particles (substances in which the velocity of light and, hence, the refractive index is the same in all directions are said to be isotropic, e.g., water or glass). The use of unpolarized light requires that τ be multiplied by $(1 + \cos^2 \theta)$.

For systems with particles size $< \lambda/20$, any observation angle may be selected since the scattering function is spherically symmetrical.* However, for experimental reasons (to avoid interference of the primary beam), one usually selects values of θ not very different from $90°$.

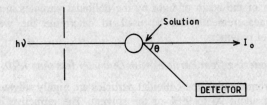

Fig. 6.12. Essential parts of a light scattering instrument.

In this simple way, through the measurement of scattered light intensity at a single angle θ, one obtains the molecular weight, but only for particle of diameters smaller than $\lambda/20$. This is true only for glycogen, a number of proteins and linear polymers of relatively low molecular weights in poor solvents.

6.5.2.1.2. Light Scattering from Particle of Diameters Larger than $\lambda/20$

For polymer molecules whose coil diameter is larger than $\lambda/20$, the scattering function is no longer spherically symmetrical. Instead, one finds a dimunition of scattered light with large particles due to interference, which is different for different observation angles. This results from the fact that for larger particles, the different scattering centres within a particle are so far from one another that the resulting scattered rays have a path difference of the order of 0 to $\lambda/20$. Since the separate scattering centres within a polymer molecule are activated by one and the same wavelength of primary light, the scattered radiation resulting from the centres is coherent and, therefore, capable of interference. The degree of dimunition resulting from the interference depends upon the path difference which in turn depends on angle θ. The scattering of light by large particles is illustrated in Fig. 6.13. Therefore, the intensity of the scattered light differs depending on the observation angle, θ). The larger the angle θ, greater is the reduction in scattering intensity $\bar{R}(\theta)$. The factor by which the scattering intensity $\bar{R}(\theta)$ (for a perpendicularly polarized incident beam undiminished by interference) is diminished by interference at the angle θ is called the scattering function $P(\theta)$ [where $\bar{R}(\theta)$ equals the scattered light in the immediate vicinity of the primary beam, i.e., with $\theta \longrightarrow 0$]. Therefore, for the scattering intensity $R(\theta)$ reduced to standard conditions and measured at the angle θ, one can write

$$\bar{R}(\theta) = \bar{R}(0)P(\theta) \tag{6.8}$$

If one substitutes this in the Debye equation (Eq. 6.7) one obtains,

$$\frac{Kc}{\bar{R}(\theta)} = \frac{1}{MP(\theta)} + \frac{2A_2c}{P(\theta)} + \cdots \tag{6.9}$$

As the macromolecules in solution are present in the form of coils, it is evident that the scattering function, $P(\theta)$ would be dependent on the scattering angle θ, the wavelength λ and the average coil diameter**$\sqrt{\bar{h}^2}$. The following approximations can

*$(R(\theta) = R(0))$
**$[\sqrt{\bar{h}^2}$ = mean end to end diameter of a coiled polymer chain].

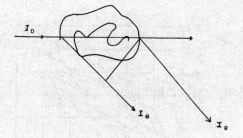

Fig. 6.13. Scattering of light by large particles

be written (Zimm),

$$\frac{1}{P(\theta)} = 1 + (8\pi^2/9\lambda^2)\bar{h}^2 \sin^2(\theta/2)$$

$$+ 2A_2c + \ldots \tag{6.10}$$

If one substitutes this in Eq. (6.9) one obtains

$$\frac{Kc}{\bar{R}(\theta)} = \frac{1}{M} + (1/M)(8\pi^2/9\lambda^2)\,\bar{h}^2 \sin^2(\theta/2) + 2A_2c + \ldots \tag{6.11}$$

If one substitutes the terms S for $[(1/M)(8\pi^2/9\lambda^2)\,\bar{h}^2]$

$$\frac{Kc}{\bar{R}(\theta)} = \frac{1}{M} + S \sin^2(\theta/2) + 2A_2c + \ldots \tag{6.12}$$

From the above equation (Eq. 6.12), it is apparent that $(Kc/\bar{R}(\theta)$ is linearly dependent on concentration c (for a constant angle of observation) and also on $\sin^2 (\theta/2)$ (provided concentration c is held constant). Numerous measurements have confirmed this prediction.

In this case, as the scattering particles were large enough in comparison to $\lambda/20$ (where λ is the wavelength of light), as is usually the case with polymer solutions, M can be replaced by $\bar{M}_w$ (for a polydisperse system). The Eq. (6.12) can, therefore, be rewritten as follows:

$$\frac{Kc}{\bar{R}(\theta)} = \frac{1}{M_w} + S \sin^2(\theta/2) + 2A_2c + \ldots \tag{6.13}$$

In a more common practice due to Zimm, all the measurements are plotted in a single diagram. The lines of low slope show the dependence of $Kc/\bar{R}(\theta)$ values on $\sin^2 (\theta/2)$ + 100c for each constant concentration, c. If one extrapolates the lines to the intercept with each corresponding 100c value of the ($\sin^2 \theta/2 = 0$) term, one obtains a series of points which together form the extrapolation of line $KC/\bar{R}(\theta) = f(c)$ at $\theta = 0$. In a similar manner, the straight lines of high slopes show the dependence of $Kc/\bar{R}(\theta)$ on $\sin^2 (\theta/2)$ + 100c at constant scattering angle θ. If one extrapolates the lines to the intercept with the corresponding abscissa values $\sin^2 (\theta/2)$ ($c = 0$), one obtains the points for the extrapolation $Kc/\bar{R}(\theta) = f(\sin^2 \theta/2)$ at $c = 0$. The two extrapolations must intercept the ordinate at the same point and this is a criterion for the validity of Eq. 6.13.

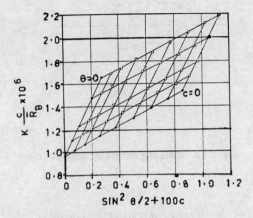

Fig. 6.14. Zimm plot for poly(styrene) ($M_w = 1,030,000$) in butanone ($A_1 = 1.29 \times 10^{-4}$ ml-mole/g; $h^2 = 460$ A, Z = 1.44)

In addition to the Zimm extrapolation procedure, there is another simple method, which in principle is not very different. This method consists of interpreting the light scattering measurements with the aid of disymmetry, Z. For this purpose one needs only to measure light scattering at two different angles, for example at 45° and 135°.

$$Z = \frac{i_{45°}}{i_{135°}} = \left(\frac{P_{45°}}{P_{135°}} \right)_{c \to 0} \tag{6.14}$$

In order to determine molecular weight from such measurements, it is necessary to know the form of the molecule, that is coil, rod or isotopic sphere. With the extrapolation procedure for the determination of M, Molecular weight on the other hand, the particle form is of no importance. Only the value of S changes (Compare e.g: Eq. 6.13) with the particle from, however, not the linear dependence of the reciprocal scattering intensity $Kc/\bar{R}$ (θ) on $\sin^2 \theta/2$.

6.5.2.2. *Light Scattering Measurements*

Developments in light scattering instrumentation have been very rapid following the availability of smaller laser radiation sources. One commercially available instrument, using laser, detects radiation scattered within a few degrees of incident beam where P (θ) is essentially unity. Knowledge of (dn/dc) and the second virial coefficient then allows calculation of $\bar{M}_w$ from a single ratio measurement. However, molecular conformational information is lost if the multiple angle study is not pursued.

Light scattering measurements traditionally require scrupulously clean, dust-free samples in order that spurious scattering may not be generated. The presence of any gel like, semi-dissolved or associated polymers causes large errors. $\bar{M}_w$ is more sensitive to their presence than to that of smaller impurities.

Q. 6.7.   What are the upper and lower limits of molecular weights between which the light scattering method can be used? What determines these values?

6.5.3. Determination of Viscosity Average Molecular Weights

The most evident distinguishing property of a polymer solution is its viscosity which is different from that of the pure solvent. Staudinger in 1930 suggested that the increase in viscosity on dissolving a polymer in a solvent may be correlated with the molecular weight of the polymer. Since then, developments in the theory and experiments have confirmed the essential basis of Staudinger's suggestion.

Determination of viscosity average molecular weight requires the measurement of the viscosity of a polymer solution relative to that of the solvent. Since only relative measurements are required, capillary viscometers are well suited for this purpose and are frequently used, the most suited viscometers being Ostwald viscometer and the Ubbelohde viscometer. In the latter, dilutions of the solution can be carried out and the measurements made at a number of concentrations.

Before proceeding further with the discussion, let us define a few common terms:

Relative viscosity $= \eta_r = \dfrac{t}{t_0}$

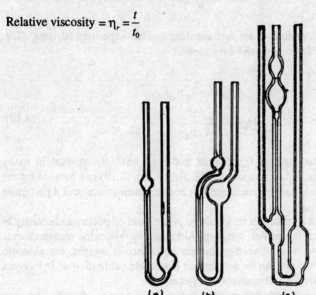

Fig. 6.15. (a) Ostwald viscometer (b) Desreux-Bischoff viscometer
(c) Ubbelohde viscometer.

Specific viscosity $\quad \eta_{sp} = \dfrac{\eta_{soln.} - \eta_{solv.}}{\eta_{solv.}} = \eta_r - 1 = \dfrac{(t - t_0)}{t_0}$

Intrinsic viscosity $\quad [\eta] = \lim\limits_{c \to 0} \dfrac{\eta_{sp}}{C} = \lim\limits_{c \to 0} \dfrac{\ln n_r}{C}$ $\qquad\qquad$ (6.15)

where $\eta_{soln.}$, $\eta_{solv.}$, η_r and η_{sp} are the solution viscosity, solvent viscosity, relative viscosity and specific viscosity, respectively at a particular temperature, c is the concentration of the polymer in g per 100 ml, t and t_0 are the times of efflux for the solution and the solvent through the viscometer.

Intrinsic viscosity of a polymer solution at a particular temperature is obtained by the use of either M.L. Huggins, (Eq. 6.16) or O. Kramer's (Eq. 6.17) equation.

$$\eta_{sp}/c = [\eta] + k_1 [\eta]^2 c \tag{6.16}$$

$$\ln \eta_r/c = [\eta] + k'_1 [\eta]^2 c \tag{6.17}$$

The current practice is to plot the results of both equations (6.16 and 6.17) using the same graph. If the higher terms of c are neglected, both lines should intersect at the ordinate to correspond to zero concentration and the constants should satisfy the condition,

$$k_1 + k'_1 = \frac{1}{2} \tag{6.18}$$

For many cases these conditions are not satisfied by the experimental data. G.V. Schulz and F. Blaschke have proposed the equation,

$$\eta_{sp}/c = [\eta] (1 + k_\eta \eta_{sp})$$

or

$$[\eta] = \frac{\eta_{sp}/c}{1 + K_\eta \eta_{sp}} \tag{6.19}$$

In which η_{sp}/c is plotted against η_{sp}. Schulz and Sing have shown that in many polymer/solvent systems, k_η has a value of 0.28. Hence, Eq. (6.19) can be used for the determination of $[\eta]$ from a single measurement at one concentration and a particular temperature.

The most important application of viscosity in the field of polymer chemistry is in the determination of molecular weights. Although by viscosity measurements alone it is only possible to observe the changes in molecular weight, the absolute value of the molecular weight can be determined if suitable calibration of $[\eta]$ versus M is available for a particular polymer/solvent system.

The original Staudinger equation relating molecular weight to intrinsic viscosity was replaced by the general relation,

$$[\eta] = KM^\alpha \tag{6.20}$$

Eq. (6.20) is known as the Mark-Howink-Sakurada equation. Neither the constant K nor the exponent α in Eq. (6.20) can be derived theoretically. The experimental method of their evaluation involves the separation of the polymer into several fractions followed by the determination of the intrinsic viscosity and absolute molecular weight ($\overline{M}_n$ or $\overline{M}_w$) of each fraction. The values of constant K and exponent α are obtained from the intercept and slope of a double logarithmic plot of intrinsic viscosity versus molecular weight. The value of the exponent α lies usually between 0.6 and 0.8 although values from 0.5 to 1.0 are possible. Further, the values of K and α vary with the solvent and temperature.

Q.6.8.   Define the following terms mathematically and explain how they are obtained experimentally (*i*) relative viscosity (*ii*) specific viscosity and (*iii*) intrinsic viscosity.

Q.6.9. What is the Mark-Howink-Sakurada equation? How can the constant K and exponent α be obtained by simple experiments and used for the determination of the molecular weight of a polymer sample.

Q.6.10. Fractions of a polymer when dissolved in an organic solvent gave the following intrinsic viscosity values at 25°C.

M (g mol^{-1}) :	34,000	61,000	130,000
$[\eta]$:	1.02	1.60	2.75

Determine K and α.

ANSWERS TO QUESTIONS

6.1. (*i*) If equal parts of the two polymers of molecular weight $M = 100,000$ and $M = 10,000$ are mixed, by definition

$$\overline{M}_n = \frac{100,000 \times 1 + 10,000 \times 10}{1 + 10} = 18181 \approx 18,200$$

$$\overline{M}_w = \frac{1 \times 10^{10} + 10 \times 10^8}{10^5 + 10^4} = 91819 \approx 92,000$$

6.2

Fraction	Weight	M_i (moles)	N_i	M_iN_i	$M_i^2N_i$
1	1000 g	1000 g/mol	1	1000	10^6
2	1000 g	10^6 g/mol	10^{-3}	1000	10^9

$$\overline{M}_n = \frac{\sum N_i M_i}{\sum N_i} = \frac{1000 + 1000}{1 + 10^{-3}} = \frac{2000}{1.001} = 2000$$

$$\overline{M}_w = \frac{\sum N_i M_i^2}{\sum N_i M_i} = \frac{10^5 + 10^9}{1000 + 1000} = 5 \times 10^5$$

$$\frac{\overline{M}_w}{\overline{M}_n} = \frac{5 \times 10^5}{2000} = 2.5 \times 10^2 = 250$$

6.3.

Sample No.	Wt. Polymer (g)	% Polymer based on total	Cumulative % polymer	DP
1	3.57	8.44	0.44	466
2	1.17	8.76	11.20	840
3	3.34	8.10	19.30	1000
4	3.74	8.84	28.14	1130
5	4.93	11.70	30.84	1270
6	3.88	9.15	48.99	1480
	15.83			
7	2.82	6.68	55.67	1480
8	4.72	11.20	60.87	1640
9	1.80	4.26	71.13	1650
10	2.78	6.86	77.69	1810
11	4.80	11.40	89.09	2420
12	4.64	10.91	100.00	2880

The integral and differential weight distribution curves can then be plotted.

6.4.
$$\overline{M}_n = \frac{\int NM\,dM}{\int N\,dM}, \overline{M}_w = \frac{\int NM^2\,dM}{\int NM\,dM}$$

If we assume that N is a constant, then

$$\overline{M}_n = \frac{M^2/2}{M} \text{ and } \overline{M}_w = \frac{M^3/3}{M^2/2}$$

Hence
$$\frac{\overline{M}_w}{\overline{M}_n} = \frac{M^3 \times 2 \times M \times 2}{3 \times M^2 \times M^2} = \frac{4}{3}$$

6.5.   This can be obtained by gel permeation chromatography. This method will give M_n, M_w and the shape of the distribution curve.

6.6.   In the measurements based on colligative properties, only the number of solute particles per unit volume of solvent is determined. Hence, if the solution contains solutes of different molecular weights, then the determination of solute molecules at a known weight concentration leads to number average molecular weight.

6.7.   The light scattering measurements are usually valid between molecular weights of 10,000 and 10,000,000. Below 10,000, there are interference effects from solvent molecules and dust particles. Above 10,000,000, there are interference effects from other polymer molecules.

6.8.

(i) $\eta_r = \dfrac{\eta_{\text{sol.}}}{\eta_{\text{solv.}}} = \dfrac{t}{t_0}$

(ii) $\eta_{sp} = \eta_r - 1 = \dfrac{\eta_{\text{sol.}} - \eta_{\text{solv.}}}{\eta_{\text{solv.}}} = \dfrac{(t - t_0)}{t_0}$

(iii) $[\eta] = \dfrac{\eta_{sp}}{c}\bigg|_{c \to 0} = \dfrac{\ln n_r}{c}\bigg|_{c \to 0}$

where t and t_0 are the times for the efflux of a measured volume of polymer solution of known concentration and the solvent respectively through the capillary of a viscometer (usually Ostwald).

6.9.   See the last para under 6.5.3.

6.10. The equation linking intrinsic viscosity with molecular weight of a polymer is,

$$[\eta] = KM^\alpha \text{ or } \log [\eta] = \log K + \alpha \log M$$

Hence, for the determination of K and α, log (η) is plotted against log M. The intercept gives the value of K and the slope the value of α. Hence, graphically $\alpha = 0.73$ and $K = 5.1 \times 10^{-4}$.

CHAPTER 7

SOME PHYSICAL PROPERTIES OF POLYMERS

7.1. INTRODUCTION

There are several factors which determine the physical properties of a polymer: the molecular weight, the secondary valency forces, polymer configuration and conformation, and the nature of chain-packing of molecules. Some of these factors have already been considered in an earlier chapter (Chapter 5). These factors influence the crystallinity of polymers, which in turn, along with other factors, influence such important physical properties as impact and tensile strength, melting point, the viscosity and solution properties of polymers.

7.2. MOLECULAR WEIGHT AND SOME MECHANICAL PROPERTIES

The physical properties of polymers notably the tensile strength and impact resistance are intimately related to the molecular weight and molecular weight distribution of the polymer. Low molecular weight compounds are liable to be brittle and of lower mechanical strength. In Fig. 7.1, the number average degree of polymerization has been plotted against a number of parameters.

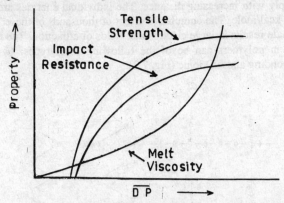

Fig. 7.1. Plot of selected properties versus $\overline{DP}$ for a hypothetical polymer.

It would be seen from this figure that both the tensile strength and impact resistance increase with increasing degree of polymerization ($\overline{DP}$) upto a point after which there is very slow increase. The melt viscosity at first increases slowly and then rapidly, after the polymer has attained a certain degree of polymerization. These generalizations must also be related to the structure of the polymer in question. The inter-chain attractions between various types of polymers vary and affect their physical properties to a large extent. For example, poly(hydrocarbons) exhibit such behaviour at a value of $\overline{DP}$ around 5000, whereas for poly(amides) the value is only 200, on account of hydrogen bonding between the different chains.

It may be pointed out that not all the physical properties of polymers are dependent on their molecular weights or for that matter on the magnitude of the intermolecular interactions. For example, the refractive index, colour, density, hardness and electrical properties are independent of the molecular weight of the polymer.

In polymer chemistry one sometimes speaks of 'threshold molecular weight', which is the minimum molecular weight a polymer must attain to develop the properties needed for a particular application. If a polymer is allowed to attain very high molecular weight, it becomes tough and intractable and cannot be easily handled. This is not desirable and the polymer chemist often needs to control the polymerization process after a certain stage depending upon the particular application.

Q.7.1.   Define the term 'threshold molecular weight.' Explain why in a poly-(hydrocarbon) such as poly(ethylene) this is much higher than that for a poly(amide)?

7.3. SECONDARY VALENCY FORCES

The secondary bonding forces present in polymers, e.g., Van der Waal and dipole-depole forces are similar to those present in small molecules. In polymers, however, many types of electrostatic forces may be present, acting among different parts of the same molecule. The strength of these forces increases with the increasing polarity and decreases sharply with increasing distance. The individual energies are low, ranging from 0.5 to 10 kcal/mole. The cumulative effect of thousands of these "bonds" along the polymer chain results in large electrostatic fields of attraction. The intermolecular bonds formed in polymers can be of the following three types: (*i*) dipole-dipole, (*ii*) hydrogen bonding and (*iii*) ionic (Fig. 7.2).

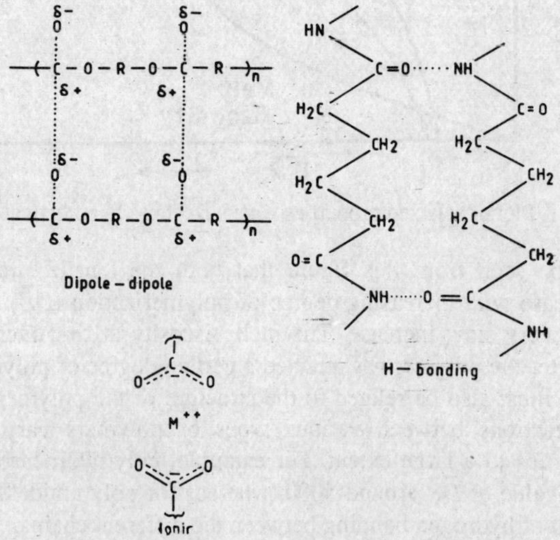

Fig. 7.2. Intermolecular bonds found in polymers

The overall bond energy of all the secondary bond forces has already been stated above. Of these, the hydrogen bond constiiutes the greatest share of the bond strength. To take an example, one may compare the physical properties of Nylon 11 with those of linear poly(ethylene). The structures of the two molecules are represented in Fig. 7.3.

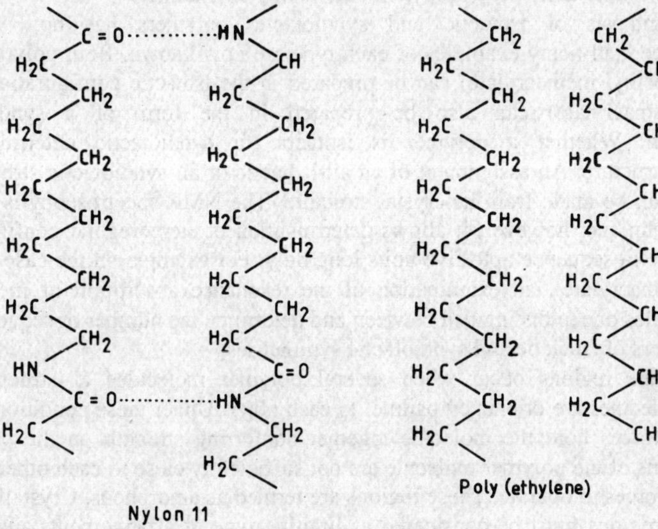

Fig. 7.3. Nylon 11 and poly(athylane).

It is evident that structurally they are similar, with the exception of the amide link in Nylon 11. A comparison of their physical properties underlines the important role that hydrogen bonding plays. Nylon 11 melts as 183-187°C and dissolves only in very strong solvents, such as formic or sulphuric acid. Linear poly(ethylene), on the other hand, melts at about 130°C and is soluble in hot aromatic solvents.

Other polymers which display hydrogen bonding are poly(vinyl alcohol) and cellulose, with their pendant hydroxyl groups, and poly(urethanes) with their carbamide linkages.

Q.7.2. What types of intermolecular bonds are present in polymers? Illustrate.

7.4. NATURE OF CHAIN-PACKING

7.4.1. Factors Affecting Crystallinity

Linear polymer chains pack both in amorphous and crystalline fashion. Small molecules are either totally amorphous and totally crystalline. Polymers, however, quite often contain both crystalline and amorphous regions. In fact polymers can be divided into two classes: those which are completely amorphous under all conditions and those which are semi-crystalline. These latter can be amorphous under certain conditions, e.g., above their melting temperature.

When vinyl monomers of the type $CH_2 = CHX$ are polymerized, the 'X' groups can be arranged along the chain in a purely random manner. Polymer with such an arrangement is known as an *atactic* polymer and is mainly amorphous. If the conditions of polymerization are so adjusted that all the 'X' groups become arranged on one side of the polymer backbone, the polymer obtained is known as *isotactic* polymer. A third type of configuration is that in which 'X' groups are arranged alternately on either side of the chain. Such a polymer is called a *syndiotactic* polymer. The subject of tacticity has been discussed at length in Chapter 5. Both *isotactic* and *syndiotactic* poly(propylene) are highly crystalline.

The synthesis of isotactic and syndiotactic polymers has now become commonplace, and many examples of each type are now known. Both poly(styrene) and poly (methyl methacrylate) can be prepared in the isotactic configuration. Poly (methyl methacrylate) can also be prepared in the form of a syndiotactic configuration. Whether a polymer is isotactic or syndiotactic determines its crystalline structure. An assignment of an all-isotactic or all-syndiotactic structure to a polymer can be made from its crystal structure. The NMR spectroscopy is a more powerful technique; however, it allows determination of stereoregular configuration of monomers in sequence upto five units long only. For example, in the case of poly (methyl methacrylate), an examination of the resonance absorption of methylene protons enables one to distinguish between and determine the number of sequences of two monomers of isotactic and syndiotactic symmetries.

Crystalline regions occur when several polymer molecules or structures of polymer molecules are orientated parallel to each other. Under these conditions weak inter-chain forces hold the molecule together conferring strength on the polymer. Other sections of the polymer molecule are not sufficiently close to each other for the inter-chain forces to operate. These regions are termed as amorphous. Crystalline and amorphous regions must be considered as limiting type of arrangements and not as distinctly different phases. For molecules to be able to approach each other sufficiently closely for the shorter range forces to operate, there must be no bulky side groups preventing close approach. If such side groups are present, they must be symmetrically arranged along the chain. Such is the case with isotactic poly(propylene).

Fig. 7.4. An unoriented crystalline polymer.

Figure 7.4 represents an unoriented crystalline polymer. It is now believed that the crystalline and amorphous demains may be represented by the switchboard model shown in Fig. 7.5, or the fringed miscelle model shown in Fig. 7.6.

The folded-chain lamella theory was put forward in 1930 when polymer single crystals were grown from polymer solutions. This theory can be applied to describe the morphology not only of single crystals grown from solution but also that of polymers crystallized from polymer melts. In commercial practice all the synthetic polymers are obtained by the latter process.

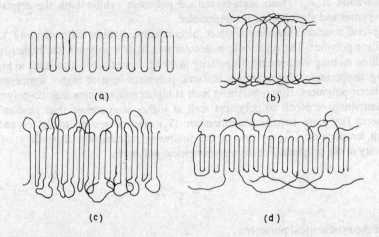

Fig. 7.5. Schematic two-dimensional representations of the fold surface in polymer lamellae: (*a*) sharp folds, (*b*) switch board model, (*c*) loop with loose folds, (*d*) a combination of these features.

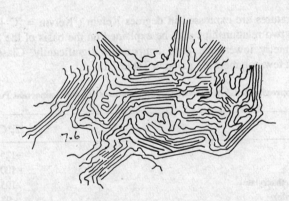

Fig. 7.6. Fringed miscelle representation of polymer crystallinity..

A plastic is mainly amorphous, but some crystallinity may be present. A fiber, on the other hand, is usually highly crystalline. A good fiber has its crystalline regions *orientated* along the length of the fiber axis. This orientation is brought about by drawing or stretching the fiber. Both crystallinity and orientation affect the physical properties of polymers.

7.4.2. Thermal Transitions in Polymers

Polymers are characterized by two types of thermal transitions, the crystalline melting temperature (T_m) and the glass transition temperature (T_g). The crystalline melting temperature is the melting temperature of the crystalline domain of a polymer sample. The glass transition temperature is the temperature at which the amorphous domain of a polymer takes on the characteristic properties of glassy state: brittleness, stiffness and rigidity. Whether a polymer sample exhibits both thermal transitions or one of them depends on its morphology. Completely amorphous polymers exhibit only T_g

and a completely crystalline polymer only T_m. Most polymers undergo only partial crystallization at T_m. These semi-crystalline polymers exhibit both the crystalline melting point and glass transition temperature.

Several structural parameters affect the crystalline melting temperature (T_m) of a crystalline polymer. The higher the molecular weight of a polymer, the higher is the crystalline melting temperature. Regularity in the chain backbone also leads to higher melting temperature. For example, isotactic polymers melt at higher temperatures than atactic polymers. Homo-polymers melt at higher temperatures than co-polymers, and alternating or block co-polymers melt at higher temperature than random co-polymers. The glass transition temperature (T_g) may be related to the crystalline melting temperature (T_m) by one of the following relationships depending on the symmetry of the polymer. For the unsymmetrical polymers,

$$T_g = \frac{2}{3} \ T_m \qquad\qquad (7.1)$$

and for the symmetrical polymers,

$$T_g = \frac{1}{2} \ T_m \qquad\qquad (7.2)$$

when the temperatures are expressed in degrees Kelvin (°Kelvin = °C + 273). The difference in the two relationships may be explained on the basis of the fact that an increase in symmetry lowers the glass temperature significantly. Glass transition temperatures of a few polymers are given in Table 7.1.

Table 7.1. Approximate Glass Transition Temperatures of Some Important Polymers

Polymer	T_g °C
Poly(ethylene)	−125
Poly(styrene)	100
Poly(methyl methacrylate)	105
Poly(acrylonitrile)	98
Poly(vinyl acetate)	30
Poly(vinyl chloride)	82
Poly(methyl acrylate)	8
Poly(propylene)	−18
cis-Poly(isoprene)	−70
Poly(dimethylsiloxane)	−123

7.4.2.1. Determination of Glass Transition Temperature and Crystalline Melting Point

The two thermal transitions are conveniently measured by measuring changes in properties such as specific volume and heat capacity. In Fig. 7.7, the changes in specific volume with temperature for completely amorphous and completely crystalline polymers are plotted.

The crystalline melting temperature (T_m) is a first order transition with a discontinuous change in specific volume at the transition temperature. The glass transition temperature is a second order transition involving only a change in the temperature coefficient of the specific volume.

More recent method of obtaining T_g and T_m are based on thermal analysis. In the differential thermal analysis method (DTA), the rate of supply of heat to a bath increases linearly with time, and a polymer sample and an inert reference material are immersed in it. A reference material is selected in such a way that its thermal capacity is close to that of the polymer sample. The temperature difference between the two samples is plotted continuously as a function of the bath temperature.

As long as the specific heat of the sample remains constant (this is below T_g), the temperature difference between the two samples is approximately zero. However, at the melting point heat absorbed is used up in the melting of the sample with the result that a decrease is observed in the temperature of the sample compound as compared to that of the reference sample. The difference in the temperatures increases till all the crystals have melted after which it decreases till a new equilibrium value of ΔT is attained (see Chapter 13).

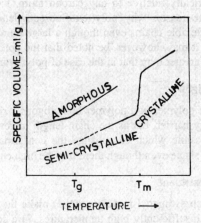

Fig. 7.7. Determination of glass transition (T_g) and crystalline melting temperature (T_m)

7.5. CHAIN FLEXIBILITY

A polymer chain is considered to be flexible if the chain segments can rotate with respect to each other with sufficient freedom. Thus, poly(butadiene) and poly(ethylene) polymers have flexible chains whereas poly(styrene) and poly(methyl methacrylate) have inflexible or rigid chains.

The flexibility of a polymer chain depends on the following factors:
  (i) Potential energy barrier to rotation
 (ii) Molecular weight of the polymer
(iii) The size of the substituents
 (iv) Cross-link density, and
  (v) Temperature

7.5.1. Potential Energy Barrier to Rotation (also see Chapter 5, Fig. 5.1)

The potential energy barrier to rotation depends on intra-molecular and inter-molecular interactions. To take an example, isolated chains of poly(acrylic acid) and poly(methacrylic acid) should be rigid due to strong intramolecular interactions of their COOH groups. In the condensed phase the intramolecular interaction of carboxylic groups lowers the potential energy barrier, the chains become flexible and assume a coiled globular structure. The reverse is also possible, when chains that are flexible in the isolated state become more rigid in the condensed phase owing to strong intermolecular interaction.

The introduction of substituetes containing polar groups in a polymer molecule intensifies the intra- and inter-molecular interactions. The most polar groups are the $-C \equiv N$ and $-NO_2$ ($\mu^* = 3.4$ D). Further, $-OH$ groups are capable of forming hydrogen bonds; hence, the energies of intra- and inter-molecular interactions are increased.

The distance between polar groups in a polymer chain has a profound effect on the flexibility of the chain. To take an example, in a co-polymer of butadiene and acrylonitrile, containing 18 parts of the latter, the flexibility of the chain is close to that of poly(butadiene). As the amount of acrylonitrile is increased in the co-polymer, the flexibility decreases and the chain becomes more rigid. If the polar substituents are arranged symmetrically relative to any carbon atom, the total dipole moment decreases. This is the reason why poly(tetrafluoroethylene) and poly(vinylidine chloride) have fairly flexible chains even though a large number of polar groups are present in the chains. It may, however, be noted that the potential energy barrier for these two polymers is greater than that in the case of poly(hydrocarbons).

7.5.2. Molecular Weight

Molecular weight of a polymer in a polymer homologous series has very little effect on the potential energy barrier, but as the chain-length increases, infinite number of conformations are possible which can change the conformation of the chain to a coiled from a rod-like shape even though there may be high energy barrier.

7.5.3. Size of the Substituent

Bulky substituents such as phenyl in poly(styrene) make the chain less flexible and consequently rigid at insufficiently high temperatures. In a co-polymer of butadiene and styrene (SBR), if styrene is randomly distributed in the chain and its percentage is small, the chain is flexible. As the amount of styrene is increased, the chain flexibility decreases. The presence of two substituents on a carbon atom decreases the chain flexibility. This is illustrated in the cases of poly(methyl methacrylate) and poly(acrylates). Poly (methyl methacrylate) contains two substituents, e.g., $-CH_3$ and $-COOCH_3$ whereas poly(methyl acrylate) has only one such substituent, a $-COOCH_3$ group. Poly(methyl methacrylate) is consequently more rigid than poly(acrylate).

7.5.4. Cross-link Density

The flexibility of natural rubber which has been vulcanized by the addition of 2-3% sulphur on the rubber content is comparable to that of unvulcanized rubber. However, if rubber is highly vulcanized, a rigid network structure is formed and the mobility of the units is decreased considerably. Consequently, the chains are said to have high cross-link density and the chains become rigid.

7.5.5. Temperature

The kinetic energy of a molecule increases with rise of temperature. As long as kT is smaller than the potential energy barrier, the units in the chain perform rotatory vibrations. When kT becomes of the order of the potential energy barrier, the units

---

$^*\mu$ = Dipole moment with the unit Debye, D.

begin to rotate freely relative to each other. Hence, the rotation becomes free with increase in temperature.

Q.7.3. What structural parameters influence the melting point of a polymer?

ANSWERS TO QUESTIONS

7.1 The threshold molecular weight of a polymer is the minimum molecular weight it must possess in order to display the properties needed for a particular application.

For polymers containing hydrogen bonding such as poly(amides), the value of this critical degree of polymerization can be low. Poly(hydrocarbons), however, must reach a degree of polymerization much higher before they attain the needed physical properties. Above this degree of polymerization, the mechanical properties increase rapidly. In the poly(hydrocarbons) like poly(ethylene), the secondary valency forces are very much smaller.

7.2 The intermolecular bonds found in polymers can be of three types: dipole-dipole, H-bonding or ionic. These are illustrated and explained in Section 7.3.

7.3 A completely amorphous polymer will not exhibit a melting point. On the other hand, a crystalline polymer will have a melting point which will be dependent on several parameters. The higher the molecular weight of a crystalline polymer, the higher will be its melting point. Regularity in the chain backbone such as that in isotactic polymers like poly(propylene) will lead to a higher melting point than that in disordered atactic polymers.

CHAPTER 8

ELASTOMERIC MATERIALS

8.1. INTRODUCTION

Polymeric materials can be classified into elastomers, fibers and plastics. Their division into these three categories is based on the chemical and geometrical structures of the individual polymer chains. Other considerations are the type of interchain arrangement within the chain aggregate at the use temperature. The present chapter deals with the preparation and structures of elastomeric materials, mainly the natural and synthetic rubber.

8.2. STRUCTURE PROPERTY RELATIONSHIPS IN POLYMERS

Elastomers—They are a group of polymers that can undergo very large reversible elongations (upto 1000%) at comparatively low stresses. The requirements for elastomer elasticity as well as the stress-strain relationships are discussed in Chapter 16. Some of the requirements for elastomers are:

(a) It should be a high polymer.
(b) It must be cross-linked and amorphous.
(c) It should have flexibility of chains.

Besides the above requirements, the modulus of the elastomer should initially be small. However, this should increase fairly rapidly with increasing elongation. Some elastomers undergo a small amount of crystallization at high elongations. For elastomers, the crystalline melting point (T_m) should be below the use temperature. These requirements are amply fulfilled by poly(isoprene) (natural rubber), which is amorphous and easily cross-linked and has a low T_g (−73°C) and T_m (28°C). The cross-linked poly(isoprene) has a low initial modulus of about 70 N*/cm^3 which increases to about 2000 N*/cm^3 at 500% elongation. The elongation is reversible upto almost the rupture point.

Fibers—The properties of fibers include high tensile strength and high modulus. These properties result when there is high molecular symmetry as well as high cohesive energy (for definition see Chapter 15) between the chains. Both require a high degree of crystallinity and the required polymers are normally linear. Poly(esters) and nylon, which are condensation polymers, exhibit these properties. If these fibers are to be drawn from melt, the T_g should be above 300°. Further, if the fabric made from these fibers is to be ironed, the T_g should be above 200°C. A small amount of cross-linking may increase the physical properties, although branching and extensive cross-linking disrupt crystallinity. Stretching of fibers improves the tensile strength enormously, as it leads to high crystallinity. The optimum values of T_m and T_g are 265°C and 50°C respectively. A part of the extensibility of fibers is reversible, a part delayed and another part permanent. The mechanical properties are not dependent on temperature.

Plastics—Plastic materials comprise a wide range of polymers, both rigid and

---

*N = Newton (unit of force) = kg m/s^2, where m = metres and s = seconds.

flexible. Their mechanical properties are intermediate between those of elastomers and fibers.

The rigid plastics are characterised by high rigidity and resistance to deformation. Examples of rigid plastics are cross-linked polymers like phenol-formaldehyde, ureaformaldehyde and melamine-formaldehyde resins. Among the linear polymers, noteworthy are poly(styrene) (T_g = 100°C) and poly(methyl methacrylate) (T_g = 105°C). The rigid plastics have high modulus, moderate tensile strength and resistance to elongation.

Flexible plastics have moderate to high crystallinity. Their T_m and T_g values vary widely. Poly(ethylene) is a typical flexible plastics. It has low modulus and tensile strength and a high elongation (500%). Other examples of flexible plastics are nylon and poly(propylene). Nylon serves as a fiber when it is highly crystalline and as a flexible plastic when it has moderate crystallinity. They have a fair range of deformability, specially at elevated temperatures. A certain part of deformability is reversible, but another part represents permanent set.

8.3. NATURAL RUBBER

8.3.1. Raw Rubber

Rubber, a poly(terpene) of formula $(C_5C_8)_n$, occurs in many tropical plants. The commercial supply comes entirely from the tree, *Heave Brasiliensis*, largely cultivated in plantations. When an incision is made through the outer bark of a rubber tree, a milky fluid called the latex oozes out. It is a colloidal suspension of rubber particles in water. Addition of dilute acetic acid or formic acid causes coagulation of rubber in a cheese like mass. In one common commercial process, this mass is passed through rollers, washing with water, to produce thin kinky sheets of light coloured pale crape rubber. In another process, it is milled into thin sheets and smoked to get amber coloured smoked rubber sheets. These two forms of raw rubber, of course, contain 95% of rubber hydrocarbon. The rubber hydrocarbon is often called *caotchouc*, a term derived from the south American words *Caa* for tears and *Ochu* for wood, thus meaning tears of wood.

8.3.2. Structure of Rubber Molecule

Many workers have contributed to the unravelling of the structure of rubber molecule. On destructive distillation of rubber, isoprene (C_5H_8) and dipentene ($C_{10}H_{16}$) were obtained. Analysis of rubber revealed that its empirical formula is $(C_5H_8)_n$. It is an unsaturated compound and adds bromine, chlorine and hydrogen halides. It was found by Pummerer that when a dilute solution of rubber in a solvent was hydrogenated in presence of a catalyst, it yielded a saturated compound $(C_5H_{10})_n$. On determining the molecular weight of rubber, the following data were obtained: *osmotic pressure* (number average), 180,000 g mol⁻¹; *viscosity*, 129,000 g mol⁻¹; molecular weight by *ultra centrifuge*, 400,000 g mol⁻¹.

Extensive investigations on the structure of rubber molecule were undertaken by C.D. Harries. On ozonolysis and hydrolysis of rubber hydrocarbon Harries obtained laevulinic acid:

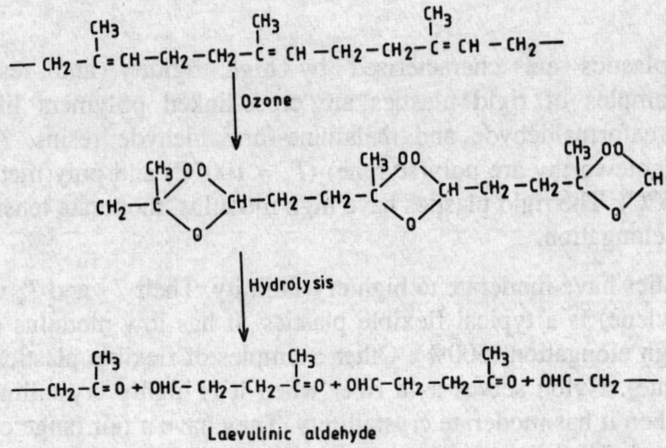

$$- CH_2 - \overset{\overset{\displaystyle CH_3}{|}}{C} = CH - CH_2 - CH_2 - \overset{\overset{\displaystyle CH_3}{|}}{C} = CH - CH_2 - CH_2 - \overset{\overset{\displaystyle CH_3}{|}}{C} = CH - CH_2 -$$

Ozone

$$CH_2 - \overset{\overset{\displaystyle CH_3}{|}}{\underset{O-O}{C}} \overset{\overset{\displaystyle O\,O}{|\,|}}{\underset{O}{CH}} - CH_2 - CH_2 - \overset{\overset{\displaystyle CH_3}{|}}{\underset{O-O}{C}} \overset{\overset{\displaystyle O\,O}{|\,|}}{\underset{O}{CH}} - CH_2 - CH_2 - \overset{\overset{\displaystyle CH_3}{|}}{\underset{O-O}{C}} \overset{\overset{\displaystyle O\,O}{|\,|}}{\underset{O}{CH}} - CH_2 -$$

Hydrolysis

$$- CH_2 - \overset{\overset{\displaystyle CH_3}{|}}{C} = O + OHC - CH_2 - CH_2 - \overset{\overset{\displaystyle CH_3}{|}}{C} = O + OHC - CH_2 - CH_2 - \overset{\overset{\displaystyle CH_3}{|}}{C} = O + OHC - CH_2 -$$

Laevulinic aldehyde

(8.1)

The rubber molecule can, therefore, be considered to be poly(isoprene), in which individual isoprene molecules are formed by 1, 4-addition.

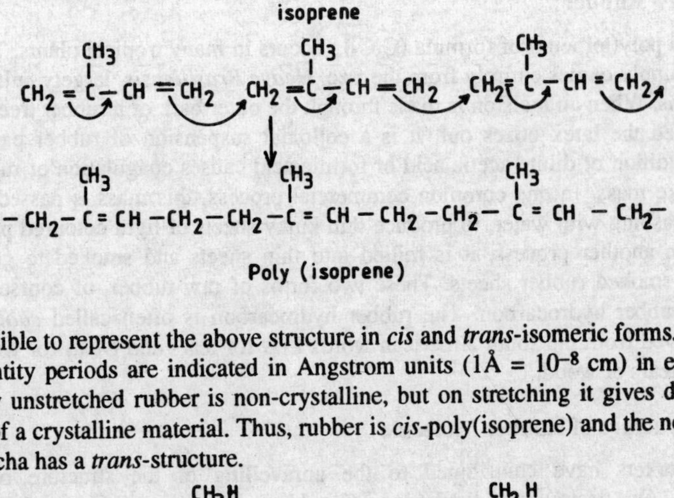

isoprene

$$CH_2 = \overset{\overset{\displaystyle CH_3}{|}}{C} - CH = CH_2 \quad CH_2 = \overset{\overset{\displaystyle CH_3}{|}}{C} - CH = CH_2 \quad CH_2 = \overset{\overset{\displaystyle CH_3}{|}}{C} - CH = CH_2$$

$$- CH_2 - \overset{\overset{\displaystyle CH_3}{|}}{C} = CH - CH_2 - CH_2 - \overset{\overset{\displaystyle CH_3}{|}}{C} = CH - CH_2 - CH_2 - \overset{\overset{\displaystyle CH_3}{|}}{C} = CH - CH_2 -$$

Poly (isoprene)

It is possible to represent the above structure in *cis* and *trans*-isomeric forms.
The identity periods are indicated in Angstrom units $(1Å = 10^{-8} cm)$ in each case. Ordinary unstretched rubber is non-crystalline, but on stretching it gives diffraction pattern of a crystalline material. Thus, rubber is *cis*-poly(isoprene) and the non-elastic guttaparcha has a *trans*-structure.

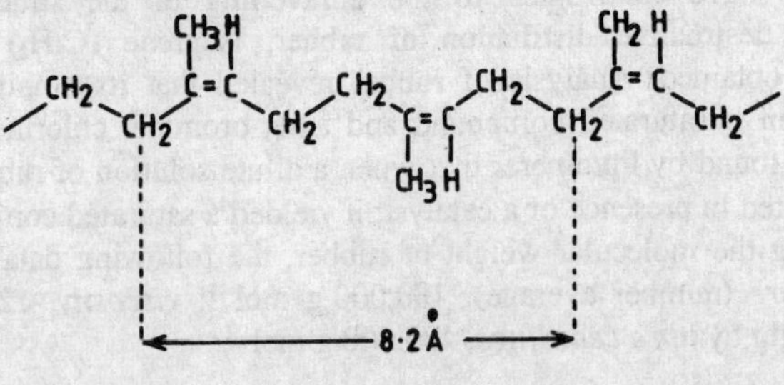

Cis Form of rubber

(1)

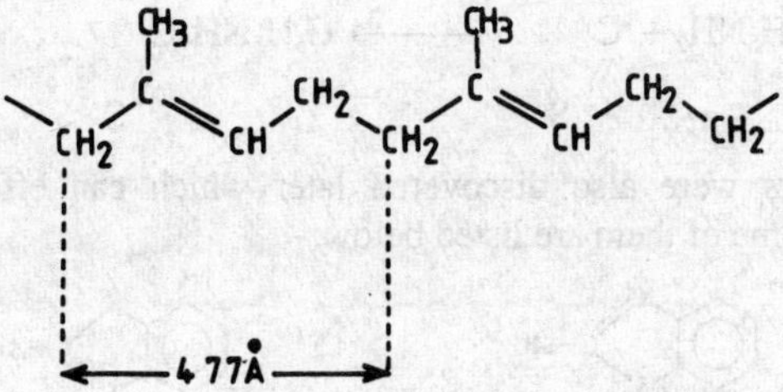

trans Form of rubber
(β – guttaparcha)
(2)

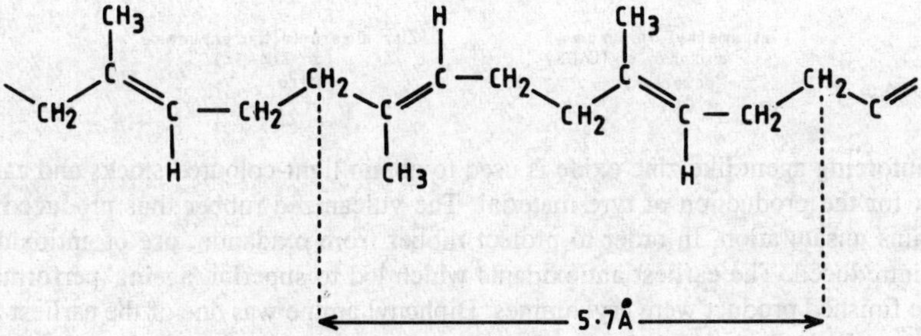

trans Form of rubber (α – guttaparcha)
(3)

8.3.3. Vulcanization of Rubber

Raw rubber is soft and tacky; its tensile strength and abrasion resistance are also low. The widespread use of rubber as an elastic material was made possible by the introduction of a process called *vulcanization*, which consists of heating rubber with sulphur. The addition of sulphur to hot rubber causes structural changes which improve the physical properties of rubber in a spectacular manner. Vulcanization establishes cross-links between linear polymer chains. Cross-linking possibly occurs both through saturation of the double bond and by way of coupling and addition reactions of C–H groups to the double bond.

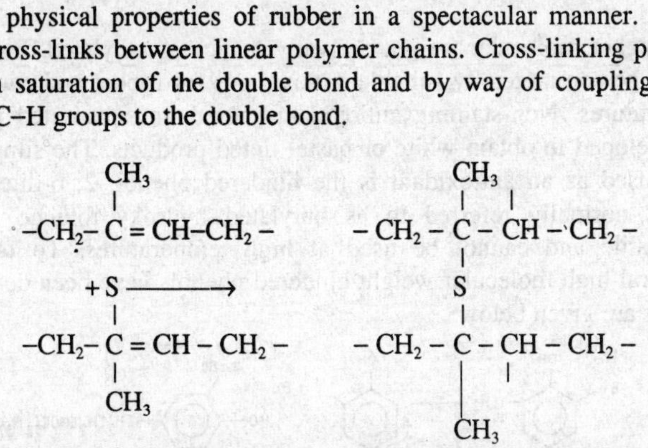

The vulcanization process is slow. It was found that it could be accelerated and vulcanized products of superior properties could be obtained if an accelerator was added. The earliest accelerator was obtained as follows:

$$C_6H_5NH_2 + \underset{S}{\overset{S}{\underset{\|}{\overset{\|}{C}}}} \longrightarrow C_6H_5NH.\underset{S}{\overset{SH}{\underset{\|}{\overset{/}{C}}}}$$

Other accelerators were also discovered later which can effect vulcanization at different rates. Some of them are listed below:

(Mercaptobenzothiazole or captax)
(medium)
(4)

(Phenylmethyldithio carbamic acid)
(medium)
(5)

$$(CH_3)_2.N.\overset{S}{\overset{\|}{C}}=SS-\overset{S}{\overset{\|}{C}}N(CH_3)_2$$

(Tetramethyl thiouram disulphide or TUADS)
(powerful)
(6)

$$(CH_3)_2N-\overset{\|}{\underset{S}{C}}-Zn-\overset{\|}{\underset{S}{C}}-N(CH_3)_2$$

(Zinc dimethyldithio carbamide or ZIMATE)
(ultra)
(7)

A reinforcing agent like zinc oxide is used to obtain light-coloured stocks and carbon black for the production of tyre material. The vulcanized rubber thus produced still contains unsaturation. In order to protect rubber from oxidation, use of antioxidants was introduced. The earliest antioxidants which led to superior 'ageing' performance of the finished product were aryl amines. Diphenyl amine was one of the earliest antioxidants ($R_1 = R_2 = H$) to be developed commercially, but it was found to be too volatile to be used in modern rubber technology.

(Substituted diphenyl amines)
(8)

(BHT)
(9)

But later its higher molecular weight derivatives such as octylated diphenyl amines were found to be commercially suitable antioxidants for rubber and could be used at higher temperatures. Non-staining antioxidants based on substituted phenols were originally developed to obtain white or pastel tinted products. The simplest member of this class used as an antioxidant is the hindered phenol, 2, 6-ditertiarybutyl-4-methylphenol, normally referred to as butylated hydroxy toluene (BHT). It is relatively volatile and cannot be used at high temperatures. To overcome this problem, several high molecular weight hindered phenols have been developed. Two such examples are given below:

(Antioxidant 2246)
(10)

(Irganox 1076)
(11)

Q. 8.1. What do you understand by the terms: (i) raw rubber, (ii) crepe rubber, (iii) vulcanized rubber, (iv) accelerator, (v) antioxidant.

Q. 8.2. What product(s) is formed by the ozonization and subsequent hydrolysis of natural rubber.

8.4. SYNTHETIC RUBBERS

During World War I, Germany was cut off from Malaya, where most of the rubber plantations were situated. The Germans, therefore, had to depend on their own ingenuity to develop rubber substitutes. Further, U.S.A. also developed its own synthetic rubbers, which could not only replace natural rubber, but in certain properties were superior to it. As a result, several synthetic rubbers were developed, a brief account of which is given in Table 10.1.

8.4.1. Buna Rubbers

Buna rubbers are based on butadiene. The name *buna* was assigned since originally these rubbers were produced by the action of sodium (Natrium) on butadiene. Butadiene can be prepared as follows:

(i) *From ethyl alcohol*

$$2\ C_2H_5OH \xrightarrow[\text{Al}_2\text{O}_3 + \text{ZnO}]{\text{Catalyst}} CH_2 = CH - CH = CH_2 + 2H_2O + H_2$$
$$\text{(Butadiene)}$$

Several by-products such as ethylene, acetylene, ethers, etc., are obtained in this process.

(ii) *Fermentation of carbohydrates*

$$(C_6H_{10}O_5)_n \xrightarrow{\text{Fermentation}} CH_3\ CH\ (OH)\ CH\ (OH)\ CH_3$$
$$\text{(Pinacol)}$$

$$\xrightarrow{\text{acetylation}} \begin{array}{c} CH_3\ .\ CH \text{——} CH - CH_3 \\ |\qquad\qquad | \\ OCOCH_3\quad OCOCH_3 \end{array} \xrightarrow{\text{hydrolysis}}$$

$$CH_2 = CH - CH = CH_2$$
$$\text{(Butadiene)}$$

(iii) *From acetylene*

$$CH \equiv CH \xrightarrow[\text{NH}_4\ \text{Cl}]{\text{Cu}_2\text{Cl}_2 +} CH_2 = CH\ .\ C \equiv CH \xrightarrow{\text{H}_2}$$
$$\text{(Acetylene)} \qquad\qquad \text{(Vinyl acetylene)}$$

$$CH_2 = CH - CH = CH_2$$
$$\text{(Butadiene)}$$

$$CH \equiv CH \xrightarrow[\text{HgSO}_4]{\text{H}_2\text{SO}_4} CH_3 CHO \xrightarrow{\text{NaOH}} CH_3CH(OH)CH_2 CHO$$

(Acetylene)　　　　　　　　　(Acetaldehyde)　　　　　　　　(Aldol)

$$\xrightarrow[\text{Ni-Al}_2\text{O}_3]{\text{H}_2} CH_3 CHOHCH_2 CH_2OH \xrightarrow[\text{catalyst}]{200°} CH_2 = CH - CH = CH_2$$

(Butane-1, 3-diol)　　　　　　　　　(Butadiene)

Table 8.1. Synthetic Rubbers

Formula of monomers	Name	Structure of polymer	Trade name	Vulcanizing agent
1. $CH_2 = CH-CH=CH_2$	1,3-Butadiene	$\{CH_2-CH=CH-CH_2\}_n$	Buna 85 Buna 115	Sulphur + accelerator
2. $CH_2=CH-CH=CH_2$ + $C_6H_5CH=CH_2$	1,3-Butadiene + Styrene	$-CH_2-CH=CH-CH_2-$ $-CH_2-CH-$ $\quad\quad\mid$ $\quad\quad C_6H_5$	Buna-S GRS SBR	Sulphur + accelerator
3. $CH_2=CH-CH=CH_2$ + $CH_2=CH-CN$	1,3-Butadiene + Acrylonitrile	$-CH_2-CH=CH-CH_2-$ $-CH_2-CH-$ $\quad\quad\mid$ $\quad\quad CN$	Buna-N Perbunan GR-N	Sulphur + accelerator
4. $CH_2=C-CH=CH_2$ $\quad\mid$ $\quad Cl$	2-Chloro-1,3-butadiene (chlorophene)	$\{CH_2-C=CH-CH_2\}_n$ $\quad\quad\mid$ $\quad\quad Cl$	Neoprene GR-M	ZnO + MgO
5. $ClCH_2CH_2Cl$ + Na_2S_4	Ethylene di-chloride + Sodium poly-sulphide	$\quad\quad S$ $\quad\quad \parallel$ $-CH_2-CH_2-S-$ $-S-CH_2-CH_2-$ $\quad\quad \parallel$ $\quad\quad S$	Thiokol-A	Zinc oxide + S as accelerator
6. $ClCH_2CH_2OCH_2CH_2Cl$ + Na_2S_4	2,2'-Dichloro diethy ether + Sodium polysulphide	$-CH_2-CH_2-O-CH_2-$ $-CH_2-S-S-$ $\quad\quad \parallel \ \parallel$ $\quad\quad S \ \ S$	Thiokol-B GR-P	ZnO + accelerator
7. $\quad CH_3$ $\quad\ \ \mid$ $CH_2=C-CH_3$	Isobutene	$\quad\quad CH_3$ $\quad\quad\ \ \mid$ $-CH_2-C-CH_2-$ $\quad\quad CH_3$ $\quad\quad\ \ \mid$ $\quad\quad -C-$ $\quad\quad\ \ \mid$ $\quad\quad CH_3$	Vistanex	Not vulcanized
8. $\quad CH_3$ $\quad\ \ \mid$ $CH_2=C-CH_3 +$ $\quad (98\%)$ $CH_2=C-CH=CH_2$ $\quad\ \ \mid$ $\quad CH_3$ $\quad (2\%)$	Isobutylene + Isoprene	$\quad\quad CH_3$ $\quad\quad\ \ \mid$ $-CH_2-C-CH_2-$ $\quad\quad\ \ \mid$ $\quad\quad CH_3$ $\quad\quad -C=CH-CH_2-$ $\quad\quad\ \ \mid$ $\quad\quad CH_3$	Butyl GR-I	Sulphur + Special accelerator

8.4.2. Buna 85 and Buna 115

These rubbers are manufactured by the action of sodium on butadiene. The names are derived from the molecular weight of the products namely 85,000 and 115,000. These rubbers are manufactured only to a limited extent. It is possible that the polymers may have some 1, 2-polybutadiene structure.

8.4.3. Buna-S (SBR, GRS) Rubber

It is a co-polymer of styrene with butadiene. Its trade names are styene-butadiene rubber (SBR) and Government regulated styrene rubber (GRS). The recipe for the preparation of this rubber is given in table 3.6. Usually, the co-polymer containing 75 parts butadiene and 25 parts styrene is suitable for rubber compositions. It is vulcanized in a similar manner as natural rubber. This rubber is superior to natural rubber with regard to mechanical strength and is used in tyre industry as it has a superior abrasion resistance.

$$- CH_2 - CH = CH - CH_2 - CH_2 - CH = CH - CH_2 - CH_2 - CH -$$
$$\overset{\mid}{\underset{C_6H_5}{}}$$

$$- CH_2 - CH = CH - CH_2 -$$
(Styrene-butadiene rubber)

8.4.4. Buna-N (Perbunan, GR-N) Rubber

This is a special purpose rubber and is a co-polymer of butadiene with acrylonitrile. It contains about 25% of acrylonitrile and 75% butadiene. This co-polymer is also vulcanized in a similar manner as the natural rubber. The product has good resistance to oils and solvents and has superior abrasion resistance. It is also superior to natural rubber towards ageing.

$$- CH_2 - CH = CH - CH_2 - CH_2 - CH = CH - CH_2 - CH_2 - CH - CH_2 - CH = CH - CH_2 -$$
$$\overset{\mid}{\underset{CN}{}}$$
(Acrylonitrile-butadiene rubber)

8.4.5. Neoprene Rubber (GR-M)

The monomer is known as chloroprene and can be synthesized from acetylene by the following steps:

$$CH \equiv CH + CH \equiv CH \xrightarrow[NH_4Cl]{Cu_2Cl_2} CH_2^4 = CH^3 . C \equiv CH^1 \xrightarrow[\text{1, 4-addition}]{HCl}$$
(Vinylacetylene)

$$\underset{\underset{Cl}{\mid}}{CH_2} - CH = C = CH_2 \xrightarrow{\text{rearrangement}} CH_2 = CH - \underset{\underset{Cl}{\mid}}{C} = CH_2$$
(2-Chlorobutadiene)
or chloroprene

The polymerization of chloroprene takes place readily and no catalyst is required. However, it is very slow in the absence of oxygen. Vulcanization of neoprene takes place in the presence of magnesium oxide. The probable mechanism is as follows:

$$- CH_2 - CH = \underset{\underset{Cl}{|}}{C} - CH_2 - CH_2 - CH = \underset{\underset{Cl}{|}}{C} - CH_2 -$$

$$+ MgO \qquad\qquad + MgO$$

$$\underset{\underset{-CH_2 - CH = C - CH_2 - CH_2 - CH = C - CH_2 -}{|}}{Cl} \longrightarrow$$

$$- CH_2 - CH = \underset{\underset{O}{|}}{C} - CH_2 - CH_2 - CH = \underset{\underset{O}{|}}{C} - CH_2 -$$

$$- CH_2 - CH = C - CH_2 - CH_2 - CH = C - CH_2 -$$

Vulcanization of neoprene

One recipe for obtaining good gum resin from neoprene is as follows: neoprene, 100 parts; magnesium oxide, 10 parts; zinc oxide, 5 parts; and wood resin, 5 parts.

In this reaction, sulphur acts as a good catalyst and tricresyl phosphate as a good softener. Neoprene is particularly noted for its resistance to chemicals, oils, light and heat and its resistance to air oxidation.

8.4.6. Polysulphide Rubbers (Thiokol)

A reference to Table 8.1 would show that the structure of these rubbers is different from that of the conventional rubber hydrocarbons. As they do not contain an ethylenic double bond, these rubbers are resistant to oxidation and solvents. Their main use is to preserve the rubber tread. A typical method for preparing thiokol is as follows:

$$Cl - R - Cl \quad + \quad Na_2S_4 \longrightarrow$$

$$\underset{\underset{S}{\parallel}}{\overset{S}{\parallel}}\hspace{-0.5em} -S - S - R \left[\underset{\underset{S}{\parallel}}{\overset{\overset{S}{\parallel}}{S - S - R}} \right]_n \underset{\underset{S}{\parallel}}{\overset{\overset{S}{\parallel}}{S - S - R}} -$$

The method of preparation of thiokol rubbers involves thorough mixing of the dichloride and the polysulphide in presence of a dispersing agent. The latex like product obtained is thoroughly washed with water and coagulated with acid to a rubber like mass. This can be further worked in a mill by adding re-inforcing agent and modifying reagents.

8.4.7. Poly(butenes) and Butyl Rubbers

Poly(butenes) are produced by the action of Friedel-Crafts catalysts like $AlCl_3$, BF_3, etc., on isobutylene.

$$n\ CH_2 = \underset{\underset{CH_3}{|}}{\overset{\overset{CH_3}{|}}{C}} \longrightarrow - CH_2 - \underset{\underset{CH_3}{|}}{\overset{\overset{CH_3}{|}}{C}} - CH_2 - \underset{\underset{CH_3}{|}}{\overset{\overset{CH_3}{|}}{C}} -$$

The polymers cannot be vulcanized as no double-bonds are present. These polymers are important as they are inert.

Another rubber which has been developed recently uses 98% isobutylene and 2% isoprene for polymerization. This polymer has occasional double bonds and can be vulcanized. A typical mixture for vulcanization contains zinc oxide (5%), stearic acid (3%), sulphur (1.5%) and an accelerator (1%). The vulcanized product shows good resistance to ageing and chemicals. Its elasticity is also good.

Q.8.3.　Write the structural formulae of the following synthetic rubbers: (i) SBR rubber, (ii) Buna-N rubber, (iii) Neoprene rubber, (iv) Thiokol-A rubber.

Q.8.4.　Which vulcanizing agents are used for the following synthetic rubbers? (i) Buna-85 and Buna-115, (ii) Neoprene (iii) Thiokol-B, and (iv) butyl rubber.

ANSWERS TO QUESTIONS

8.1. (i) The latex from the rubber tree is coagulated by the addition of dilute acetic acid. The cheese like natural rubber obtained after washing with water is the raw rubber.

　(ii) The raw rubber when passed through rollers and washed with water and consolidated in the form of sheets of pale yellow colour is called 'crepe rubber.'

　(iii) Raw rubber is soft and tacky and has poor mechanical properties. When heated with sulphur, sulphur cross-links are introduced between different chains. This improves the properties of rubber. This process is called vulcanization.

　(iv) The vulcanization process of heating raw rubber with sulphur is slow. To accelerate this process certain additives, like zinc dimethyldithio carbamide are mixed with the rubber-sulphur material. The vulcanization is achieved in minutes rather than hours.

　(v) Rubber stocks deteriorate on standing due to the action of ozone. For protection against ageing an antioxidant like phenyl-β naphthyl amine is added to rubber.

8.2. Please see Section 8.2.2.

　(i) $-CH_2-CH=CH-CH_2-CH_2-CH=CH-CH_2-CH_2-$ CH$-CH_2-CH=CH-CH_2-$

　　　　　SBR rubber　　　　　　　　　　　C_6H_5

　(ii) $-CH_2-CH=CH-CH_2-CH_2-CH=CH.CH_2-CH_2-$ CH$-CH_2-CH=CH-CH_2-$

　　　　　Buna-N rubber　　　　　　　　　　CN

　(iii) $\{CH_2-CH=C-CH_2\}_n$
　　　　　　　　|
　　　　　　　Cl　Neoprene rubber

　　　　　　　S
　　　　　　　‖
　(iv) $\{CH_2CH_2-S-S\}_n$
　　　　　　　　　‖
　　　　Thiokol-A　S

8.4. (i) Sulphur + accelerator　　　　　(iii) ZnO+sulphur

　(ii) MgO+ZnO　　　　　　　　　　(iv) Sulphur+special accelerator

CHAPTER 9

FIBER-FORMING POLYMERS

9.1. INTRODUCTION

The development of synthetic fiber industry has been intimately associated with the growth of plastics as a whole. The synthetic fibers (See Table 9.1) can be broadly divided into the following classes: (i) rayons, (ii) proteins, (iii) nylons, (iv) poly(esters) (terylene), and (v) vinyls. Table 9.1 gives a list of the important synthetic fiber types, their sources and trade names.

9.2. RAYONS

The important processes for the preparation of rayons are given below:

9.2.1. Viscose Process

Cellulose in the form of high quality wood pulp or cotton linters, or a mixture of these materials is used in this process. The first step consists of 'Steeping' 150–200 parts of pulp in the form of sheets in 18% caustic soda solution. This step requires from 30 to 90 minutes at 18°C. The excess caustic soda is removed and the sheets pressed until the weight of the sheets is three times their original weight. After pressing, the wet sheets are removed and transferred to powerful shredders where they are milled and disintegrated into a fluffy condition. The temperature is maintained at 20°C.

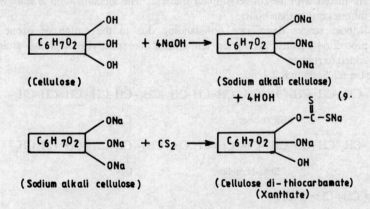

Fig. 9.1. Rayon viscose

The fluffy alkali cellulose is stored at a temperature of 25–30°C for 48 hours so that the chain-length becomes shorter. The alkali cellulose is removed to revolving drums and reacted with carbon disulphide whereby thiocarbamate of cellulose is formed. The reaction is allowed to proceed for four hours at 25–30°C. The reactions that take place can be represented as in Fig. 9.1.

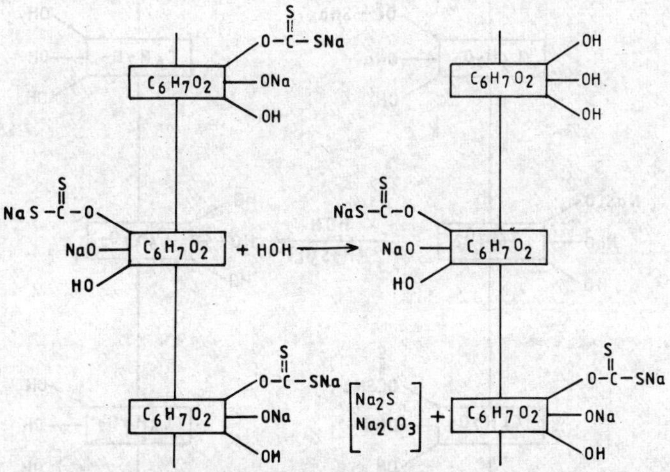

Fig. 9.2. Rayon viscose ripening

The xanthate is now transferred to water jacketed mixers where it is mixed with dilute NaOH at 17°C. The resulting orange coloured liquid is known as 'viscose. After thorough mixing, the viscose solution is transferred to large storage tanks, dissolved air removed by evacuation, filtered to remove dissolved particles and allowed to 'ripen', until ready for spinning operation. Hydrolysis, which continues during the ripening process, is controlled by controlling the temperature and time of ripening.

The viscose solution, on reaching the spinning machine, is filtered and extruded by pressure through spinnerets in a precipitation bath. One popular type of bath contains an aqueous solution of sulphuric acid, sodium sulphate, zinc sulphate and glucose. The sulphuric acid, which is mainly responsible for regeneration in most cases, is about 8–12%. The acid in the coagulating and regenerating bath precipitates the cellulose xanthate and further breaks down the xanthate, releasing cellulose. This cellulose is known as the 'regenerated cellulose.'

After spinning, the cakes or bobbins of rayon are given a wash to remove the acid. The final treatment involves removal of sulphur compounds and bleaching.

9.2.2. Cuprammonium Process

In the cuprammonium process cotton linters are used as the source material for cellulose. Into a large mixer are added measured amounts of (i) ammonium hydroxide and (ii) copper hydroxide. After mixing, cotton linters are added and with proper agitation, dissolution of cellulose can be accomplished.

After dissolving cotton, the solution is removed to a second mixing tank and water added until the cellulose concentration is about 4%. After deaeration, the solution is filtered and pumped to a storage tank for spinning.

Spinning can be accomplished by extruding through spinnerets into sodium hydroxide solution (4% or stronger) maintained at room temperature. The thread is given a slight stretch before winding. Both ammonia and copper are recovered in the process.

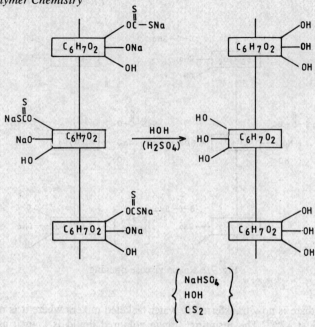

Fig. 9.3. Regeneration in rayon production.

Table 9.1. Synthetic Fibers

Type	Source	Trade name
Regenerated cellulose	Natural cellulose	Viscose
Cellulose acetate	Partial acetylation of cellulose	Dicel
	Total acetylation of cellulose	Tricel
Poly(amides)	Condensation of diamines with dicarboxylic acids	Nylon
	Condensation polymers produced from cyclic amides	Perlon
Poly(esters)	Condensation of methyl esters of dicarboxylic acids and dihydric alcohols	Terylene, Dacron
Poly(urethanes)	Condensation polymer of poly(ethylene glycol) and diisocyanate	Lycra
Poly(acrylonitrile) co-polymers with acrylonitrile	Free radical emulsion polymerization of monomers	Orlon, Acrylan, Courtelle
Co-polymer of vinyl chloride with vinylidine chloride	Free radical emulsion polymerization of monomers	Saran
Poly(propylene)	Obtained by use of Ziegler-Natta catalyst	Ulstron
Poly(ethylene)	High pressure polymerization of ethylene	Courlene
Co-polymer of vinyl chloride with vinyl acetate	Free-radical polymerization of a mixture of monomers	Vinyon

9.2.3. Cellulose Acetate Rayon

Cellulose acetate rayon differs from both viscose and cuprammonium types in that it is cellulose ester and not regenerated cellulose. The cellulose acetate required for rayon should contain 54.5% of combined acetic acid (corresponding to the average degree of acetylation of slightly over two hydroxyl groups).

The primary acetate or the triacetate is treated with water in the form of dilute acetic acid and allowed to remain for some time under controlled temperature conditions. When the hydrolysis proceeds to around 54.5% combined acetic acid, the entire mass is precipitated by diluting with water and agitating vigorously. An acetate solution is obtained, which is soluble in acetone and is termed as 'secondary acetate'.

The concentration of cellulose acetate in solution generally varies between 18 and 25% depending on the viscosity desired. The acetate flake is dissolved in the solvent (acetone) in large closed mixers. When dissolution is complete, the charge is blended with several other batches to ensure uniformity. After de-aeration, it is transferred to a feed tank.

The spinning operation consists of extruding the viscous acetate through fine-holes (0.002 to 0.005 inch in diameter) in a nickel or some other non-corrosive metal spinneret into the 'spinning cell' through which warm air is circulated to evaporate the solvent. As the filaments leave the cell, they are drawn together and led to some form of winding device.

9.3. PROTEIN FIBERS

The first synthetic fiber of this type was developed in Italy in an attempt to find a substitute for wool. Casein is used as the protein and the extruded filaments of the protein are hardened by treatment with formaldehyde. Formaldehyde reacts with the free amino groups of the protein and makes it less hygroscopic. The casein used for fibers is obtained from skimmed milk, which should be free from fats, sugars etc. Rennet, the enzyme found in the stomach of pigs, is added to slightly warmed skimmed milk. Within quarter of an hour, curds are formed. The curds formed in this way are broken up, the whey is drained off, and the curds washed. They are then subjected to hydraulic pressure to get rid of water. After being finely powdered, they are then dried at about 40°C. The casein is now in a form fit for undergoing the next operation.

An alkaline solution of casein is prepared by keeping its 10% solution at a temperature of 50°. To strengthen the fibers, a small amount of a calcium, barium or aluminium salt is added. The solution is de-aerated and allowed to ripen as in the case of viscose solution used for the viscose rayon. Ripening (hydrolysis) is carried out to a point till the cross-links in the side chains are broken.

The ripened solution is then extruded through the rough spinneret into a coagulating bath. The coagulating bath may contain from 2 to 20% of the acid chosen. Coagulating bath containing (i) 20% H_2SO_4 – 27% Na_2SO_4 or 20% H_2SO_4 – 14% $Al_2(SO_4)_3$ – 5% NaCl may be used.

Protein fibers have certain characteristics of wool, but lack in wet strength. The greatest use of felt fibers is in the manufacture of felt hats. Casein fibers mixed with other fibers, principally wool, are spun into yearn for making clothing.

Q.9.1.   Give the equations outlining the main steps in the manufacture of viscose rayon.

Q.9.2.   Explain the terms 'primary' and 'secondary' acetate.

Q.9.3.   What is understood by the term 'ripening' in the manufacture of casein fibers?

9.4. NYLONS

Nylon is the generic name of the synthetic linear poly(amides) obtained by the condensation polymerization of a diamine and a diacid or by the self condensation of an amino acid like ε-aminocaproic acid. The preparation of Nylon plastics has already been described earlier (see Section 2.5.2).

For the manufacture of fibers, the material is re-melted prior to spinning. The molten material is forced through spinnerets. After extrusion, the filaments immediately cool and solidify. The filaments are conveyed and wound on bobbins. In this form the yarn is unsatisfactory for most uses, but by using a relatively small force it is cold drawn to as much as seven times its original length. During this operation the molecules are orientated parallel to the fiber axis and the yarn becomes tough and elastic. Nowadays it is mostly used for making ladies' stockings and hosiery. Nylon ropes are as strong as steel ropes.

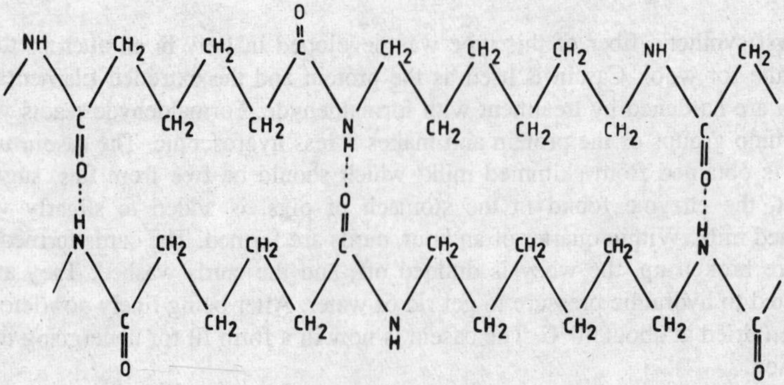

Fig. 9.4. Hydrogen bonding in Nylon.

In order that the Nylon 6,6 fibers be strong, the molecule should have from 50 to 90 molecules of each variety in the combination. Further, the Nylon molecules are inter-linked with other similar molecules by weak hydrogen bonds as shown above (Fig. 9.4.):

The hydrogen bonds are weak Van der Waal's forces individually but since their number is large, these forces assume importance.

The preparation of the polymer from ε-aminocaproic acid has already been described earlier (Section 2.5.2). Perlon fiber is also obtained by a similar method.

Q.9.4.   What is cold drawing and how does it affect the properties of Nylon?

5. How does hydrogen bonding affect the properties of Nylon fiber?

9.5. POLY (ESTER) FIBERS

The most important poly(ester) fiber is Terylene. This name was given to it, as it is the condensation product of *tere*phthalic acid and ethy*lene* glycol. The preparation of Terylene polymer has already been described earlier (see Section 2.5.1). The method of spinning of Terylene yarn is similar to that described for Nylon. Terylene mixed with cotton or wool is widely used for the fabrication of clothing, which is crease-resistant.

9.6. VINYLS

9.6.1. Acrylonitrile Fibers(Orlon)

Poly(acrylonitrile) is prepared by the free-radical polymerization of acrylonitrile monomer. Its structure is not known definitely, but from its reactions, it appears to have the head-to-tail structure as given below:

$$- CH_2 - CH - CH_2 - CH - CH_2 - CH -$$
$$\quad\quad | \quad\quad\quad\quad | \quad\quad\quad\quad |$$
$$\quad\quad CN \quad\quad\quad CN \quad\quad\quad CN$$

(I)

The molecular weight of poly(acryonitrile) is between 40,000 and 70,000 and its density is in the range 1.13–1.16. It decomposes at 220° and is soluble in dimethylformamide. It is resistant to heat and has good mechanical properties. It is more light-resistant than most other polymers.

Poly(acrylonitrile) fibers resemble wool in some of the properties, but they are very difficult to dye. Co-polymerization of acrylonitrile with amines such as vinyl pyridine gives the following co-polymers which can be easily dyed.

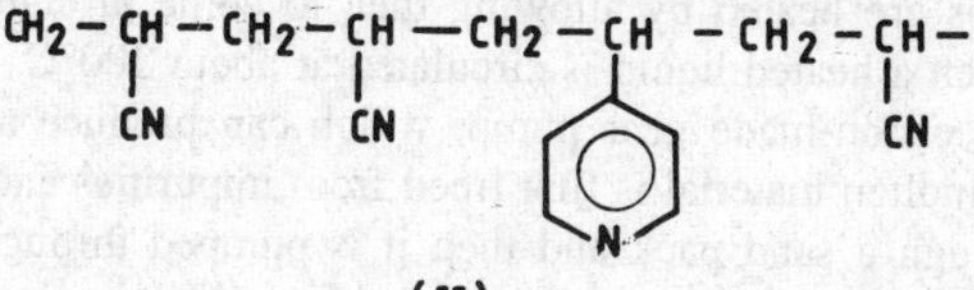

$$CH_2 - CH - CH_2 - CH - CH_2 - CH - CH_2 - CH -$$
$$\quad\quad | \quad\quad\quad\quad | \quad\quad\quad\quad\quad\quad\quad\quad\quad\quad |$$
$$\quad\quad CN \quad\quad\quad CN \quad\quad\quad\quad\quad\quad\quad\quad\quad CN$$

(II)

9.6.2. Vinyon

This fiber is obtained by the co-polymerization of vinyl chloride and vinyl acetate. The co-polymer is linear and has the probable structure (random) shown below.

$$- CH_2 - CH - CH_2 - CH - CH_2 - CH - CH_2 - CH -$$
$$\quad\quad | \quad\quad\quad\quad | \quad\quad\quad\quad | \quad\quad\quad\quad |$$
$$\quad\quad Cl \quad\quad\quad Cl \quad\quad\quad OCOCH_3 \quad Cl$$

(III)

As the formula for the co-polymer suggests, it is made up of vinyl chloride and vinyl acetate. The co-polymer used for rayon manufacture consists of 77–90% of vinyl chloride and 10–12% of vinyl acetate.

The dry spinning process described in the case of Nylon is used in the spinning of vinyon yarns also. During stretching the molecules are orientated in a direction parallel to the axis of the fiber. The drawing operation increase its tensile strength.

9.6.3. Saran

Saran or vinylidine co-polymer is a random co-polymer of vinylidine chloride with vinyl chloride. The composition and accordingly properties of the co-polymer may be varied over a wide range from a flexible, moderately soluble resin, having a softening point of around 70°C to a hard, tough thermoplastic, having a softening point of at least 180°C.

9.7. SPINNING

All fibers, whether natural or synthetic, consist of large molecules. The first job in the production of a synthetic fiber is to convert it into fine filaments. There are three methods of carrying out this operation. Some plastic materials such as Nylon and Terylene melt to a viscous liquid at temperatures around 300°C. This molten mass can be extruded through fine holes in the form of filaments, which cool and solidify. This is the method of melt spinning. In the second and third methods, called the dry spinning process and the wet spinning process, the material is dissolved in some suitable solvent and the viscous solution is extruded through orifices. In the dry spinning process, the extruded mass comes in contact with hot air. The solvent evaporates on coming in contact with hot air and is subsequently recovered. The extruded filaments solidify. In the wet spinning process, the extruded mass is precipitated in a precipitating bath.

The choice of the process to be adopted depends on the nature of the filaments.

9.7.1. Melt Spinning

The method of melt spinning is explained below for the typical case of Nylon. Nylon in the form of small chips is fed into a hopper. The hopper in turn feeds a pot in which the melting occurs in an inert atmosphere. It is necessary to surround the melting Nylon with an inert atmosphere of nitrogen to avoid oxidation of the material. The chips are heated by allowing then to come in contact with a coil of pipes through which a heated liquid is circulated at about 300°C. The molten Nylon then goes to a precision-made gear pump, which can produce a pressure of 1000 1bs/sq. inch. The molten material is first freed from impurities and non-molten chips by passing it through a sand pack and then it is pumped through a spinneret. The spinneret consists of a flat disc of steel about 1/4 inch thick having a diameter of 2–3 inches, perforated with a large number of evenly spaced holes. The filaments solidify when they attain a temperature of 260°C. At this point the yarn is free from moisture and it is necessary to condition it by passing it through rolls immersed in water and a lubricant. The yarn is then wound on a bobbin. A similar process is carried out for the melt spinning of Terylene.

9.7.2. Dry Spinning

In this process, used for spinning cellulose acetate yarn, the material is dissolved in acetone to obtain a viscous solution of requisite consistency. It is freed from impurities and insoluble matter. For this purpose, the solution is pumped through a porcelain filter of cylindrical shape. A pump forces the solution through a spinneret into an enclosed heated chamber through which hot air is blown in the same direction as the filaments. The solvent evaporates in the upper part of the apparatus and the multi-filament yarn is withdrawn from the bottom without any further treatment. It is very important to recover the solvent. A 95% recovery of the solvent (acetone in this case) can be achieved by absorbing the solvent on solid carbon or dissolving it in water from the air charged with the solvent. The recovered solvent is used again. For quick evaporation of the solvent, the spinning cell is usually kept at a temperature of around 100°C.

9.7.3. Wet Spinning

In this process the solution is passed through a spinneret into a coagulating bath which removes the solvent and precipitates the yarn. This method of spinning is particularly relevant to the manufacture of *viscose* yarns. The chemistry and technology of producing viscose fiber has already been described earlier (see Section 9.2.1).

9.7.4. Cold Drawing

The filaments spun in melt spinning are completely unorientated. Consequently, a separate drawing operation is necessary to produce the orientation of the crystallites which is necessary for optimum physical properties. During earlier times two separate steps were necessary. Now they are combined in one single continuous operation. The drawing step utilizes two rolls, one to feed the undrawn yarn from a supply package at a velocity, S_1 and the other moving four times as fast to collect the drawn yarn at velocity S_2. The filaments may pass over a metal pin between the twin sets of rolls; drawing is localized in the neighbourhood of the pin. The yarn is then collected on a strong metal bobbin.

ANSWERS TO QUESTIONS

9.1 See the steps given under Section 9.2.1.

9.2 There are three available hydroxyl groups in the anhydroglucose units joined together to give a cellulose chain. When cellulose is completely acetylated, all these three hydroxyls are acetylated. This is called a 'primary acetate'. When, however, the primary acetate is partially hydrolysed by allowing it to stand in open pans in presence of dilute acetic acid, it is called a 'secondary acetate.' In the secondary acetate, approximately 2.0 hydroxyl groups per molecule remain in acetylated condition.

9.3 Casein contains cross-links in the side chain. When treated with alkali and allowed to stand, these cross-links are hydrolysed and chains are also shortened in length. This is called the process of 'ripening'.

9.4 Before stretching of the Nylon fiber, the molecules are distributed haphazardly. On drawing the filament, the molecules arrange themselves in a direction parallel to the axis, thus giving it immense tensile strength.

Unorientated   Cold drawing →   Orientated

9.5 Hydrogen bonding in Nylon chains is illustrated (Fig. 9.4) in which the oxygen atom of the carbonyl group of one chain is hydrogen bonded to the hydrogen of the NH group of the second Nylon chain. These forces are small individually, but since a large number of these hydrogen bonds are involved, the net force assumes importance.

CHAPTER 10

FABRICATION OF POLYMERS

10.1. INTRODUCTION

If a substance during its stage of manufacture has passed through a plastics condition (is capable of flowing), it is called a *plastic*. *Plastics* are, therefore, an arbitrary group of artificially prepared substances of organic origin which sometimes during the stage of their manufacture could be caused to flow. Plastics can be divided into two classes: (i) thermoplastics and (ii) thermosetting.

Thermoplastic resins are those which under the influence of heat and sometimes of pressure become plastic (can flow) and can be moulded into different shapes under this condition. They retain their shapes on cooling. They can be remoulded into different shapes on heating. Examples are poly(styrene) and cellulose acetate.

Thermosetting resins are those that under the influence of heat and pressure become soft and can be moulded into different shapes. On further heating, they become hard and infusible on account of chemical change. They retain their shapes and cannot be remoulded. Examples are phenol-formaldehyde and urea-formaldehyde resins.

10.2. MOULDING OF PLASTICS

10.2.1. Moulding of Thermosetting Resins

The simplest method of moulding consists of placing a weighed quantity of moulding powder between the platens of a highly polished steel mould and heating under pressure till the powder flows into a homogeneous mass. The mould is allowed to cool and the object ejected from the mould. A diagram of the lay out of such a laboratory press and its method of working are shown in Fig. 10.1.

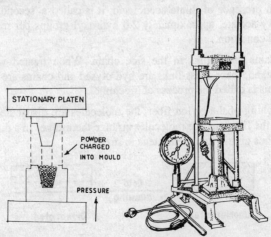

Fig. 10.1 (a) Compression moulding principle and (b) hand-operated laboratory compression press.

The platens are heated either by steam or electrically. These heated platens convey the heat to the mould. There are several types of moulds in use but one known as semi-positive mould is usually preferred. This is illustrated in Fig. 10.2.

In order to obtain a satisfactory moulding, a slight excess of the moulding powder is placed in the mould so that a thin skin of material, flash, is extruded from the mould. Temperature of moulding depends on the chemical nature of the concerned plastics and ranges upto 200°C. In order to get uniform mouldings free from defects, pressure in the range of 2-3 tons/sq. inch is applied. The cycle is known as curing time and depends on the thickness of the moulding desired.

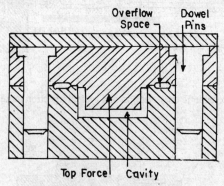

Fig. 10.2. Semi-positive mould.

This process is useful for those resins which harden or thermoset on the application of heat. Cross-linking results on account of chemical change and a network polymer results.

10.2.2. Moulding of Thermoplastics

Thermoplastic materials are usually moulded by injection moulding. A machine suitable for moulding resins such as poly(styrene) is shown in Fig. 10.3. The moulding powder is fed through a hopper to an electrically or steam-heated cylinder where it becomes fluid. A mechanically operated plunger then pushes the molten material through a nozzle into a cold mould which consists of two parts. A torpedo is provided for thorough mixing of the molten mass. After the moulding operation, the plunger is withdrawn and simultaneously the mould opens and the moulded object is ejected. Temperatures similar to those used in compression moulding are used but pressures applied are upto 15 tons/sq. inch.

The cycle may consist of about 25 seconds. This cycle is repeated and the moulding of several objects can be carried out simultaneously.

10.2.3. Extrusion Moulding

For many purposes plastics material may be required in the form of films, tubes or for coating wires. The process of production has got to be continuous and the cross-section uniform. This is achieved best by an extrusion machine. Products such as poly(vinyl chloride), which may decompose in an injection moulding machine, are handled by this method of processing. The principle of operation of an extrusion machine is shown in Fig. 10.4. The extruder is essentially a screw conveyer carrying cold plastic pellets forward and compacting them in the compression section with heat being supplied by external heaters. The pressure is highest just before the plastic enters the die which shapes the extrudates. When thermoplastics are used, it is necessary to cool the extruded material for the sake of dimensional stability.

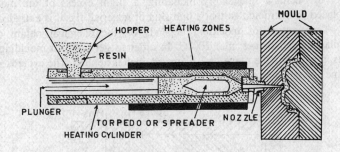

Fig. 10.3. Schematic cross section of a typical injection moulding machine.

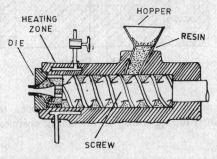

Fig. 10.4. Cross-section of a typical extruder.

With some plastics like poly(ethylene) the tube is blown into a bigger cylinder which is then slit to prepare sheets.

Q. 10.1. Which of the following materials are thermoplastic and which are thermosetting?
(i) Poly(methyl methacrylate), (ii) urea-formaldehyde, (iii) phenol-formalde-hyde, (iv) poly(styrene), (v) Terylene, (vi) Nylon, (vii) natural rubber, (viii) poly(vinyl acetate), (ix) cellulose acetate, and (x) styrene-butadiene rubber.

Q. 10.2. Which of the moulding procedures--compression, injection or extrusion--is applicable for the following polymers?
(i) Poly(styrene), (ii) poly(vinyl chloride), (iii) urea-formaldehyde, (iv) poly(ethylene), (v) poly(methyl methacrylate), (vi) phenol-formaldehyde.

10.3. CASTING OF FILMS

Films of safety materials such as cellulose acetate are widely used these days for photographic films. It is essential that they should have dimensional stability and be uniform. For this, a simple casting machine is used. The plastic material such as cellulose acetate is dissolved in a solvent (acetone in this case) and fed through a slit onto a travelling metal band. The two types of machines commonly used are illustrated in Fig. 10.5. In the first there is a drum about 15 feet in diameter upon which the film is cast. The quality of the film depends upon the perfectness of the surface of the cylinder. The big size of the drum and its high speed of rotation allow the solvent to evaporate. In the second type of machine a continuously moving metal belt is used. The machines are usually put in enclosures so that the solvent can be recovered.

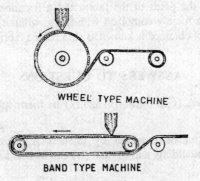

WHEEL TYPE MACHINE

BAND TYPE MACHINE

Fig. 10.5. Film-forming machine.

10.4. CALENDERING

The calendering operation consists of producing the materials like rubber or poly(vinyl chloride) in sheet form or combining these with fabric. There are three general types of calendering: (i) sheet calendering, (ii) frictioning, and (iii) skin coating. The calender is usually designed in the form shown in Fig. 10.6 (a) and (b).

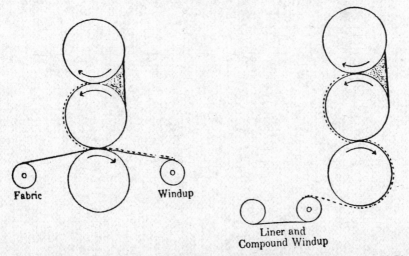

Fig. 10.6. (a) Skin coating showing combined-compound and fabric being wound, and (b) sheet calendering showing continuous film being wound with fabric interliner to prevent sticking of adjacent plies.

The operation consists of first warming the composition and then placing it in the nip of the middle and top rolls. By the proper regulation of the speed of rolls and temperature the sheet formed between the rolls may be caught on the middle roll and then passed between the middle and bottom rolls. The film may be cooled and rolled if sheet calendering is required. In the case where frictioning or skin coating is desired, a thin sheet of fabric is passed between the middle and bottom rolls along-with the rubber or plastics composition. When the compound is sufficiently plastic, it may be squeezed into the pores of the fabric and a frictioned product is obtained. When the compound is merely combined with fabric without filling up of the pores appreciably, the product obtained is known as skin coated fabric.

ANSWERS TO QUESTIONS

10.1.   (i), (iv), (v), (vi), (vii), (viii), (ix) and (x) are thermoplastics. (ii) and (iii) are thermosetting.

10.2.   (i) and (v) can be moulded by injection moulding, (iii) and (vi) by compression moulding, and (ii) and (iv) can be best handled by the extrusion technique.

CHAPTER 11

POLYMER DEGRADATION AND STABILIZATION

11.1. INTRODUCTION

Degradation of a polymer can be defined as the undesirable change that occurs in a polymer affecting its physical and chemical properties. Examples of degradation would include the wearing of tyres, loss of a plasticizer from a system by evaporation or migration or separation of a polymer from its fillers, leaving a void at the interface. The degrading agencies may be heat, radiation, chemicals and mechanical energy. In practice, all of them may act together. The term weathering is sometimes used to describe the action of ultraviolet radiation, heat, moisture and oxygen acting together.

To overcome these effects, polymer scientists and technologists have developed heat stabilizers, anti-fatigue agents and antiozonants. Further, stabilizers have been developed which increase (polymer) life and performance. The investigation of polymer degradation has been carried out at two levels. The polymer technologist has been mainly concerned with the empirical study of change in the physical properties with time. On the other hand, the polymer scientist has been more concerned with the study of reasons for these changes.

11.2. PHYSICAL METHODS USED IN THE STUDY OF DEGRADATION PROCESSES

There are four important physical methods which can be used in the study of polymer degradation processes:

  (i)   Determination of molecular weight during degradation
  (ii) Thermal analysis
  (iii) Spectroscopy
  (iv) Chromatography

11.2.1. Molecular Weight

The molecular weight of a polymer can be determined by viscosity measurements osmometry, light scattering or ultra-centrifuge. The molecular weight determination can provide evidence as to whether a polymer is degrading by random scission, stepwise at the chain ends or it is breaking along the chain leading to 'unzipping'.

11.2.2. Thermal Analysis

The methods most commonly employed in thermal analysis of polymers are:

(a) Thermogravimetric analysis (TGA) in which the loss of weight of polymer is recorded as a function of temperature.
(b) Differential thermal analysis (DTA) in which thermal changes within the polymer due to physical or chemical processes are recorded.
(c) Thermal volatilization analysis (TVA) which records the evolution of volatile products.

11.2.3. Spectroscopy

Spectroscopic techniques such as UV, IR, NMR, mass and electron spin resonance (ESR) spectroscopy are used to study the structural changes in the polymers on degradation.

11.2.4. Chromatography

Chromatographic techniques such as gas liquid chromatography (GLC) and thin layer chromatography (TLC) have been used in the separation, identification and quantitative determination of the products of polymer degradation.

11.3. THERMAL DEGRADATION OF POLYMERS

The thermal degradation of polymers is more complicated than that of small molecules. Many different kinds of degradation reactions may be introduced thermally in polymers. They can be divided into two main classes, namely, depolymerization and substituent reactions. In depolymerization reactions the main polymer backbone is broken in such a way that the products of degradation are similar to the parent material in the sense that the monomer units are still distinguishable. Examples of this type are the thermal degradation reactions of poly(methyl methacrylate) and poly(ethylene).

In the substituent reactions, the substituents attached to the polymer backbone are involved. The chemical nature of the repeating unit is changed although the chain structure of the backbone remains intact. An example of this type is the dehydrohalogenation of poly(vinyl chloride).

11.3.1. Depolymerization Reactions

11.3.1.1. *Radical Depolymerization Reactions*

11.3.1.1.1. *Poly(methyl methacrylate)*—In the depolymerization of poly(methyl methacrylate) a quantitative yield of the monomer is obtained. A random scission along the chain takes place and depropagation or 'unzipping', which is the opposite of propagation, is envisaged as shown in Scheme 11.1.

11.3.1.1.2. *Poly(ethylene)*—The visible products obtained from poly(ethylene) by thermal degradation are still more complex. A rapid decrease in molecular weight accompanied by intermolecular and intramolecular transfers takes place. The most abundant olefins formed are hex-1-ene and propene, as shown in Scheme 11.2.

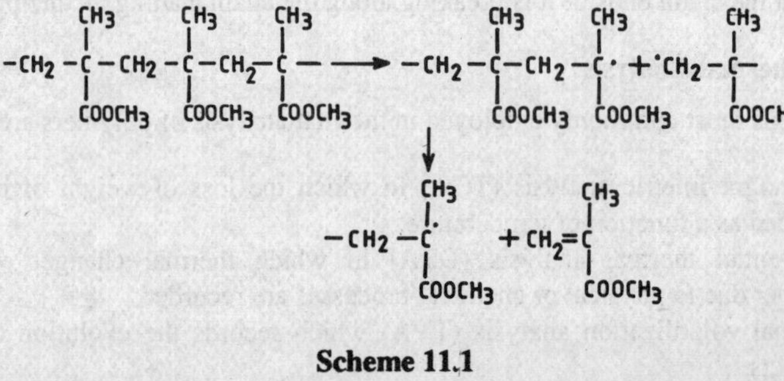

Scheme 11.1

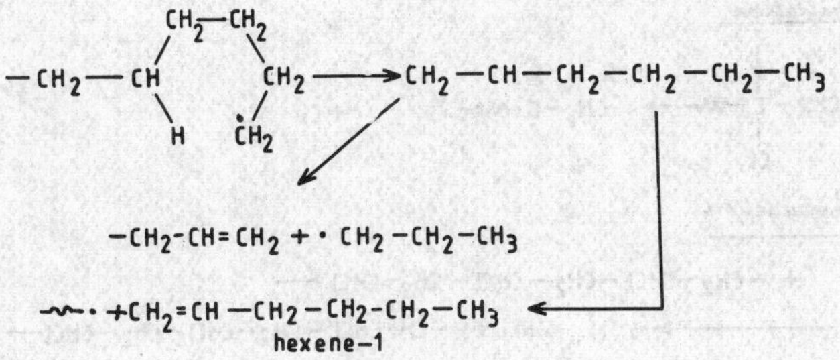

$$-CH_2-CH \overset{CH_2-CH_2}{\underset{H \quad \dot{C}H_2}{\diagdown CH_2}} \longrightarrow CH_2-CH-CH_2-CH_2-CH_2-CH_3$$

$$-CH_2-CH=CH_2 + \cdot CH_2-CH_2-CH_3$$

$$\text{w}\cdot + CH_2=CH-CH_2-CH_2-CH_2-CH_3 \longleftarrow$$
$$\text{hexene-1}$$

Scheme 11.2

11.3.1.2. *Non-radical Depolymerization Reactions*

Several polymers undergo depolymerization by non-radical processes. Notable examples of such polymers are poly(esters), poly(urethanes) and poly(siloxanes). A typical example of such a polymer is terylene whose reactions are shown in Scheme 11.3.

A large number of products are obtained and molecular weight decreases by random scission along the chain.

11.3.2. Substituent Reactions

Poly(vinyl chloride)

The thermal degradation of poly(vinyl chloride) yields hydrogen chloride and a conjugated system. Both radical and molecular mechanisms have been proposed for the dehydrohalogenation reaction. The radical mechanism, which envisages the occurrence of tertiary carbon atom at the end of some poly(vinyl chloride) chains, can trigger the reaction as shown in Scheme 11.4.

The molecular mechanism may operate as shown in Scheme 11.5:

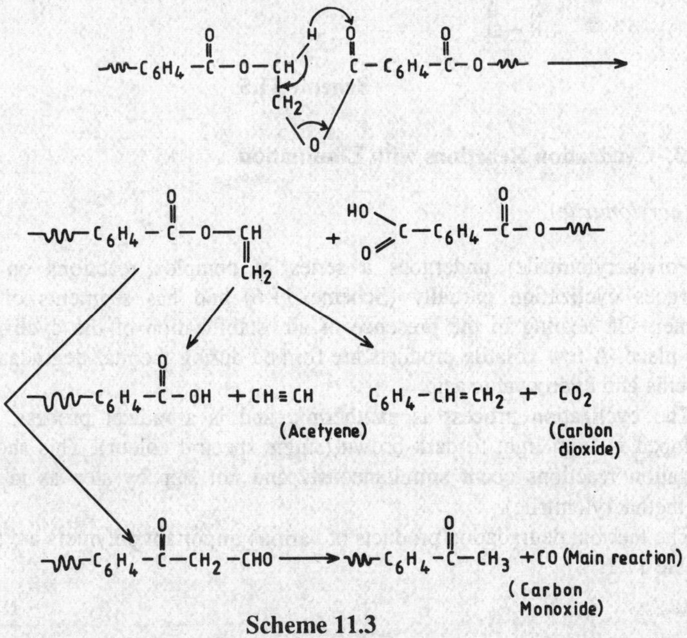

$$\text{w}-C_6H_4-\overset{O}{\overset{\|}{C}}-O-CH \overset{H \quad O}{\diagdown} C-C_6H_4-\overset{O}{\overset{\|}{C}}-O-\text{w} \longrightarrow$$
$$\underset{CH_2}{\overset{|}{}} \diagdown_O$$

$$\text{w}-C_6H_4-\overset{O}{\overset{\|}{C}}-O-CH \quad + \quad \overset{HO}{\underset{O}{\diagup}} C-C_6H_4-\overset{O}{\overset{\|}{C}}-O-\text{w}$$
$$\underset{CH_2}{\overset{|}{}}$$

$$\text{w}-C_6H_4-\overset{O}{\overset{\|}{C}}-OH + CH\equiv CH \qquad C_6H_4-CH=CH_2 + CO_2$$
$$\text{(Acetyene)} \qquad\qquad \text{(Carbon dioxide)}$$

$$\text{w}-C_6H_4-\overset{O}{\overset{\|}{C}}-CH_2-CHO \longrightarrow \text{w}-C_6H_4-\overset{O}{\overset{\|}{C}}-CH_3 + CO \text{ (Main reaction)}$$
$$\text{(Carbon Monoxide)}$$

Scheme 11.3

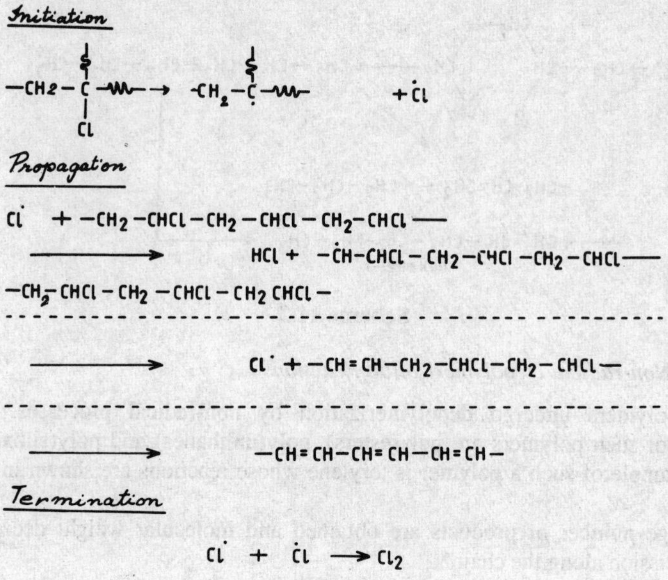

Initiation

$$-CH_2-\overset{\underset{|}{\xi}}{\underset{CL}{C}}-\text{\Large\char"29B}\longrightarrow \quad -CH_2-\overset{\underset{|}{\xi}}{\overset{}{\dot{C}}}-\text{\Large\char"29B} \qquad +\dot{C}l$$

Propagation

$$\dot{C}l \quad + \quad -CH_2-CHCl-CH_2-CHCl-CH_2-CHCl-$$

$$\longrightarrow \qquad HCl + \ -\dot{C}H-CHCl-CH_2-\dot{C}HCl-CH_2-CHCl-$$

$$-CH_2-CHCl-CH_2-CHCl-CH_2\,CHCl-$$

- -

$$\longrightarrow \qquad Cl^{\cdot} + -CH=CH-CH_2-CHCl-CH_2-CHCl-$$

- -

$$\longrightarrow \qquad -CH=CH-CH=CH-CH=CH-$$

Termination

$$Cl^{\cdot} + \dot{C}l \longrightarrow Cl_2$$

Scheme 11.4

$$-\text{\Large\char"29B}-CH_2-\underset{\underset{Cl}{|}}{CH}-\underset{\underset{H}{|}}{CH}-CHCl- \longrightarrow -\text{\Large\char"29B}-CH_2-CH=CH-CHCl +HCl$$

or alternately

$$-\text{\Large\char"29B}-CH_2-\underset{\underset{\underset{H-Cl}{H}}{|}}{\underset{|}{CH}}-\underset{\underset{H}{|}}{CH}-CHCl \longrightarrow -\text{\Large\char"29B}-CH_2-CH=CH-CHCl +2HCl$$

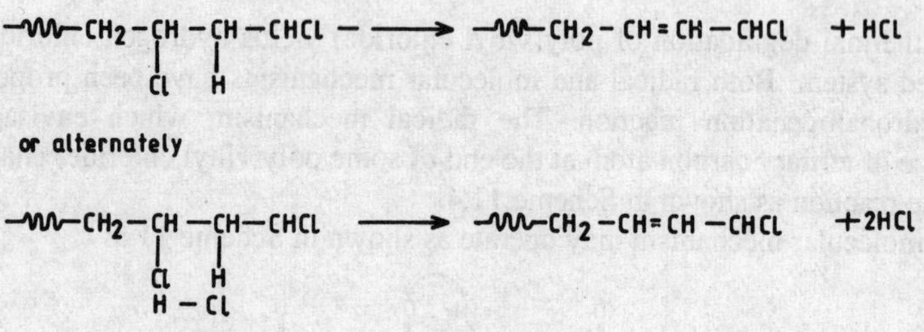

Scheme 11.5

11.3.3. Cyclization Reactions with Elimination

Poly(acrylonitrile)

Poly(acrylonitrile) undergoes a series of complex reactions on heating. It undergoes cyclization partially (Scheme 11.6) and has segments of uncyclized polymer. On heating in the presence of air, stabilization of the cyclized structure takes place. A few volatile products are formed during thermal degradation, such as ammonia and hydrocyanic acid:

The cyclization process is exothermic and is a radical process. The colour developed is from light to dark brown (single spectral colour). This shows that the cyclization reactions occur simultaneously and not step by step as in the case of poly(methacrylonitrile).

The thermal degradation products of various important polymers are summarized in Table 11.1.

Table 11.1. Thermal Degradation Products of Polymers

Polymer	Monomer unit	% Monomer	Degradation products
Poly(methyl methacrylate)	$-CH_2-\overset{\displaystyle CH_3}{\underset{\displaystyle COOCH_3}{C}}-$	100	Monomer
Poly(methyl acrylate)	$-CH_2-\underset{\displaystyle COOCH_3}{CH}-$	traces (less than 1%)	Chain fragments
Poly(styrene)	$-CH_2-\underset{\displaystyle C_6H_5}{CH}-$	42	Monomer, dimer, trimer, tetramer etc.
Poly(α-methyl styrene)	$-CH-\overset{\displaystyle CH_3}{\underset{\displaystyle C_6H_5}{C}}-$	100	Monomer
Poly(acrylonitrile)	$-CH_2-\underset{\displaystyle CN}{CH}-$	Traces (less than 1%)	Hydrocyanic acid, ammonia
Poly(methyacrylonitrile)	$-CH_2-\overset{\displaystyle CH_3}{\underset{\displaystyle CN}{C}}-$	100	Monomer
Poly(ethylene)	$-CH_2-CH_2-$	(Less than 1%)	Chain fragments
Poly(vinyl chloride)	$-CH_2-\underset{\displaystyle Cl}{CH}-$	Nil	Hydrogen chloride and conjugated products
Poly(vinylidene chloride)	$-CH_2-\overset{\displaystyle Cl}{\underset{\displaystyle Cl}{C}}-$	Nil	—
Poly(tetrafluoro ethylene)	$-\overset{\displaystyle F}{\underset{\displaystyle F}{C}}-\overset{\displaystyle F}{\underset{\displaystyle F}{C}}-$	100	Monomer
Poly(isoprene)	$-CH_2-\underset{\displaystyle CH_3}{C}=CH-CH_2-$	12	Monomer
Poly(butadiene)	$-CH_2-CH=CH-CH_2-$	1.5	—

Q. 11.1 Give the mechanism of the thermal degradation of poly(styrene) into monomer and oligomers.

Q. 11.2 In the cyclization of poly(methacrylonitrile), a colour develops gradually from yellow through orange to red and black. Explain.

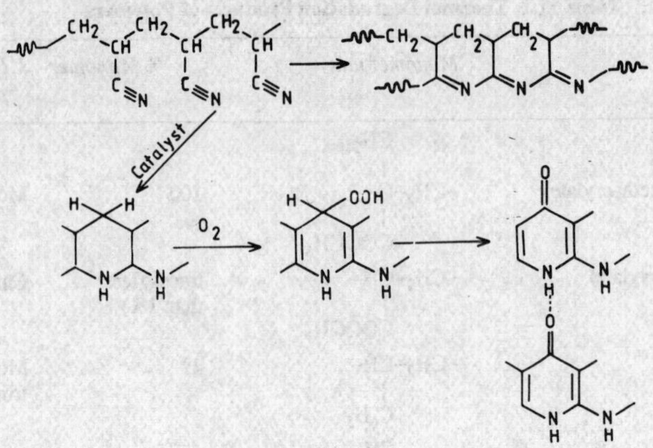

Scheme 11.6

11.4. OXIDATIVE DEGRADATION OF POLYMERS

11.4.1. Autoxidation

Oxygen in the ground state occurs as a diradical $\cdot O-O\cdot$. Oxygen normally reacts with organic compounds in a radical chain reaction involving the ground state diradical:

$$R^\bullet + O_2 \longrightarrow ROO^\bullet \tag{11.1}$$

$$ROO^\bullet + RH \longrightarrow ROOH + R^\bullet \tag{11.2}$$

Since reaction (11.1) is a radical pairing reaction, it takes place readily as it has a low energy of activation. The driving force for reaction (11.2), on the other hand, is the stability of radical $R^\bullet$, usually due to resonance. The termination can take place by one of the following reactions:

$$R^\bullet + R^\bullet \longrightarrow R-R \tag{11.3}$$

$$ROO^\bullet + ROO^\bullet \longrightarrow ROOR + O_2 \tag{11.4}$$

$$R^\bullet + ROO^\bullet \longrightarrow ROOR \tag{11.5}$$

Although initiation can take place by any free-radical generator (azobisis-obutyronitrile or benzoyl peroxide), it usually occurs by the thermal or photochemical decomposition of the hydroperoxide initially formed.

$$2ROOH \xrightarrow{\Delta H} RO^\bullet + {}^\bullet OH + H_2O \tag{11.6}$$

$$ROOH \xrightarrow{h\nu} RO^\bullet + {}^\bullet OH \tag{11.7}$$

Transition metal ions can also generate free-radicals

$$ROOH + M^+ \longrightarrow RO^\bullet + OH^- + M^{2+} \tag{11.8}$$

$$ROOH + M^{2+} \longrightarrow ROO^\bullet + H^+ + M^+ \tag{11.9}$$

11.4.2. The Role of Additive on Oxidation of Polymers

Poly(styrene)

In the oxidative degradation of poly(styrene), hydroperoxides are formed in the initial stages of the reaction. The decomposition of the hydroperoxide may then result in the rupture of the chain, as shown in Scheme 11.7.

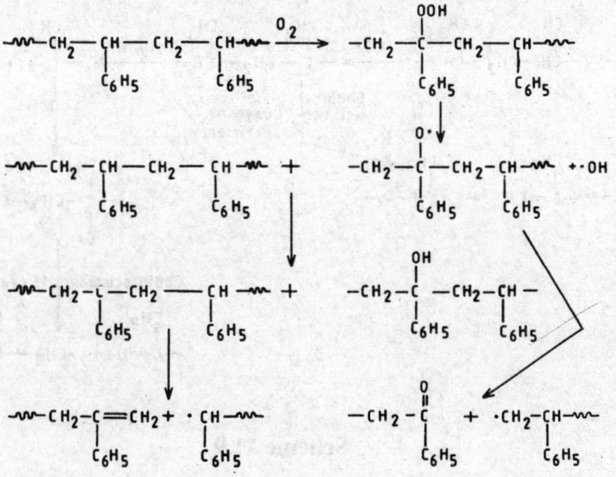

Scheme 11.7

Poly(ethylene)

The oxidative chain fission in poly(ethylene) can be represented as:

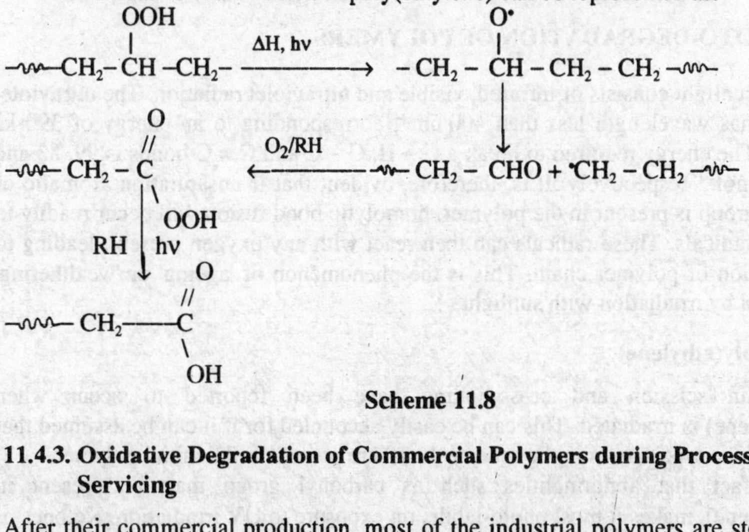

Scheme 11.8

11.4.3. Oxidative Degradation of Commercial Polymers during Processing and Servicing

After their commercial production, most of the industrial polymers are stored in the dark at ambient temperatures. Under these conditions, they do not suffer any notable deterioration. However, most of the thermoplastics are moulded by injection or extrusion moulding (see Chapter 10), where they are subjected to temperatures between 150 and 300°, and shear. Further, during servicing, specially at high temperatures, they undergo degradative oxidation.

11.4.3.1. *Degradation during Melt Processing*

Poly(propylene)

Scheme 11.9 summarises the degradation of poly(propylene) during high temperature processing.

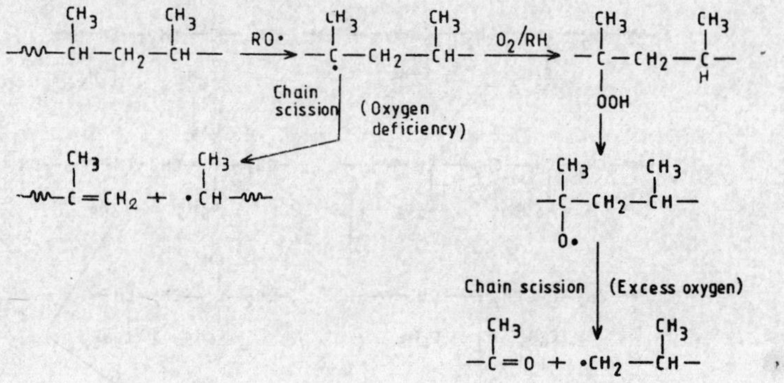

Scheme 11.9

11.4.3.2. *Degradation during Servicing*

Most commercial rubber products are manufactured with sulphur vulcanization. The sulphur cross-link further activates the rubber molecule. However, the tensile strength of vulcanized rubber is completely destroyed by as little as 3% by weight of oxygen and even 1% of oxygen renders it technologically useless.

11.5. PHOTO-DEGRADATION OF POLYMERS

Ordinary sunlight consists of infrared, visible and ultraviolet radiation. The ultraviolet radiation has wavelength less than 400 nm[+], corresponding to an energy of 390 kJ mole^{-1}[*]. The energy required to break a C – H, C – C and C = C bonds is 99, 83 and 145 kcal mol^{-1} respectively. It is, therefore, evident that if unsaturation aromatic or carbonyl group is present in the polymer, homolytic bond fusion will occur readily to give free-radicals. These radicals can then react with any oxygen present, leading to the oxidation of polymer chain. This is the phenomenon of 'ageing' or 'wealthering' occasioned by irradiation with sunlight.

11.5.1. Poly(ethylene)

Both chain scission and cross-linking have been reported to occur when poly(ethylene) is irradiated. This can be easily accounted for if it can be assumed that radicals of the type shown in the scheme 11.10 are present in the initial polymer.

The fact that abnormalities such as carbonyl group may be present in poly(ethylene), makes it most photo-labile, on exposure to UV irradiation. Ketones of this type undergo the following types of reaction.

[+]1 nm = 10 Å = 10^{-7} cm = 10^{-9} m

[*]1 cal = 4.184 J

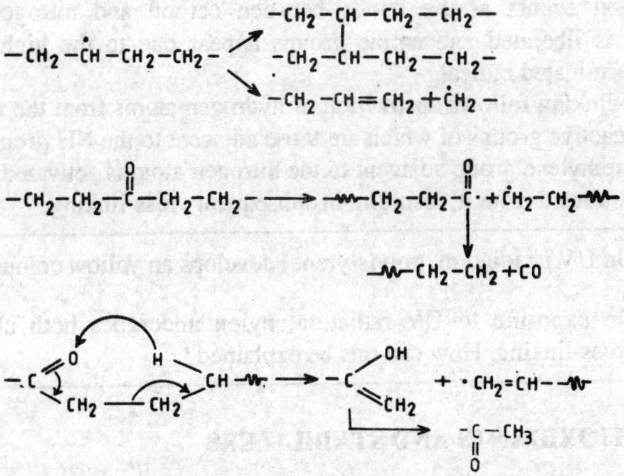

$$-CH_2-\overset{.}{C}H-CH_2-CH_2- \quad
\begin{cases}
-CH_2-CH-CH_2-CH_2- \\
\qquad\quad | \\
-CH_2-\overset{.}{C}H-CH_2-CH_2- \\
-CH_2-CH=CH_2+\overset{.}{C}H_2-
\end{cases}$$

$$-CH_2-CH_2-\overset{O}{\overset{||}{C}}-CH_2-CH_2- \longrightarrow \text{\small www}CH_2-CH_2-\overset{O}{\overset{||}{C}}\cdot+\overset{.}{C}H_2-CH_2\text{\small www}$$

$$\downarrow$$

$$\text{\small www}CH_2-CH_2+CO$$

$$-C\overset{O}{\underset{CH_2}{\diagdown}}\overset{H}{\underset{CH_2}{\diagdown}}CH\text{\small www} \longrightarrow -C\overset{OH}{\underset{CH_2}{\diagdown}} + \cdot CH_2=CH-\text{\small www}$$

$$\downarrow$$

$$-\overset{}{\underset{O}{\overset{||}{C}}}-CH_3$$

Scheme 11.10

In the first case chain scission occurs adjacent to the carbonyl group to form radicals directly. The second is a non-radical reaction, which results in chain scission by way of a six membered ring transition state, yielding a methyl ketone and vinyl unsaturation scheme 11.10 illustrates the degradation of a copolymer of ethylene and carbon monoxide.

11.5.2. Poly(amides)

In poly(amides) both cross-linking and chain scission occur although it has been reported that this is wavelength dependent. Thus although cross-linking usually predominates, chain scission can become predominant if short wavelengths (< 300 nm) are filtered out of the incident radiation.

Scission

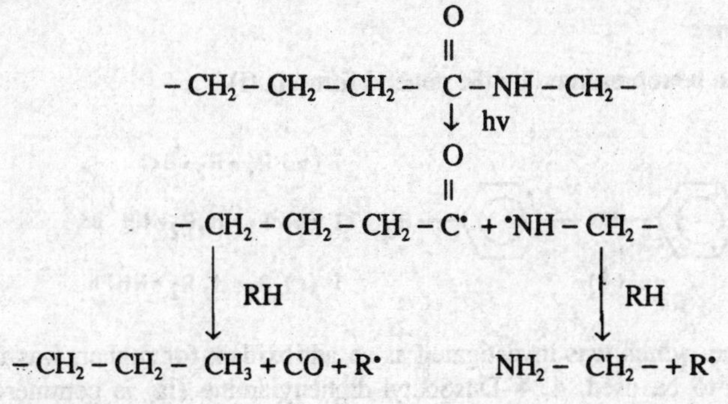

$$- CH_2 - CH_2 - CH_2 - \overset{O}{\overset{||}{C}} - NH - CH_2 -$$

$$\downarrow \text{ hv}$$

$$- CH_2 - CH_2 - CH_2 - \overset{O}{\overset{||}{C^\bullet}} + {}^\bullet NH - CH_2 -$$

$$\Big\downarrow RH \qquad\qquad\qquad \Big\downarrow RH$$

$$- CH_2 - CH_2 - CH_3 + CO + R^\bullet \qquad NH_2 - CH_2 - + R^\bullet$$

Cross–linking

$$-CH_2-CH_2-CH_2-\overset{O}{\overset{||}{C}}-NH-CH_2-CH_2- \overset{hv}{\longrightarrow} - CH_2-CH-CH_2-\overset{O}{\overset{||}{C}}-NH-CH-CH_2-$$

$$- CH_2-CH_2-CH_2-C-NH-CH-CH_2-$$
$$\overset{||}{O}$$

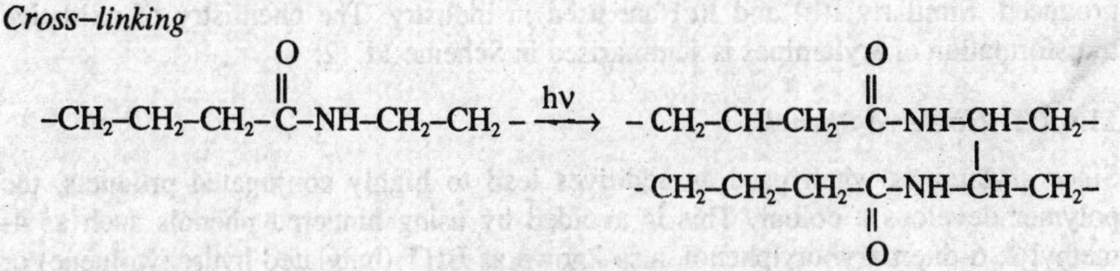

Scheme 11.11

Scission occurs at the bonds between carbon and nitrogen atoms. Carbon monoxide is liberated and amine groups appear due to the high reactivity of the nitrogen terminated radical.

Cross-linking follows abstraction of hydrogen atoms from the methylene groups, the most reactive groups of which are those adjacent to the NH groups.

The methylene group adjacent to the nitrogen atom is activated and the hydrogen atom is abstracted from it, resulting in subsequent cross-linking.

Q. 11.3 On UV irradiation, poly(styrene) develops an yellow colour. Explain.

Q. 11.4 On exposure to UV radiation, nylon undergoes both chain scission and cross-linking. How can this be explained?

11.6. ANTIOXIDANTS AND STABILIZERS

Polymer materials can suffer deterioration by thermal, photo and oxidative degradation, leading to 'ageing', 'weathering' and fatigue. Additives which counteract these effects are known as antioxidants. A large amount of work has been done to counter chain breaking and peroxodylitic reactions. A large number of antioxidants and UV stabilizes have been developed. The detailed mechanism of their action is beyond the scope of the present book. However, a few important members will be briefly discussed.

11.6.1. Chain Breaking Antioxidants

The main kinetic chain breaking mechanism, known as electron donor mechanism (CB-D), is given below:

$$ROO^\bullet \xrightarrow{\ +e\ } [RO\bar{O}] \xrightarrow{\ +H^+\ } ROOH$$

These antioxidants are of two types namely, arylamines and hindered phenols.

11.6.1.1. *Arylamines*

The arylamines can be represented by the general formula (I)

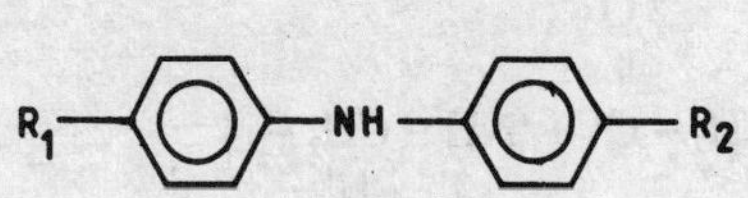

I (a) $R_1 = R_2 = $ Oct

I (b) $R_1 = H, R_2 = NH$ pr

I (c) $R_1 = H, R_2 = NHPh$

Diphenylamine, which was investigated as an antioxidant for rubber, was found to be too volatile to be used. 4, 4'-Düsooctyl-diphenylamine (Ia) is commercially produced. Similarly I(b) and I(c) are used in industry. The chemistry of oxidative transformation of arylamines is summarised in Scheme 11.12:

11.6.1.2. *Hindered Phenols*

Since arylamines when used as additives lead to highly conjugated products, the polymer develops a colour. This is avoided by using hindered phenols such as 4-methyl 2, 6-ditertiarybutylphenol, also known as BHT (butylated hydroxytoluene) or Topanol OC.

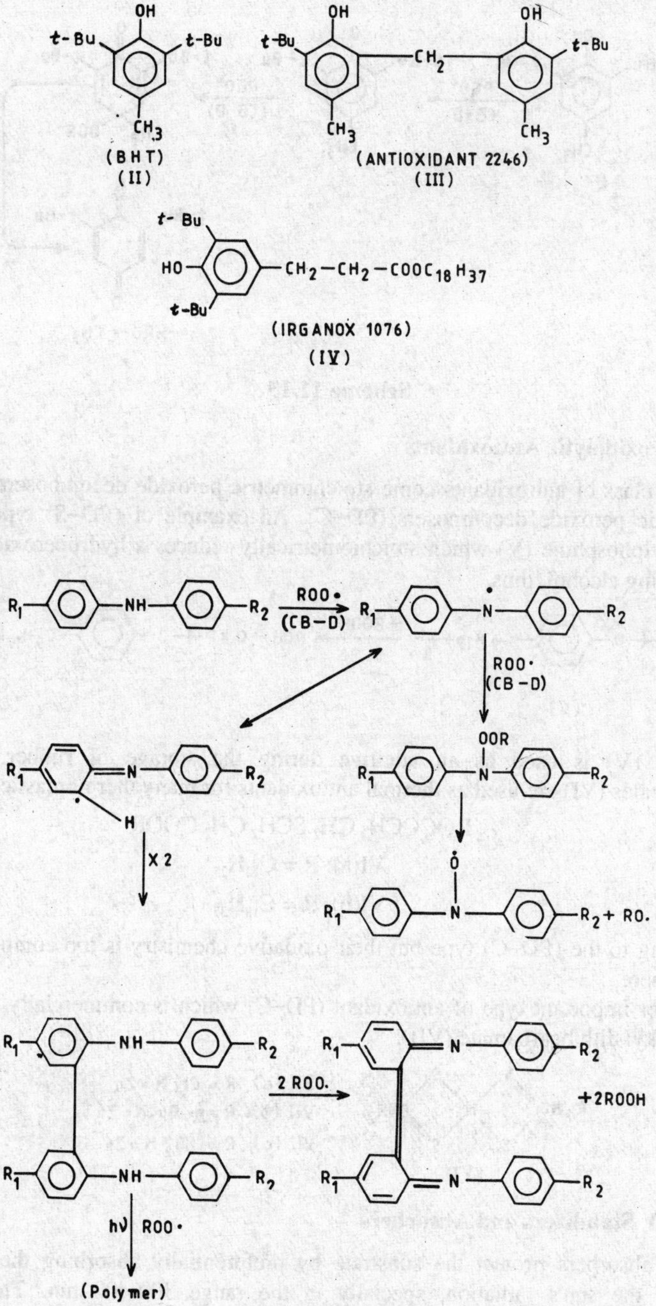

(BHT)
(II)

(ANTIOXIDANT 2246)
(III)

(IRGANOX 1076)
(IV)

Scheme 11.12

BHT (II) is a highly volatile antioxidant and hence it is not used as such at high temperatures. Instead, higher molecular weight hindered phenols such as (III) and (IV) are used in modern rubber technology. Some oxidative transformations of BHT are given in Scheme 11.13.

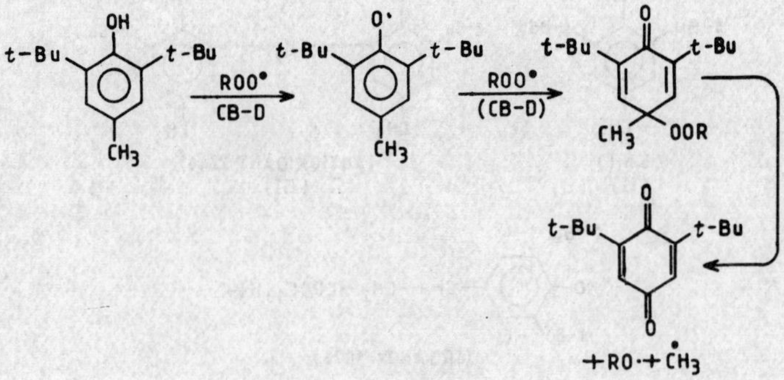

Scheme 11.13

11.6.2. Peroxidolytic Antioxidants

Under this class of antioxidants come stoichiometric peroxide decomposers (PD–S) and catalytic peroxide decomposers (PD–C). An example of (PD–S) type is tris-nonylphenylphosphine (V) which stoichiometrically reduces a hydroperoxide to the corresponding alcohol, thus,

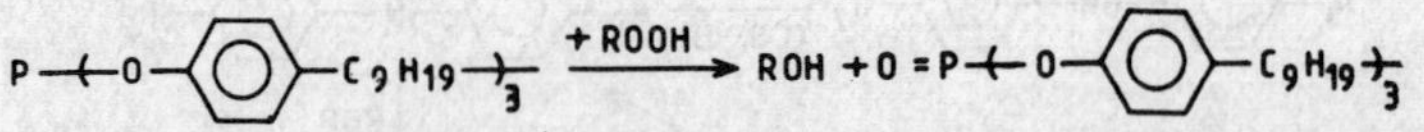

(V)

Compound (V) is used as an additive during the storage of rubber. Dialkyl thiopropionates (VI) are used as thermal antioxidants for many thermoplastics.

$$ROCOCH_2\ CH_2\ SCH_2\ CH_2\ COOR$$

$$VI(a): R = C_{12}H_{25}$$

$$VI(b): R = C_{18}H_{37}$$

They belong to the (PD–C) type but their oxidative chemistry is too complex to be discussed here.

Another important type of antioxidant (PD–C) which is commercially available is zinc dialkyl dithiocarbamate (VII).

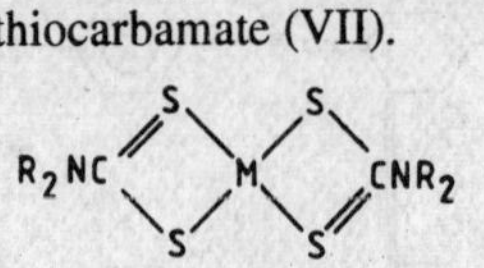

VII (a) R = Et, M = Zn
VII (b) R = n–Bu, M = Zn
VII (c) R = ^{1}NO, M = Zn

(VII)

11.6.3. U V Stabilizers and Absorbers

Most UV absorbers protect the substrate by preferentially absorbing the harmful portion of the sun's radiation, specially in the range 300–400 nm. The earth's atmosphere filters out most of the shorter wavelength radiation. To be useful, the absorber should dissipate the absorbed energy by transfering it to its surroundings as heat or by remitting it at higher wavelengths.

Carbon black can be tolerated in a number of formulations when colour is not a criterion of choice. It not only absorbs light, it reacts with free-radicals that may be formed. Two absorbing dyestuffs, which are colourless in visible light, have been discovered for plastic materials.

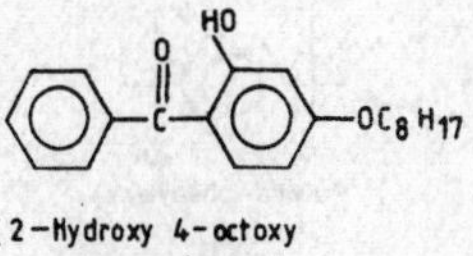

(2 - Hydroxy 4 - octoxy
benzophenone)
(VIII)

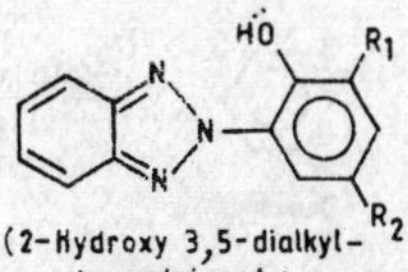

(2 - Hydroxy 3,5 - dialkyl -
benzotriazole)
(IX)

IX (a) $R_1 = H$, $R_2 = CH_3$
IX (b) $R_1 = R_2 = $ ter.Bu

Both (VIII) and (IX) show strong absorbance in the region around 330 nm. This is associated with the hydrogen-bonded interaction between the phenyl hydroxyl group and adjacent double bonded groups (C = O and C = N).

The nickel complex (X) functions primarily as an antioxidant.

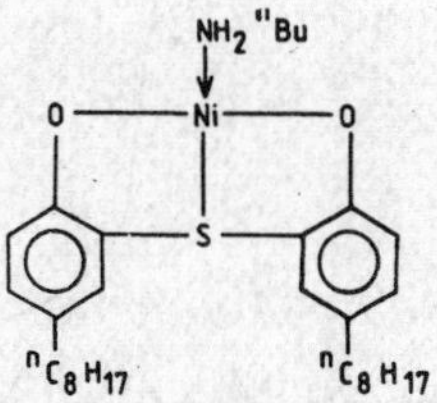

(X)

11.7. THERMALLY STABLE POLYMERS

The major impetus for the growth of interest in the field of temperature resistant polymers comes from their application in the aeronautic and air space industry.

11.7.1. Requirements for Stability

The polymers most resistant to degradation should have high degrees of crystallinity and high melting points. In crystalline state, the polymers are more closely packed and intermolecular forces are more difficult to overcome.

The following chemical factors also affect the stability:

(i) High bond dissociation energy.
(ii) Structural features which do not allow degradation by low energy process.
(iii) Formation of inter-or intra-molecular bonds on heating, resulting in cross-linked structure.
(iv) The polymeric structure must be chemically inert to oxygen, moisture, bases and other substances to which it might be exposed.

In the following account some important heat resistant polymers have been discussed.

11.7.2. Polymers Containing Aromatic Units

Polymers containing aromatic structures, e.g., poly(*p*-phenylene) are known to have high bond energy, low degree of reactivity and rigidity. Poly(*p*-phenylene) is prepared as follows (oxidative coupling):

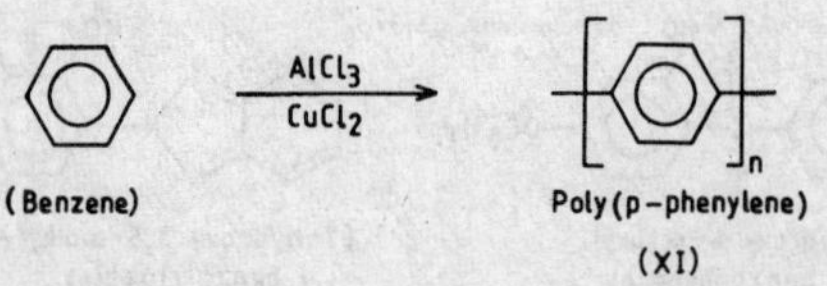

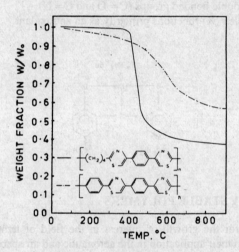

The polymer (XI) is an insoluble and infusible material.

Poly(thiazoles) containing alternating phenyl and thiazole rings are also quite stable thermally. Thermogravimetric analysis (Fig. 11.1) shows that the polymer with aliphatic units degrades rapidly at 350°-400°C in nitrogen, but the wholly aromatic polymer is stable from 500°-600° and loses only 40% of its weight upto 900°C.

Fig. 11.1 Thermogravimetric curves of poly(thiazoles).

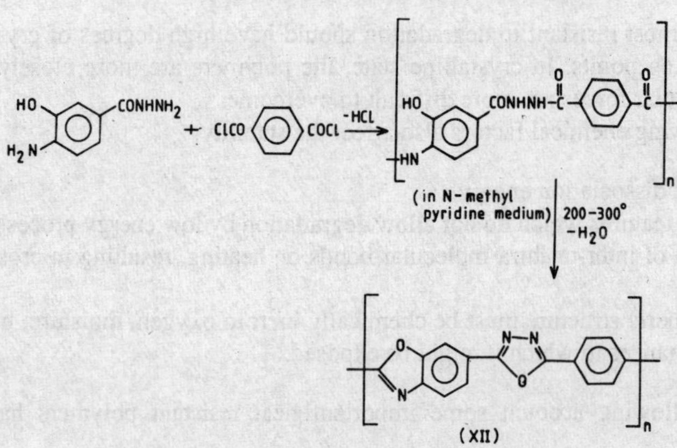

Scheme 11.14

The poly(thiazoles) are soluble only in sulphuric and trifluoroacetic acids. Their use is limited as they are not high molecular weight compounds and do not give films.

Oxadiazoles (XII), prepared according to Scheme (11.14), show excellent thermal stability in air as well as in nitrogen upto 520°C. Above 550°C, polymer degradation occurs, which is more rapid in air.

Poly(pyromellitimide) (XIII) fibers, made up of poly(imide) are extremely heat-stable in addition to being flame-, radiation- and oxidation-resistant. Their preparation has been described earlier (Scheme 2.15)

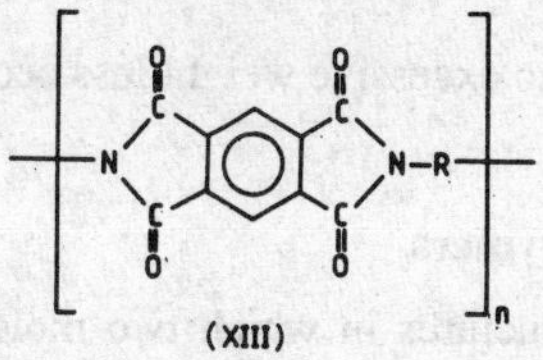

(XIII)

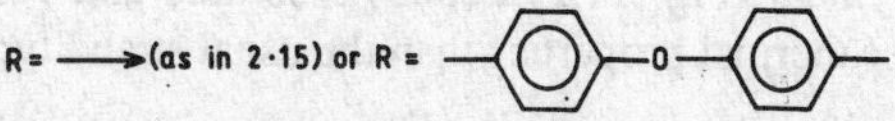

$R = $ (as in 2·15) or $R = $

High molecular weight film and fiber forming benzimidazole polymers (XIV) and (XV) have been prepared. Thermogravimetric (TGA) studies show that they do not begin to lose weight below ~500°C.

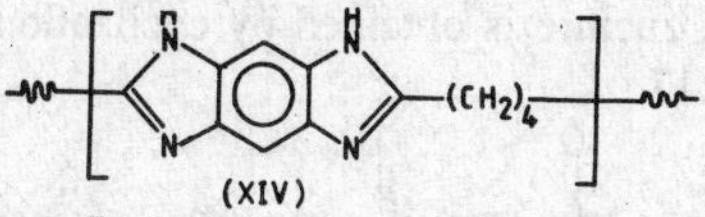

(XIV)

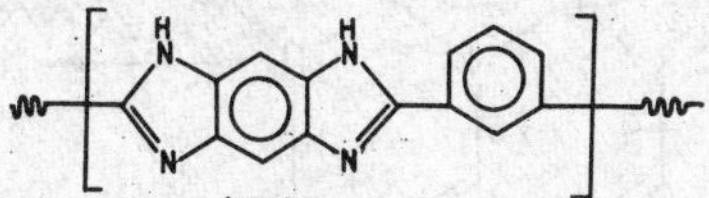

(XV)

An interesting commercial polymer is poly(phenylene oxide) (XVI) derived from 2, 6-dimethylphenol. It is prepared by a catalytic oxidative coupling reaction (Scheme 11.15).

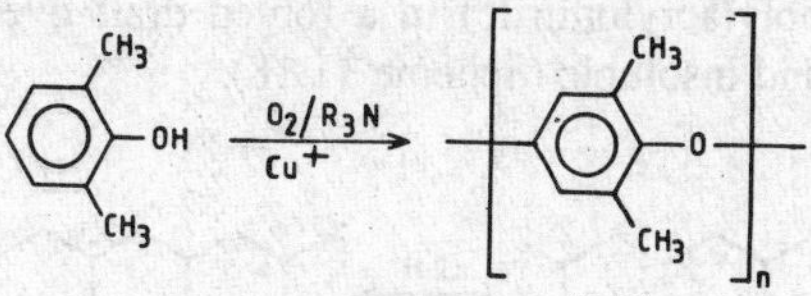

(XVI)

Scheme 11.15

The polymer has a high molecular weight, a useful temperature range and is chemically inert.

A similar polymer, poly(phenylene sulphide) (XVII) with a degree of polymerization ($\overline{DP}$) equal to ~ 100 has been obtained by the following reaction:

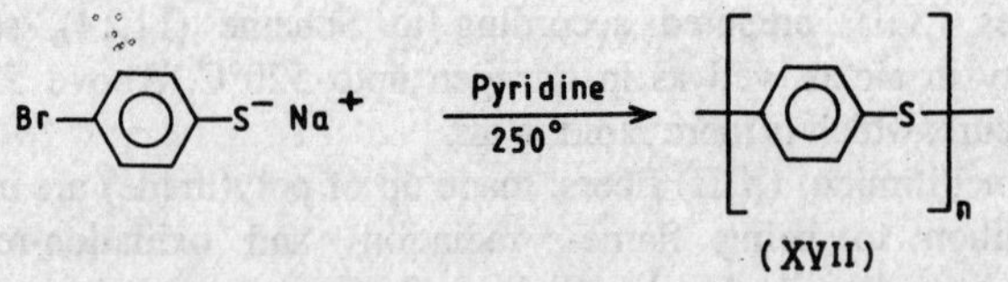

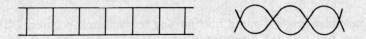

Scheme 11.16

TGA studies have shown that no excessive weight loss occurs upto 400°C in either N₂ or air.

11.7.3. Ladder and Spiro Polymers

These are non-cross-linked structures in which two molecular strands are joined to each other. Such polymers (Fig. 11.2) could be soluble and fusible and would be expected to have better thermal properties than the usual strand type polymers.

Fig. 11.2. Representation of ladder and spiro polymers.

A ladder polymer structure is obtained by cyclization of poly(vinyl isocyanate) as shown in Scheme 11.17.

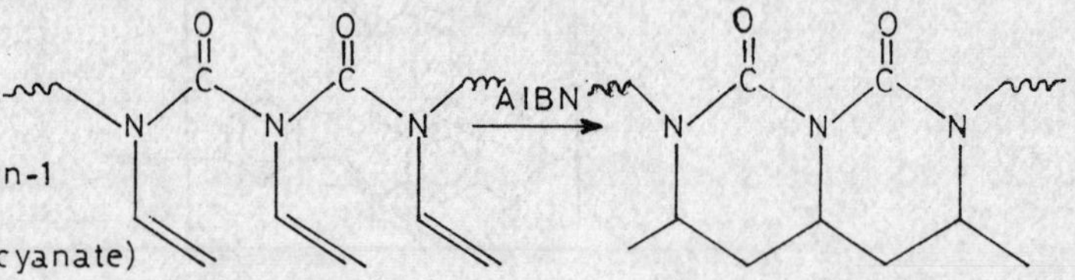

N-vinyl Nylon-1

poly(vinylisocyanate)

Scheme 11.17

The above mentioned Nylon-1 derivative decomposes above 300°C. The ladder polymer obtained by treatment with AIBN (azobisisobutyronitrile) decomposes at 385°-390°C. Another useful ladder structure known as Black Orlon or fiber AF is prepared by heating poly(acrylonitrile) in a forced draft oven at 250°-275°C. The final product is black and insoluble (Scheme 11.18).

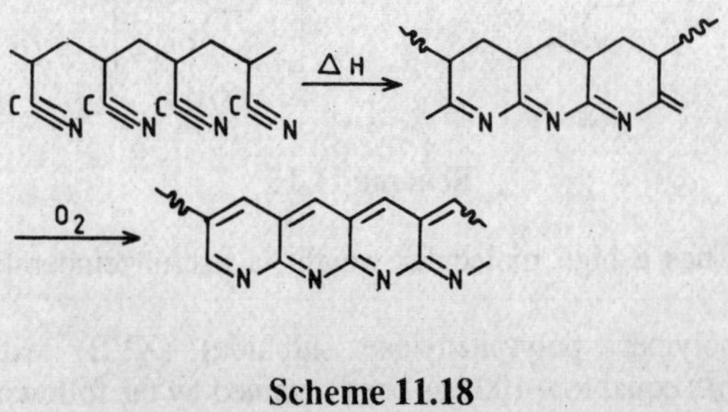

Scheme 11.18

It does not burn when placed directly in flame. It has tenacity of 1.0 to 1.5 g/den and is suitable for the fabrication of flame-proof clothing.

Poly(benzimidazopyrolones) (XVIII) and poly(quinoxalines) (XIX) are two more such ladder polymers which have been prepared but these seem to have little or no advantage in stability over other polymers described in Section (11.7.3).

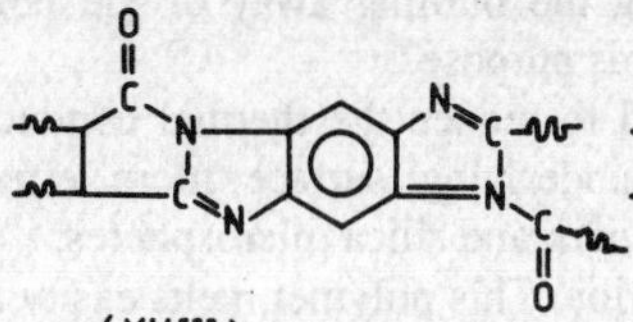

(XVIII 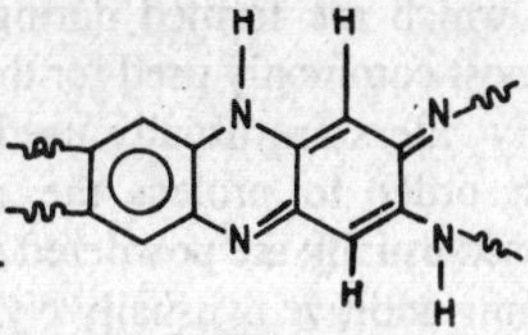)   (XIX)

Spiropolymers differ from the ladder polymers in having attachment at one point between the rings. An example of spiropolymers is the product (XX) obtained by the condensation of 1, 4-cyclohexadione and pentaerythritol.

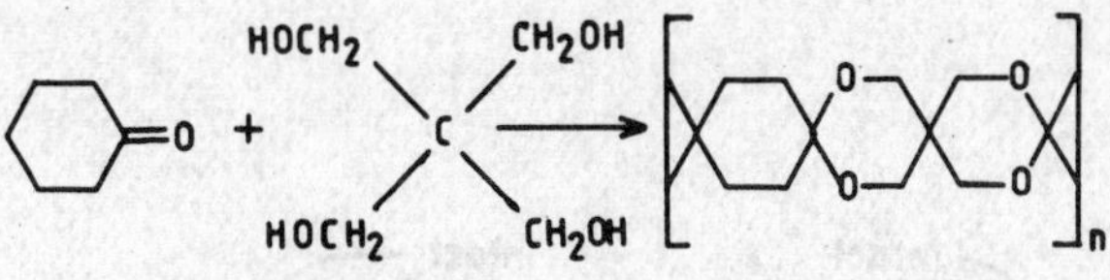

(X X)

Q.11.5 What are the main requirements of a high temperature polymer?

Q.11.6 Distinguish between a spiro and a ladder polymer. Give one example of each.

11.8. ABLATION

When a space-craft re-enters the earth's atmosphere at high velocity, even the most refractory material will be vaporised by the intense heat liberated due to the conversion of the kinetic energy of the material into heat. This conversion of the kinetic energy into heat will occur due to the slowing down of the space-craft by the friction of the atmosphere. Temperatures of around 4000 Kelvin are encountered at which no temperature resistant material can be of any avail. This problem was solved by applying the principle of ablation or burning away.

Polymers have been used in ablative systems for a combination of requirements. The polymer should have high heat absorption and dissipation and should have excellent thermal insulation which eliminates or reduces the need for an additional internal cooling system. It should have useful performance in a wide variety of hypothetical environments. The materials should be of light weight and have automatic temperature control of surface by self regulated ablative degradation. They should be of low cost and simply designed. Further, the materials should be capable of being tailored by varying the individual material components and construction to give the performance needed under a particular situation.

In accordance with the above requirements, the ablative mixture is composed of four different kinds of ingredients:

(i) A cross-linked polymer such as phenol-formaldehyde resin is used. It decomposes on heating to give a char with requisite properties.

(ii) A fiber reinforcement is used to give additional strength to the char and reinforce the cracks which are formed during the burning away of the polymer. Silica fibers are most commonly used for this purpose.

(iii) Low density fillers are usually used to reduce the thermal conductivity of the material in order to protect the underlying surface more effectively. The materials used usually are powdered cork and silica microspheres.

(iv) The subliming additive is usually Nylon. This polymer melts easily and degrades to give large volume of gaseous products which results in the formation of a char of requisite porosity.

The various stages of heating, charring and melting are shown in Fig. 11.3 for two composites.

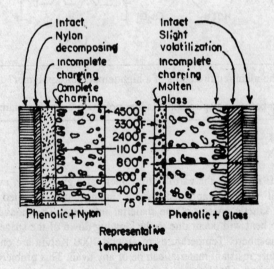

Fig. 11.3. Temperature distribution in two plastics during ablation.

ANSWERS TO QUESTIONS

11.1.

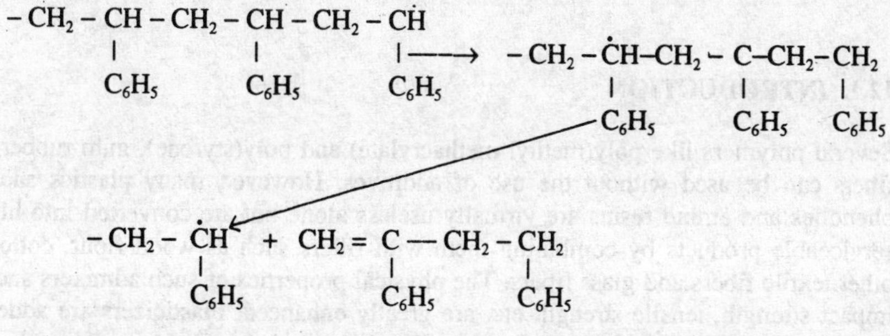

$$-CH_2-CH-CH_2-CH-CH_2-\overset{\cdot}{C}H \longrightarrow -CH_2-\overset{\cdot}{C}H-CH_2-C-CH_2-CH_2$$

with C_6H_5 groups on the first structure's three CH centres, and C_6H_5 groups on the product.

$$-CH_2-\overset{\cdot}{C}H \quad + \quad CH_2=C-CH_2-CH_2$$

with C_6H_5 substituents.

Poly(styrene) degrades thermally in one step to give the principal volatile product styrene (40%) and oligomers.

11.2 The reason for this is that cyclization of poly(methacrylonitrile) occurs in steps, leading to conjugation and thus colour development. The spectral colour, therefore, develops into darker colours as the cyclization proceeds.

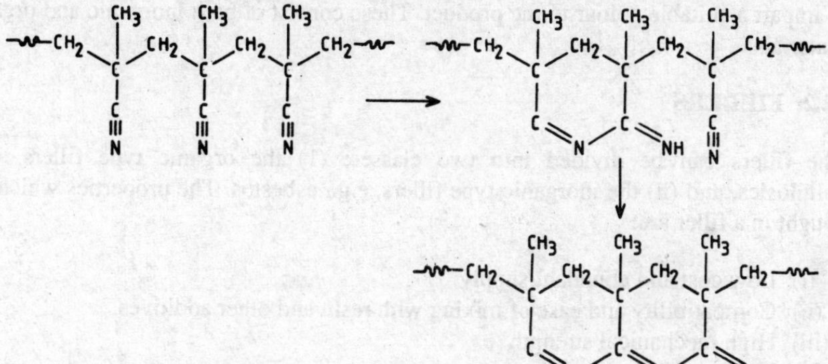

11.3 Abstraction of a hydrogen atom from the carbon atom adjacent to the radical centre would be energetically favourable and would result in unsaturation, in conjugation with the benzene ring. A conjugated chain results, leading to yellow colour.

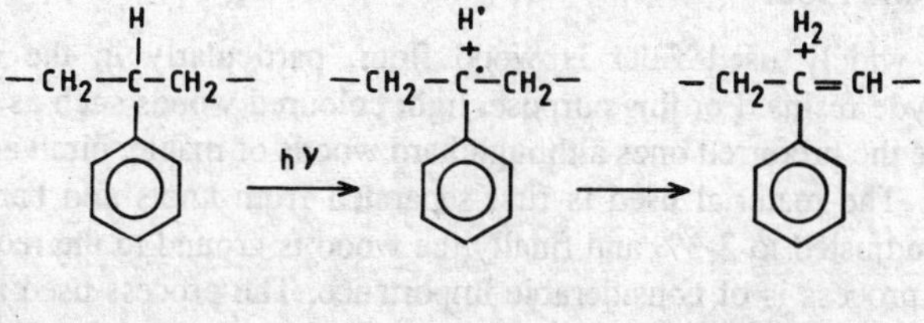

11.4 Please see Scheme (11.11).

11.5 Please see Section (11.7.2).

11.6 Please see concluding part of section (11.7.4).

CHAPTER 12

POLYMER ADDITIVES

12.1. INTRODUCTION

Several polymers like poly(methyl methacrylate) and poly(styrene), gum rubber and fibers can be used without the use of additives. However, many plastics such as phenolics and amino resins are virtually useless alone but are converted into highly serviceable products by combining them with fillers such as wood flour, cotton or other textile fibers and glass fibers. The physical properties of such admixers such as impact strength, tensile strength etc. are greatly enhanced. Plasticizers are added to improve flow properties and reduce the brittleness of the products. Several other additives such as antioxidants, flame retarders and stabilizers are also added to improve the performance of the plastics material. The role of anti-oxidants such as substituted diaryl amines and hindered phenols is to prevent oxidation of polymers. A flame retarder usually antimony trioxide is added to improve the fire resistance of the polymers. In addition to antioxidants and flame retarders, certain polymers require stabilizers to achieve and maintain stability. An important example is the stabilization of vinyl chloride by metallic salts of lead, barium, tin etc. Finally a colourant is added to impart a suitable colour to the product. These consist of both inorganic and organic materials.

12.2. FILLERS

The fillers can be divided into two classes; (1) the organic type fillers, e.g., cellulosics, and (ii) the inorganic type fillers, e.g., asbestos. The properties which are sought in a filler are:

(i) Low cost and abundant supply
(ii) Compatibility and ease of mixing with resin and other additives.
(iii) High mechanical strength
(iv) Low moisture absorption
(v) High heat resistance
(vi) Good electrical characteristics
(vii) Ease of moulding
(viii) Absence of abrasive or chemical reaction on the mould.

12.2.1. Wood Flour

The most widely used filler is wood flour, particularly in the case of phenol-formaldehyde resins. For this purpose, light coloured woods such as pine, spruce and popular are the preferred ones although hard woods of maple, birch and oak have also been used. The material used is first separated from knots and bark. The moisture content is adjusted to 2-3% and finally the wood is ground to the required mesh size. The latter process is of considerable importance. The process used is called attrition grinding. Although the wood is ground to a fine mesh size, the fibers are left in an undamaged condition.

Wood flour has low specific gravity. Its value in extending the resin is due to the greater volume of moulded products formed per kilogram of the base. Its general characteristics also favour its wide-spread use. Some of these are its excellent mouldability, good shock resistance, high tensile strength, desirable electrical characteristics and low heat conductivity. The moulded objects can be easily machined. However, there are some drawbacks as well. The moisture resistance is poor and it suffers from shrinkage in service. The temperature resistance is not high enough (135°C), which restricts its use. When pastel shades are required, for example in urea-formaldehyde resins, a special grade of cellulose known as alpha cellulose is used in place of wood flour.

12.2.2. Cotton and Other Textile Fabrics

Where high strength is required, cotton flocks or finely ground cotton is generally used. Use of cotton flocks imparts high impact and tensile strength to moulded objects. Fabric scraps, clippings and rags are bleached, purified and shredded and then used as fillers. However, there are certain limitations to the use of fabric fillers. The moisture absorption of moulded pieces is high and machinibility is more difficult because of the exposure of fabrics at the surface. Further, the mouldability is decreased because of the high bulk ratio of the moulding powder.

12.2.3. Asbestos

Asbestos is a mineral filler which is the hydrated form of magnesium silicate. In its natural form, it consists of small fibers admixed with some impurities. This leads to some loss of strength in its action as a filler compared to cotton fabrics, but this is compensated by the improvement in other properties. The moisture absorption of mouldings is very low and its electrical characteristics are excellent. The plastics employing asbestos as a filler such as asbestos filled phenolics have excellent heat resistance and can withstand temperatures upto 200°C. The mouldings are also resistant to chemicals.

Besides the loss of strength, when asbestos is used as a filler, there is abrasion of the mould and the fibers show at the surface in a discoloured form.

12.2.4. Graphite

Graphite is used as a filler to a limited extent, because of its lubricating value. The lubricating influence is evident during the moulding operation. It can be very easily removed from the mould. Due to this property, it finds use in self-lubricating bearings.

12.2.5. Mica

One grade of mica used in plastics is a finely wet-ground, white viscovite. In cases where the electrical characteristics and heat resistance are important, it is widely used; for example with phenolics, shellac and alkyd resins. The advantages in the use of mica are that it has prominent electrical characteristics and high heat resistance. It is easily wetted by resins and dyes and does not absorb moisture. It has a low specific gravity of 2.26. It has no abrasive effect on moulds and its light colour is retained at elevated temperatures in the presence of chemicals. The main drawback is that mica has a tendency to cause peeling or sticking to the mould as phenol does not coat the mica sufficiently.

12.2.6. Jute Fibers

Jute fibers, yarn and fabrics have been used for reinforcement in phenolic compositions. These fabrics have low strength but this deficiency is overcome in part by using parallel jute fibers.

Besides the fillers mentioned above, there are miscellaneous mineral fillers, containing mainly silica, like *Vermiculite*. The electrical characteristics and heat resistance of the mouldings are considerable but they tend to be brittle.

Table 12.1. Comparison of Various Fillers in Phenol-Formaldehyde Mouldings

	Filler	Characteristic displayed	Important applications
1.	Wood flour	General charac-teristics	Radio cabinets
2.	Wood flour (vacuum dried)	Electrical charc-teristics	Electrical components, ignition coil housings
3.	Cotton fibers	Good impact strength, good surface appearance	telephone handsets
4.	Graphite	Lubricating properties	Self lubricating bearings,
5.	Asbestos	Heat and chemical resistance	brake linings, plugs
6.	Mica	Low loss electrical properties	Coil housings, mechanical support for radio parts

Q.12.1 Which fillers are suitable for

    (i)   General purposes mouldings
    (ii)   Pastel shade urea-formaldehyde mouldings
    (iii)   Heat resistant mouldings
    (iv)   Mouldings for use in ratio parts

Q.12.2 Which filler is most suitable for mouldings with high mechanical strength and why?

There is a tendency to use fibrous reinforcing agents to enhance the mechanical properties and sometimes the thermal and chemical properties also of the polymers themselves. While the leading fibrous material has always been glass, there are new high performance speciality fibers and inorganic single crystals of micro-dimensions and high aspect (length/diameter) ratio. These are increasingly referred to as high performance composites and they are used frequently in aerospace and automotive industries and conventional consumer applications.

The properties of phenol-formaldehyde resins compounded with the various types of fillers have been compared in Table 12.1.

12.3. PLASTICIZERS

Most of the synthetic resins and cellulose derivatives are obtained in the form of white powders or horny tough materials. In these forms their plastic properties are latent and they do not flow appreciably under the action of heat and pressure and thus cannot be processed. As a result, the applications of these pure materials are rather limited. Most thermoplastics are, therefore, modified with the addition of plasticizers, which are generally liquids of high boiling points. The addition of a plasticizer brings out the best characteristics of a synthetic resin and, in fact, transforms it into a workable plastic.

A plasticizer can be defined as a high boiling liquid, or in rare cases a solid with low volatility, used to toughen and flexibilize a plastics base or to soften it at workable temperatures. The plasticizing effect is considered to be due to the solubilization action and an accompanying reduction in inter-molecular forces to permit freer movement of molecules relative to one another.

The two main functions of plasticizers are thus:

(i) Production of resilient elastic characteristics, and

(ii) Development of ease of fabrication.

When cellulose derivatives and vinyl resins are used in moulding, casting, extrusion or as lacquers, a suitable plasticizer must be added. The plasticizer becomes an integral part of the resin mixture and gives cohesive plastics material. The properties imparted are flexibility, impact strength, elasticity and toughness.

12.3.1. Characteristics of Plasticizers

Plasticizers have been described as 'non-volatile solvents.' They have the essential characteristics of high boiling point and low volatility. These properties are important because if the plasticizer evaporates from a plastic or a film then it will revert to its original brittle condition. As a consequence, the boiling point of the plasticizer should be above 300°. In general, the plasticizer must be chemically inert, fast to light, resistant to moisture, non-toxic and non-fuming. If the plastics material is to be used for packing food material or beverages, it should, in addition to the above properties, have no taste or smell. Besides, the plasticizer should have some basic properties, e.g., it should be compatible with the polymer, impart flexibility, show permanent retention and be water insoluble and stable in its presence.

12.3.2. Some Important Plasticizers

The properties of a few important plasticizers, their important uses and their drawbacks are discussed below:

Tricresyl phosphate

It is a high boiling liquid, having a specific gravity of 1.18 at 20°C and has low volatility. It freezes at −35°C and its loss on evaporation is negligible. It is almost insoluble in water and is moisture resistant. It is non-inflammable and confers fire-resistant properties to polymers. It is compatible with most of the synthetic resins. It is the most suitable plasticizer for poly(vinyl chloride). Tricresyl phosphate finds wide use in the field of synthetic rubber. However, it is reputed to be toxic and care has to be exercised in its use.

Dimethyl Phthalate

Dimethyl phthalate [$CH_3OOC. C_6H_4. COOCH_3$ (*o*)]. is a clear liquid with a specific gravity of 1.20 at 20°C and a b.p. of 285°C. The solubility in water is 0.3 per cent. It is soluble in a number of organic solvents. Dimethyl phthalate is compatible with most of the plastic resins and oils. It is widely used as a plasticizer for cellulose acetate, for which it is a good solvent, inspite of its volatility it is fast to light and is non-toxic. It finds use in the preparation of cellulose acetate dopes and lacquers and also films.

Dibutyl Phthalate

Dibutyl phthalate [C_4H_9 OOC. C_6H_4 COOC$_4H_9$ (*o*)]. is perhaps the most widely used plasticizer. It is a clear liquid, b.p. 340°C. Its specific gravity is 1.05 at 20°C and the freezing point is − 35°C. It is miscible with all organic solvents and oils and its solubility in water is only 0.001 per cent at 25°C. It is widely used as a plasticizer for almost all plastics, notably for cellulose acetate, cellulose nitrate and poly(vinyl chloride). It is fast to light and is non-toxic.

Triphenyl Phosphate

Triphenyl phosphate, available in white flakes, is a low melting solid (m.p. 45° and b.p. 390°C). It is compatible with tricresyl phosphate and has low volatility. It is resistant to moisture and its solubility in water is only 0.001 per cent at 25°C. It is compatible with oils and resins and is soluble in organic solvents. It is the plasticizer of choice for cellulose acetate films and sheets, and gives clear and tough products.

Camphor

Until recently, camphor was obtained exclusively from the camphor tree. At present, it is produced synthetically from α-pinene which is a constituent of turpentine oil. It is a colourless, transparent material with a characteristic odour. It melts at 175°C and boils at 209°C. Its specific gravity is 0.92. It is extensively used as a plasticizer for the manufacture of celluloid. It is also used in explosives and for therapeutic purposes.

In Table 12.2 are given the solubility parameters of some typical plasticizers.

Table 12.2. Solubility Parameters of Some Typical Plasticizers

Plasticizer	Solubility parameter
Paraffin oil	7.5
Tricresyl phosphate	8.4
Triphenyl phosphate	8.6
Dimethyl phthalate	10.7
Diethyl phthalate	10.0
Dipropyl phthalate	9.7
Dibutyl phthalate	9.3
Dioctyl phthalate	7.9
Dihexyl phthalate	8.9
Dibutyl sebacate	9.2
Glycerol	16.5

Q. 12.3 What is a plasticizer and how does it function?

Q. 12.4 Compare and contrast the advantages and disadvantages in the use of tricresyl phosphate and dibutyl phthalate.

Q. 12.5 What plasticizer would you use with (i) poly(vinyl chloride), (ii) nitrocellulose and (iii) cellulose acetate?

12.4. ANTIOXIDANTS AND THERMAL STABILIZERS

A detailed description of antioxidants and thermal stabilizers has been given in Sections 11.6.1, 11.6.2 and 11.6.3 (pages 160-162).

12.5. U.V. STABILIZERS AND ABSORBERS

U.V. Stabilizers and absorbers are described in Section 11.6.4 (pages 162-163).

12.6. FIRE RETARDANTS

From times immemorial, the building materials for the construction of houses have been wood, textiles and paper, which are different forms of cellulose. In case of fires, during the fire fighting one had to contend against smoke and occasionally carbon monoxide. Modern houses are built and decorated with synthetic plastics in various forms. On burning, these materials evolve toxic gases and fire fighting becomes a very hazardous business.

To minimize fire hazards, fire-retardant chemicals can be added to polymers as additives. These additives are usually incorporated into the polymer during processing. Six elements are particularly associated with fire retardance. These are boron, aluminium, phosphorus, antimony, chlorine and bromine. Inorganic phosphates like ammonium phosphate have been used for a long time for the fire protection of cellulose used in the form of wood, paper and cotton. In more recent times, tricresyl phosphate and tris (2-chloroethyl) phosphate $[(ClCH_2CH_2O)_3P = O]$ have come into use, and they have been found to give a degree of protection to various polymeric materials.

The action of ammonium poly(phosphate) as a fire retardant on various types of poly(urethanes) has been studied. Poly(phosphates) are precursers of phosphoric acid which has the fire retarding property. Antimony trioxide (Sb_2O_3 or more correctly Sb_4O_6) is very widely used as a flane retardant in conjunction with halogens. It is reported that 3% Sb_2O_3 and 5% of bromine by weight are as effective as 13-15% by weight of bromine with epoxy resins. With poly(vinyl chloride), 3-5% by weight of Sb_2O_3 is quite effective. Aluminium oxide trihydrate ($Al_2O_3.3H_2O$) is widely used for fire-retardation in polymer technology. A mixture of borax and boric acid is used as a fire-retardant additive for textiles. The reasons for using the mixture are two fold. In the first place, the mixture does not crystallize out and thus a cohesive coating is maintained on the textile. Secondly, the mixture is more effective as a fire-retardant.

Q. 12.6 What is the function of a fire-retardant? Name a few fire-retardants being used commercially.

12.7. COLOURANTS

Some polymers like poly(methyl methacrylate) are used as such without the addition of any colourants. But, most of the polymers are dark and require the use of colourants to obtain desired colours. Where dyes are used as colourants, a polar group must be present in the polymer molecule. Titanium dioxide is a pigment most widely used for imparting a white colour to polymers. Various colour shades may be obtained by the use of other pigments.

Pigments are insoluble coloured materials. They can be inorganic or organic compounds. The inorganic pigments are chromates (yellow), ferrocyanides (blue), sulphides, oxides and silicates etc. Aluminium silicates can give colours from red to blue. Many colour modifications are possible by co-precipitation of various metallic salts.

The organic colourants include phthalocyanins with colours ranging from blue to green. Another important colourant is 6-chloro-4-toluidine-3-sulphonic acid (permanent Red 23). Another organic colourant is qurnacridine which can be used alone or mixed with pigments.

Organic dyes are usually more compatible with polymers than pigments. They can be used to give transparent colours to polymers. If opaque material is required, their mixture with titanium dioxide is used.

Q. 12.7 Name a few pigments and organic colourants used as polymer additives. What colours they impart to the polymer?

ANSWERS TO QUESTIONS

12.1.   (i) Wood flour, (ii) alpha cellulose, (iii) asbestos (iv) mica.

12.2.   Cotton and other textile fabrics are the most suitable materials imparting high mechanical strength. The fibers in cotton textiles impart high strength as fillers. As these fibers are long, strong and compatible with the resin, they impart the high mechanical strength.

12.3.   A plasticizer can be defined as a high boiling liquid (and in rare cases a low melting solid) with low volatility used to toughen and flexibilize a plastics base or to soften it at workable temperatures. The two main functions of a plasticizer are:
(i) Production of resilient elastic characteristics in a resin, and
(ii) Development of ease of fabrication.

12.4.   Tricresyl phosphate is a high-boiling non-volatile liquid. The evaporation from a resin is of very low order. It is compatible with most resins and is not attacked by moisture and is insoluble in water. However, it is toxic and cannot be used in materials used in for packing food and beverages. Dibutyl phthalate has a b.p. of 340°. It is soluble in organic solvents and is more volatile than tricresylphosphate and thus susceptible to migration. It is, however, non-toxic.

12.5.   (i) Tricresyl phosphate, (ii) camphor, (iii) dibutyl phthalate, and (iv) dimethyl phthalate.

12.6.   The function of a fire retardant is to retard the flammability of a polymer with which it is compounded. A few common fire retardants are:
(i) Tricresyl phosphate
(ii) Tris-(trichloroethyl) phosphate, $(ClCH_2CH_2O)_3P = O$
(iii) Ammonium polyphosphate
(iv) Antimony trioxide, Sb_4O_6 etc.

12.7.   *Pigments*
Titanum dioxide - white colour
Chromates - yellow colour
Ferrocyanides - blue colour
Aluminium silicate - red to blue colour
Organic colourants
Phthalocyanins - blue to green colour
6-Chloro-4-toluidine 3-sulphonic acid - permanent red etc.

CHAPTER 13

IDENTIFICATION AND CHARACTERIZATION OF POLYMERS

13.1. INTRODUCTION

Identification of an unknown sample of a plastic, rubber or a fiber may be required for a variety of reasons. Many of these polymers carry their trade name only and their detailed structure is not disclosed. In this age of competition, a manufacturer may like to develop a product similar to that of another producer. In the context of a developing country like India, an up-and-coming manufacturer may not be provided with the detailed nature of the products being used in the fabrication of a component, for example in the field of rocketry or aircraft industry. In such cases it may be necessary to analyse the polymer sample and identify it. The value of molecular weight determinations in the characterization of polymers has already been discussed in detail in Chapter 6. Besides this, the other methods of characterisation may be broadly divided into the following categories: (i) testing, (ii) spectroscopic examination and (iii) thermal analysis.

13.2. PRELIMINARY TESTS

To begin with, a range of (around 40) representative samples of common types of polymers should be available for comparison. The preliminary tests include visual examination, cutting, effect of heat and burning, solubility and fusion tests. Besides these, the melting point of a thermoplastics material (T_g or T_m) can also give valuable information regarding the probable identity of the product.

13.2.1. Visual Examination

A visual examination of the sample may reveal useful information. For example, the presence of a hard, inflexible flash line would indicate a thermoset moulded material. Similarly, the presence of a gate scar would indicate an injection moulded material. However, this should not be confused with the gate scar on a thermoset transfer moulded material. The physical form of the sample, i.e., whether granules, film, sheet or fiber and its flexibility or rigidity would give some indication of its identity. The colour of the product can also be used for identification of a thermoset material. For example, a pastel shade would rule out a phenolic and probably indicate that it is a urea or melamine-formaldehyde plastics material.

13.2.2. Cutting

The cutting of a plastics material with a pen knife can also provide some information. For example, one can differentiate between crystal clear cellulose acetate and poly(styrene). Thermoset and thermoplastic material can also be differentiated in this way. The ivory like cut of casein and cast phenolics is noteworthy.

13.2.3. Heating Tests

A small amount of the material to be tested is taken in a spoon type of spatula and heated on a small bunsen flame. The ease of burning, whether the burning continues after removal from the flame, colour of flame and so on, all give an indication of the possible identity of the material. If the material explodes or burns away rapidly, it is possibly a cellulose nitrate composition.

A second heating test can be conducted with the help of a clean copper wire. The wire is first heated in a clear bunsen flame and then touched with a small quantity of the material. It is heated again and the colour of the flame is noted. Blue and green colours indicate the presence of halogens in the composition, that is, chlorine, fluorine and rarely, bromine. Presence of poly(vinyl chloride) or its co-polymers, poly(vinylidine chloride), poly(tetrafluoroethylene), chlorinated rubber, rubber hydrochloride or cellulose acetate containing a plasticizer like tricresyl phosphate is indicated by this test.

A third heating test is carried out by heating a small sample in a hard glass tube. The gas evolved is condensed in another tube and very carefully smelled. This can then be compared with the gas evolved from a known polymer.

13.2.4. Fusion Test

The metallic sodium fusion test, can show the presence or absence of nitrogen and halogens. Similarly, the potassium nitrate/potassium carbonate fusion test will indicate the presence or abscence of phosphorus in the material under examination. Thus, one can easily distinguish between Nylon and Terylene.

13.3. IDENTIFICATION OF POLYMERS

From the preliminary tests, a great deal of information can be gathered regarding the possible identity of the unknown polymer. If the polymer does not show rubber like elasticity, an elastomer is ruled out. Further, if on heating it does not melt or flow, a thermosetting polymer is indicated. If it does melt, a thermoplastic polymer is indicated.

13.3.1. Carbon, Hydrogen and Nitrogen Analysis

If the fusion test indicates the presence of nitrogen, a quantitative analysis of the compound for C, H and N is carried out and if it tallies with that of a known polymer the identity of the polymer material is known with some amount of certainty. If other elements besides C, H and N are present, such as P or Cl, they have also to be taken account of in the calculations.

13.3.2. Solubility

For thermoplastic polymers, a scheme of identification based on their solubility behaviour has been put forward (Fig. 13.1). The chart shows six different examples where at least two polymers are found by the same solubility route. Other similar examples will certainly be found if this chart is further generalized. In case of ambiguity, the solubility test has to be complemented by calorimetric and/or spectroscopic analyses. For example, the different nylons can be distinguished on the basis of their T_m (crystalline melting point) values and poly(isoprenes) can be distinguished by NMR spectroscopy.

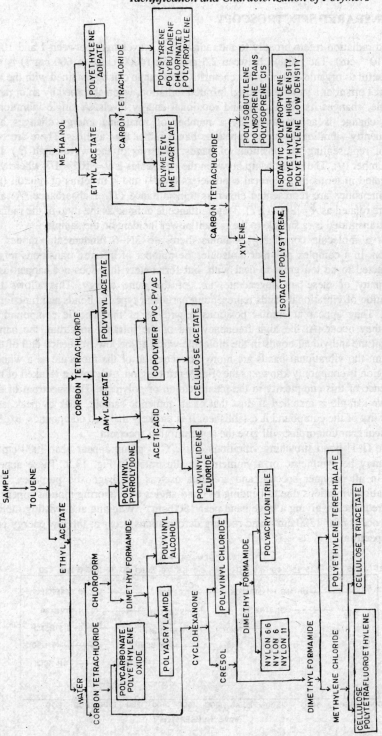

Fig. 13.1. Solubility Chart: the soluble polymers follow the arm at the right of the solvent considered while the insoluble ones follow the arm at the left.

13.4. INFRARED SPECTROSCOPY

Infrared radiation refers broadly to the radiation of wavelength between 1 and 100µ ($1\mu = 10^{-4}$ cm). The portion between 2.5 and 15µ (4000 cm^{-1} to 600 cm^{-1}) is the most useful to organic and polymer chemists. This range can be obtained with the aid of a NaCl prism or a suitable grating. Infrared radiation, when absorbed by an organic molecule, changes its vibrational and rotational energy levels. A single vibrational energy change is accompanied by a number of rotational energy changes and consequently vibrational spectra appear as bands rather than as lines. There are two ways of representing the positions of bands: in terms of the wavelength (λ) and wavenumber (ν). The relationship between the two scales is $\nu = 10^4 \times 1/\lambda$ where ν is represented in terms of reciprocal centimeters (cm^{-1}) and λ in terms of micron (μ). Band intensities are represented either as transmittance (T) or absorbance (A) and these are related as $A = \log_{10}(1/T)$. Transmittance is defined as the ratio of the radiant power transmitted by a sample to the radiant power incident on the sample.

For a molecule containing N atoms there are 3N–6 fundamental modes of vibration. In a complex polymer molecule, the number of infrared transitions might be expected to be too great to deal with, but fortunately this does not happen as a great many of these are degenerate, i.e., of the same energy. This allows the recognition of vibrational bands representing particular types of bonds and functional groups. They appear at similar positions regardless of the specific compound in which they occur. At the high frequency end of the infrared spectrum, the bands represent the individual bonds in the molecule while at the low frequency end of the spectrum, the vibrational bands are more characteristic of the molecule as a whole. This region is commonly known as the `finger print region.' Advantage is taken of the occurrence of this complexity in the identification of polymers. The spectrum of the unknown sample is matched against that of an authentic sample, peak by peak, and the identity of the compound is established. It is unlikely that two compounds, except the optical enantiomorphs, will give the same infrared spectrum.

The C–H bond stretching vibrations of the CH$_2$ group appear near 3000 cm^{-1}, which may be symmetric or asymmetric as illustrated in Fig. 13.3. These appear nearly in all polymer spectra, and are thus useless for diagnostic purposes. The deformation vibrations due to bending of bond angles or scissoring motion appear at lower frequencies, giving a large band near 1500 cm^{-1}. Wagging and twisting modes of CH$_2$ occur near 1300 cm^{-1} and rocking deformations occur at the low energy end of the spectrum.

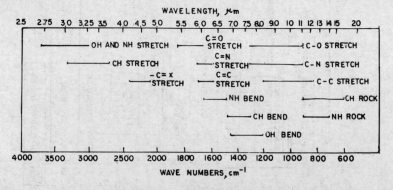

Fig. 13.2. Infrared bands of interest in polymers.

In Fig. 13.2. are shown a number of other characteristic vibrational bands and their frequency ranges. The carbonyl (C=O) stretching band near 1700 cm^{-1}, the C=C stretching band near 1600 cm^{-1} and the olefinic C–H bending bands between 900 and 1000 cm^{-1} are specially to be noted.

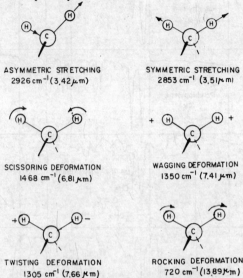

ASYMMETRIC STRETCHING
2926 cm^{-1} (3.42μm)

SYMMETRIC STRETCHING
2853 cm^{-1} (3.51μm)

SCISSORING DEFORMATION
1468 cm^{-1} (6.81μm)

WAGGING DEFORMATION
1350 cm^{-1} (7.41μm)

TWISTING DEFORMATION
1305 cm^{-1} (7.66μm)

ROCKING DEFORMATION
720 cm^{-1} (13.89μm)

Fig. 13.3. Stretching and deformation vibrational modes of the methylene group.

The sample for analysis can be prepared in various forms. It can be in the form of a thin film, a pellet of the powdered compound with KBr, a nujol mull (a hydrocarbon) or as a solution in chloroform or carbon tetrachloride held in a NaCl cell. In a double beam infrared spectrophotometer having matched cells, one can compensate for the solvent also, if a solution has been used for recording the spectrum.

In Fig. 13.4 are shown the transmission spectra in the infrared region of a few commercially important polymers. A few of the spectra may now be considered in detail. In the IR spectrum of poly(ethylene) (Fig. 13.4) one observes the presence of a C–H stretching band (2940 cm^{-1}), C–H bending band of CH$_2$ groups (1470 cm^{-1}), stretching modes of CH$_2$ groups (shoulder at 1380 cm^{-1} on an amorphous band on 1370 cm^{-1}) and CH$_2$ rocking mode at 720 cm^{-1} of the sequence of CH$_2$ groups in paraffin structures.

Poly(styrene) has a repeating unit – (– CH$_2$ – CH C$_6$H$_5$) consisting of 16 atoms. It has no symmetry and hence all vibrations are active. It has (3 n – 6) or 42 possible fundamental vibrations and 3 degrees of rotational and 3 degrees of vibrational freedom. The spectrum of poly(styrene) shown in Fig. (13.4) is used as a standard for calibrating IR spectrophotometers. The spectrum shows typical C–H stretching vibrations at 3030, 2940 and 2898 cm^{-1}, and C–C stretching vibrations at 1612 and 1492 cm^{-1}. The out-of-plane C–H bending vibrations of the benzene ring are evident at 910 and 700 cm^{-1}. The bands at 1150 and 1030 cm^{-1} are in the 'finger-print region' of the spectrum and as such difficult to explain.

13.5. NUCLEAR MAGNETIC RESONANCE SPECTROSCOPY

Nuclear magnetic resonance (NMR) spectroscopy makes use of the fact that certain nuclei possess nuclear magnetic moment. What does one understand from nuclear

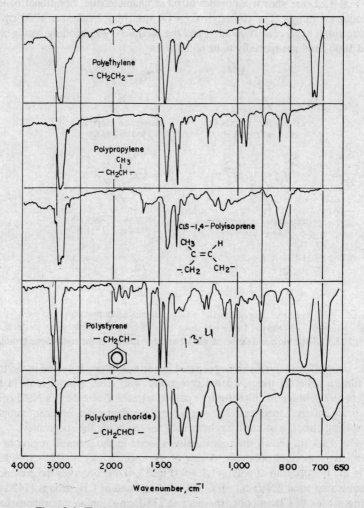

Fig. 13.4. Typical transmission spectra in the infrared region.

magnetic moment and which nuclei have this property? Certain nuclei behave like tiny bar magnets as they possess both charge and spin. Of course, all nuclei have charge but few have spin. Those nuclei which have magnetic moment, have odd number of protons or odd number of neutrons or odd number of both.

The commonly occurring isotopes of carbon (^{12}C) and oxygen (^{16}O) contain an even number of protons and neutrons and so have no nuclear magnetic moment. If a nucleus with magnetic moment such as proton (^{1}H) is placed in a magnetic field, it can align itself either with the field or against it. These two orientations have different energies. The number of orientations for a particular nucleus is determined by its nuclear spin quantum number, (I). I may have integral or half-integral values. The value of I for both ^{1}H and ^{13}C is 1/2. The number of orientations is given by the expression $2I + 1$.

In addition to spin quantum number, another feature is of importance in NMR spectroscopy. This is the magnitude of the nuclear magnetic moment (μ) and also the magnetic flux density (B) of the applied magnetic field. The energy difference between the two arrangements is $2\mu B$ (see Fig. 13.5). This is the amount of energy required in a magnetic field of strength B to change the nucleus from its lower energy state to its high energy state, that is to change the direction of its magnetic moment. The basis of NMR spectroscopy is to supply exactly this quantum of energy and monitor the absorption by the nuclei. Energy is supplied in the form of electromagnetic radiation. It is known that for radiation of frequency ν, the associated energy is given by $h\nu$,

$$\Delta E = h\nu \tag{13.1}$$

where h is the Planck's constant.

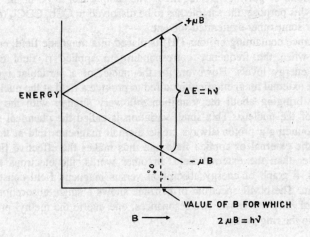

Fig. 13.5. Effect on the energies of the two states of increasing applied magnetic field.

From the condition of absorption of energy,

$$h\nu = 2\mu B \tag{13.2}$$

When a large number of nuclei of a particular magnetically active isotope, such as ^1H, reach thermal equilibrium, more nuclei will be found in the lower energy state than those in the upper energy state. The relative number in the two states will correspond to the Boltzman distribution,

$$\frac{N_u}{N_1} = \frac{\text{number in upper state}}{\text{number in lower state}} = e^{-\Delta E/kT}$$

For the ordinary magnetic fields that are experimentally available ΔE is so small compared (at room temperature) to kT, that there are almost as many nuclei in the upper state as in the lower state.

As in ultraviolet and infrared spectroscopy, transitions or jumps between two energy states can occur if the nuclei are irradiated with electromagnetic radiation, the energy of which is given by,

$$\Delta E = h\nu$$

Since according to Eq. (13.2), the difference in the energies of the two levels depends upon the magnetic field experienced by the nuclei, the frequency of the electromagnetic radiation that will cause transition between the two levels will depend on the strength of the magnetic field.

For example, for protons experiencing a field of 1.41 T (T = Tesla) (14,100 Gauss), the frequency of electromagnetic radiation required is 60 MHz (magahertz) or 60 megacycles/sec. This radiation occurs in the radiofrequency (rf) part of the electromagnetic spectrum and the corresponding magnetic field is often referred to as an rf field.

The basic features of a continuous wave spectrometer include a transmitter coil which introduces the rf field and a coil surrounding the sample which detects the energy absorbed by the sample. The sweep coil permits variation of the magnetic field experienced by the sample. Thirty to forty milligrams of the sample of the organic compound or polymer are required. The sample has to be in the form of a liquid and for this purpose, the sample has to be dissolved in CCl_4, $CDCl_3$ (deuterated chloroform) or some other deuterated solvent.

If a substance containing protons (^1H) is placed in a magnetic field, energy will be absorbed when the frequency corresponding to applied rf field equals the difference in energy levels. However, for the nuclei of a particular type of the strength of the external magnetic field required to provide a field at the nucleus of the right size for bringing about the transition will vary slightly with the structural environment of the nucleus. This small variation is called the chemical shift. The electrons surrounding a proton always create a small magnetic field at the nucleus that opposes the external or applied field and thus makes the effective field at the nucleus smaller than the external field. In other words, the electrons shield the nucleus. Thus, a graph of energy absorption versus magnetic field constitutes the NMR spectrum. The NMR spectrum of benzene shows a single absorption whereas the spectrum of *p*-xylene shows two resonances, one due to the methyl protons and the other due to the ring protons.

(Benzene)   (p –Xylene)   (Tetramethyl silane)
(I)   (II)   (TMS)
                                            (III)

Since it is difficult to measure the absolute value of the external field to the required accuracy of 1 part in 100,000,000, resonances are measured with respect to that of a standard reference substance. This is usually tetramethylsilane (TMS), which is added to the sample whose NMR spectrum is being determined. The chemical shift scale, δ, is given by the relation,

$$\delta = \frac{\Delta B}{B} \times 10^6 \qquad (13.3)$$

where $\Delta B = B_{standard} - B_{sample}$, for a constant frequency.

Also, $\delta = \frac{\Delta v}{v} \times 10^6$ for a constant magnetic field.

Another scale τ is sometimes used, where $\tau = 10 - \delta$.

A list of chemical shifts (δ) for protons in organic molecules is given in Table 13.1.

Since the degree to which energy is absorbed by any particular type of magnetically active nuclei is independent of its structural environment, the total area or integral of an absorption peak is proportional to the number of protons responsible for the absorption. Thus, in the low resolution NMR spectrum of methanol (CH_3OH), the peaks under CH_3 and OH give, on integration, integrals in the proportion 3:1. These correspond to 3 protons of CH_3 and 1 of the OH group. Similarly, the low resolution NMR spectrum of bromoethane (CH_3CH_2Br) gives two peaks with integrals in the ratio of 3:2, corresponding to 3 protons of CH_3 and 2 protons of CH_2.

Table 13.1. Some Values of Chemical Shift for Protons Presenting Different Environments in Organic Molecules

Group	Compound type	Chemical shift (δ)
$CH_3 - C$	Alkane	0.9 – 1.0
$CH_3 - \overset{\overset{O}{\|\|}}{C}\diagdown$	Ketone	2.1 – 2.6
$CH_3 - O -$	Ether	3.3 – 3.9
$\diagup^{H}C = C\diagup$	Alkene	4.5 – 8.0
$H-\langle\bigcirc\rangle$	Arene	6 – 9
$H - \overset{\overset{O}{\|\|}}{C}\diagdown$	Alkanal	9.7
$CH_3 - \overset{\overset{O}{\|\|}}{\underset{O}{C}}$	Ester, acid	2 – 2.2
$CH_3 - N\diagup\diagdown$	Amine	2.2
$H-\langle\bigcirc\rangle^R$	Substituted benzene	6.5 – 8.5
$H - O - \overset{\overset{O}{\|\|}}{C} -$	Carboxylic acid	10 – 13

In a high resolution ^1H NMR spectrum of 1, 1, 2, 2-tetrachloroethane only a singlet is obtained as the two protons experience the same molecular environment.

```
    Cl Cl                      Br  Cl
    |  |                       |   |
 H - C -C - H              H - C - C - H
    |  |                       |   |
    Cl Cl                     Br   Cl
(1,1,2,2-Tetrachloroethane)  (1,1,-Dibromo-2,2-dichloroethane)
        (IV)                          (V)
```

On the other hand, the NMR spectrum of 1, 1-dibromo-2, 2-dichloroethane shows four lines, two for each proton. The resonance of each proton has been split into a doublet by its single neighbour.

The splitting pattern (relative intensities) is given by the following Pascal triangle according to the number of protons causing splitting.

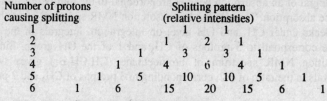

Number of protons   Splitting pattern
causing splitting   (relative intensities)

```
1                         1        1
2                      1      2      1
3                    1    3      3    1
4                  1   4    6    4   1
5                1   5   10   10   5   1
6              1   6   15   20   15   6   1
```

Fig. 13.6. Splitting pattern.

In the example of 1, 1-dibromo-2, 2-dichloroethane where the neighbouring protons are on adjacent carbon atoms connected by a single bond, i.e., free to rotate, the magnitude of splitting is about 7 Hz. The usual notation for splitting is J, i.e., $J = 7$ Hz.

Thus, the following information can be obtained from a high resolution NMR spectrum:

Integration trace: The chemical environment or type of each group of protons,

Splittings: (i) Number of protons on atoms adjacent to the group which contains the proton(s) whose resonance is being recorded, (ii) The measurement of splittings and their dependence on structure.

Sometimes the NMR spectra of certain compounds are too complex to be deciphered or else the splittings are too close to be measured, specially with a 60 MHz spectrophotometer. To obviate these shortcomings, recourse is taken to the following:

(1) Spectra are recorded on 100 and 200 MHz instruments. The individual peaks become very distinct and measurements can be easily carried out (see Fig. 13.7).

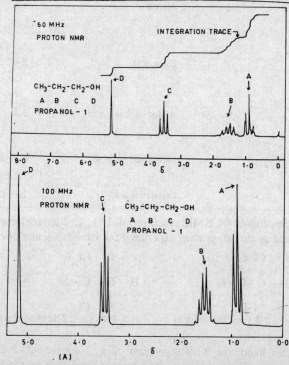

Fig. 13.7. (a) Effect of applied rf frequency on resolution of NMR spectrum.

(2) Shift reagents are used in cases where complex spectra are obtained and over-lapping takes place. Examples of this type are amino and hydroxy compounds. A typical shift reagent is tris-(2, 2, 6, 6-tetramethylheptane 3, 5-dionato)europium. Addition of a shift reagent improves the spectra distinctly.

(3) In decoupling experiments the magnetic field is adjusted for the resonance of one proton. The second proton in the neighbourhood undergoes rapid transitions between its two spin states, induced by incident radiation of the appropriate frequency.

C – 13 NMR

^{13}C occurs only in 1% natural abundance (^{12}C occurring in 99% natural abundance). Since the relative abundance of ^{13}C is so low, it is improbable that a particular nucleus in a molecule will have a second ^{13}C nucleus as an immediate neighbour. Therefore, the splitting of a ^{13}C resonance by coupling to a neighbouring ^{13}C nucleus is unlikely. If the protons in the molecule are spin-decoupled, the ^{13}C NMR spectrum of a substance will consist of a series of single peaks. In a favourable case, the number of structurally different carbon atoms in a molecule can be determined simply by counting the peaks in its protons-decoupled ^{13}C NMR spectrum.

The magnetic moment of ^{13}C is approximately one-fourth that of the proton. This reduces the intensity of the NMR signal. The lower sensitivity is overcome by the use of pulsed Fourier transform NMR in which a high power microsecond pulse of radio frequency energy sets all the carbon nuclei into resonance simultaneously, eliminating the need to sweep the frequency or the magnetic field. The data are recorded as the subsequent decay of the resonances with time, which is the Fourier transform of the desired spectrum.

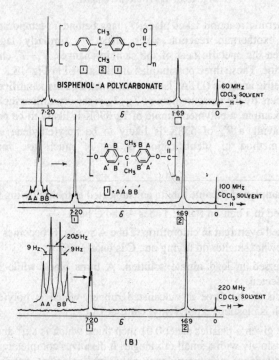

Fig. 13.7 (b) Effect of applied rf frequency on resolution of NMR spectrum.

13.6. THERMAL ANALYSIS OF POLYMERS

The physical and chemical changes in polymers may be investigated by differential thermal analysis (DTA) and thermo-gravimetric analysis (TGA). The DTA technique is based on the occurrence of a difference in temperatures (ΔT) of a polymer sample and an inert reference material when both are heated at a programmed rate in an inert atmosphere. The reference material is often powdered alumina. Any change in the sample's specific heat as at T_g, any structural change which is endothermic or exothermic as at T_m or chemical reactions will change the temperature differences between the sample temperature T_s and the reference temperature T_r. In a simple instrument thermocouples are inserted in the sample and the reference. The sample and the reference material are heated in a metal block at a controlled temperature T_o. As T_o is raised (a rate of 0.5° to 50°/ min is used), T_s and T_r follow, perhaps by as little as 0.1°C.

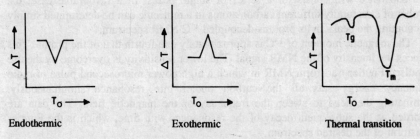

Endothermic   Exothermic   Thermal transition

Fig. 13.8. DTA thermographs.

If an endothermic reaction takes place, T_s lags behind T_r temporarily. If on the other hand an exothermic reaction takes place, T_r similarly lags behind T_s temporarily. When the specific heat of the sample changes, $T_s - T_r$ changes with a shift in the base line. These three possibilities are illustrated in Fig. 13.8.

The calorimetric analysis (DTA) leads to an unambiguous identification of semi-crystalline polymers by the measurement of T_m (T_m is the crystalline melting point for polymers). For example, a polymer sample of T_m 396K is likely to be poly(ethylene), while another having a T_m of 538K is likely to be poly(ethylene terephthalate). However, this method of identification is not of much use for amorphous thermoplastics.

Q. 13.1 Three samples of flexible tubing are subjected to the following tests:

(a) Heated in a bunsen flame. Tubes A and C burn.

(b) Soaked overnight in chloroform. Tube A swells, B becomes very stiff and somewhat smaller on drying out, C is unaffected.

(c) Immersed in lead nitrate solution. A turns black while B and C are unaffected.

Which of the three is vulcanized rubber, which is poly(ethylene) and which is plasticized poly(vinyl chloride).

Q. 13.2 You are given a plastics sheet 0.01 inch thick which is stiff and transparent. It burns slowly with a smell of vinegar. It dissolves completely in acetone. Is it poly(ethylene), poly(propylene), poly(vinyl chloride), poly(vinyl acetate), nylon or cellulose diacetate?

Q. 13.3 How would you identify whether a sample is poly(vinyl acetate) or poly(vinyl alcohol) on the basis of IR spectroscopy.

ANSWERS TO QUESTIONS

13.1. The samples are identified as follows:

A. Vulcanized rubber. It will burn, swell in chloroform and turn lead nitrate black, as it has sulphur in the vulcanized form.

B. Plasticized poly(vinyl chloride). It will not burn; on extraction with chloroform the plasticizer will be dissolved away leaving unplasticized poly(vinyl chloride) which will be stiff.

C. Poly(ethylene).

13.2. The material is cellulose acetate; on heating the smell of vinegar (acetic acid) will be evident. It will burn slowly as it is a cellulose derivative. It will dissolve in acetone.

13.3. Poly(vinyl acetate) will show the characteristic group frequencies due to (C=O) stretching at ~1700 cm^{-1} and C−O−C stretching vibrations at ~1240, 1010–1040 cm^{-1} while poly(vinyl alcohol) will show OH stretching vibration at 3200–3550 cm^{-1}.

CHAPTER 14

POLYMER REACTIONS AND POLYMER REACTANTS

14.1. INTRODUCTION

In earlier chapters, an account of polymer syntheses by a variety of reactions involving step- and chain-polymerizations has been given. In addition, new polymers can be synthesized by modifying the structure of known polymers by such simple reactions as cyclization, halogenation, cross-linking, etc. Another important development is the discovery of polymer supports for the synthesis of poly(peptides) and proteins. Syntheses of a few polymer reactants has also been described in Section 14.7.

14.2. CYCLIZATION REACTIONS

The most well known example of this type is the cyclization of poly(acrylonitrile) (see Section 11.7.4.):

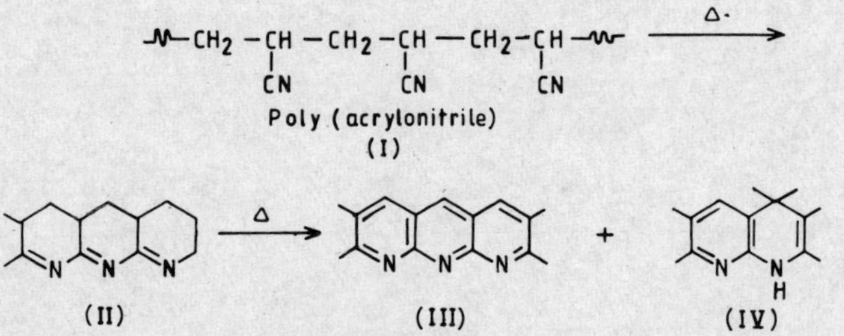

$$\text{(14.1)}$$

Cyclization takes place (II) on heating (I) and on further heating, a mixture of completely aromatized poly(quinizarine) (III) and hydrogenated poly(quinizarine) (IV) is obtained. On heating at 1500-3000°, a carbon fiber is obtained.

One of the best sequences of polymer reactions is probably that of changing poly(vinyl acetate) to poly(vinyl acetal).

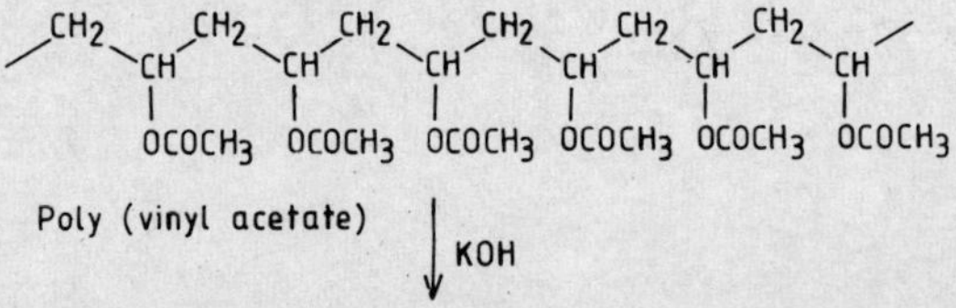

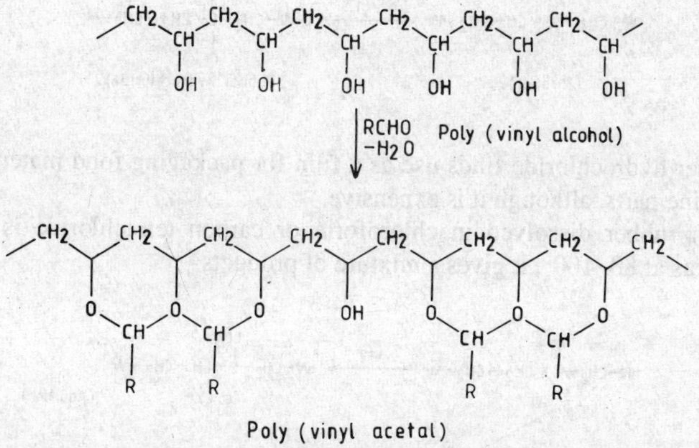

$$\begin{array}{c}\text{(structure of poly(vinyl alcohol) chain)}\end{array}$$

$$\downarrow \begin{array}{l}\text{RCHO}\\ -\text{H}_2\text{O}\end{array} \quad \text{Poly (vinyl alcohol)}$$

$$\text{(structure of poly(vinyl acetal) chain)} \tag{14.2}$$

Poly (vinyl acetal)

This reaction is interesting because it brings out a point stressed by Flory, namely, that in a chain like poly(vinyl alcohol) which has groups that may react in pairs, some of the groups should become isolated by chance in such a reaction. When natural rubber is treated with a strong protonic acid or Lewis acid, a carbonium ion is first formed. The cation then attacks the neighbouring double bond to give a cyclic product.

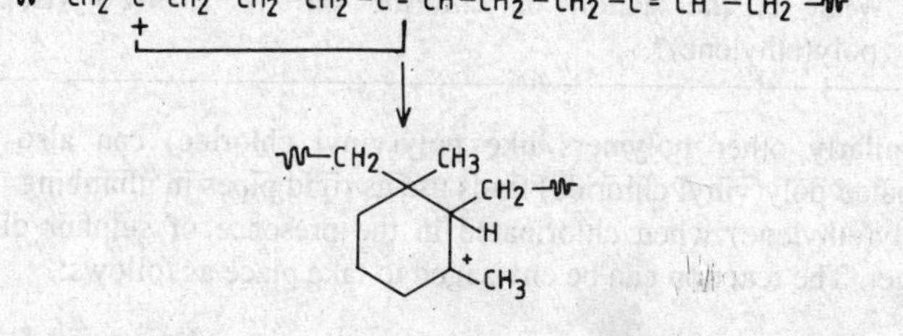

$$-\text{CH}_2-\overset{\overset{\text{CH}_3}{|}}{\text{C}}=\text{CH}-\text{CH}_2-\text{CH}_2-\overset{\overset{\text{CH}_3}{|}}{\text{C}}=\text{CH}-\text{CH}_2-\text{CH}_2-\overset{\overset{\text{CH}_3}{|}}{\text{C}}=\text{CH}-\text{CH}_2-$$

Poly (isoprene)

$$\downarrow \text{H}^+$$

$$-\text{CH}_2-\overset{\overset{\text{CH}_3}{|}}{\underset{+}{\text{C}}}-\text{CH}_2-\text{CH}_2-\text{CH}_2-\overset{\overset{\text{CH}_3}{|}}{\text{C}}=\text{CH}-\text{CH}_2-\text{CH}_2-\overset{\overset{\text{CH}_3}{|}}{\text{C}}=\text{CH}-\text{CH}_2-$$

$$\downarrow$$

$$\text{(cyclic carbonium ion product structure)} \tag{14.3}$$

Q.14.1 Show by equations how the cyclization of 1,4-poly(isoprene) takes place?

14.3. HALOGENATION

In the commercial preparation of rubber hydrochloride, raw rubber is dissolved in benzene and reacted with hydrogen chloride at 10°C. The addition takes place according to Markownikoff's rule.

$$\text{-W-CH}_2 - \overset{\overset{\displaystyle CH_3}{|}}{C} = CH - CH_2 \text{-W-} \xrightarrow{H^+ Cl^-} \text{-W-CH}_2 - \overset{\overset{\displaystyle CH_3}{|}}{\underset{\underset{\displaystyle Cl}{|}}{C}} - CH_2 - CH_2 \text{-W-}$$

(Rubber)   (Rubber hydrochloride)

$$(14.4)$$

Rubber hydrochloride finds use as a film for packaging food material, machines and machine parts, although it is expensive.

When rubber dissolved in chloroform or carbon tetrachloride is treated with chlorine gas at 80–100°, it gives a mixture of products.

$$(14.5)$$

$$\text{-W-CH}_2 - \overset{\overset{\displaystyle CH_3}{|}}{C} = CH_2 - CH_2 \text{-W-} \xrightarrow{Cl_2} \text{-W-CH}_2 - \overset{\overset{\displaystyle CH_3}{|}}{\underset{\underset{\displaystyle Cl}{|}}{C}} - \overset{}{\underset{\underset{\displaystyle Cl}{|}}{CH}} - CH_2 \text{-W-}$$ (Addition)

$$\overset{HCl}{\searrow}\,Cl_2 \qquad \text{-W-CH}_2 - \overset{\overset{\displaystyle CH_3}{|}}{\underset{\underset{\displaystyle Cl}{|}}{C}} = CH - CH \text{-W-}$$ (Allylic substitution)

Chlorine addition to the double bond and substitution at the allylic position both take place. Because of its resistance to moisture and chemicals, chlorinated rubber finds use in the corrosion resistant paints, lacquers and adhesives.

Poly(ethylene) can be chlorinated in the presence of an initiator, U.V. light or on heating. It is a free-radical process.

$$\text{-w-CH}_2\text{-w-} + \dot{C}l \longrightarrow \text{-w-}\dot{C}H\text{-w-} + HCl$$

$$\text{-w-}\dot{C}H\text{-w-} + Cl_2 \longrightarrow \text{-w-}CHCl\text{-w-} \qquad \dot{C}l$$

$$(14.6)$$

Q.14.2 What is the action of chlorine on (i) 1, 4-Poly(isoprene) and (ii) poly(ethylene)?

Similarly other polymers like poly(vinyl chloride) can also be chlorinated. Chlorinated poly(vinyl chloride) finds use as rigid pipes in plumbing.

Poly(ethylene) when chlorinated in the presence of sulphur dioxide yields an elastomer. The reaction can be envisaged to take place as follows:

$$\text{-w-CH}_2\text{-w-}CH_2\text{-w-} \xrightarrow{Cl_2 \cdot SO_2} \text{-w-}\overset{\overset{\displaystyle Cl}{|}}{CH}\text{-w-}\overset{\overset{\displaystyle SO_2Cl}{|}}{CH}\text{-w-} \qquad (14.7a)$$

The number of chlorines and SO_2Cl groups is small. Vulcanization across the SO_2Cl groups of the product initially obtained can be carried out with the help of metal oxides like lead peroxide.

$$\overset{|}{\underset{|}{CH}} - SO_2Cl \; + \; PbO_2 \; + \; ClSO_2 - \overset{|}{\underset{|}{CH}} \longrightarrow \overset{|}{\underset{|}{CH}} - SO_2 - O - Pb - O - SO_2 - \overset{|}{\underset{|}{CH}} \qquad (14.7b)$$

Vulcanization of chlorosulphonated rubber

The chlorosulphonated rubber is resistant to temperature, chemicals and oxidation. However, its high cost discourages its use in gaskets.

14.4. REACTIONS OF CELLULOSE

Cellulose is a linear poly(saccharide) made up of recurring glucose units, joined by 1, 4-glucosidic linkages:

The hydroxyl groups, the hydroxymethyl group and the ether linkages are all in equatorial positions.

Viscose

The chemical reactions involved in the manufacture of viscose rayon have been discussed in detail in Chapter 9 (Section 9.2.1). The reactions in short can be written as follows:

$$R - OH + CS_2 + NaOH \longrightarrow R - O - \overset{\displaystyle S}{\overset{\displaystyle \|}{C}} - S\,Na + H_2O \qquad (14.8a)$$

Cellulose   Sodium xanthate

The cellulose is then regenerated by the action of acid,

$$R-O-\overset{\displaystyle S}{\overset{\displaystyle \|}{C}}-SNa + H^+ \longrightarrow ROH + CS_2 + Na^+ \qquad (14.8b)$$

Cellulose

Cellophane

For the manufacture of cellophane, the viscose solution is extruded in the form of a film, which is precipitated from a bath containing sodium and ammonium sulphates and sulphuric acid. The film which is in the form of regenerated cellulose is washed, bleached and desulphurized. It is finally passed through a bath of glycerol, which also acts as a plasticizer and finally dried.

Cellophane films are very thin but have fair physical properties. Cellophane finds wide use as a wrapping and packaging material.

Cellulose Acetate Rayon

The manufacture of cellulose acetate rayon fibers has been described in detail in Chapter 9 (Section 9.2.3).

Cellulose Acetate Plastics

The usual cellulose acetate, used for injection moulding contains about 50–55% acetyl by weight, corresponding to between a diacetate and a triacetate. It is usually plasticized with a phthalate plasticizer such as dibutyl phthalate. For the preparation of rods and tubes injection, extrusion and compression mouldings can be used.

The sheets are cast from solution onto highly polished metal surfaces as described in Chapter 10 (Section 10.3). These are used for making photographic and X-ray films and for making bags and boxes for packing.

Cellulose Acetate-Butyrate

The usual mixed cellulose ester contains 13% acetyl and 37% of butyral content and is an excellent injection moulding material. It can be more easily plasticized than cellulose acetate. It finds wide use for making films for motion pictures and in automobile parts. It has high impact strength, good dimensional stability and higher moisture resistance. It is soluble in several organic solvents.

Cellulose Nitrate

The use of cellulose nitrate as a plastics material has decreased due to its heat sensitivity, moisture absorption and dimensional instability. But, cellulose nitrate is still widely used in lacquers.

For the manufacture of cellulose nitrate, the Du Pont process is used. The nitration mixture consists of HNO_3 (25 per cent), H_2SO_4 (55 per cent) and water (20 per cent). Purified cotton linters are reacted with the acid mixture at a temperature of 95-105°F and the nitration is allowed to proceed for twenty minutes. The spent acid is spun off by centrifugation. The material still retains a lot of acid which is removed by dipping the material in water. From the drowning tank, the slurry is pumped into other tanks, where it is stabilized by admitting live steam. During these operations, the cellulose nitrate is partially hydrolysed to a dinitrate stage. The product is now bleached with chlorine or sodium hypochlorite.

The water content of the whizzed cellulose nitrate must be substantially reduced. This is done by treating the wet material with denatured alcohol and expelling it by applying pressure. Alcohol-wet cellulose nitrate is converted into celluloid by kneading it with camphor until a uniform dispersion or dough is formed; this is filtered under hydraulic pressure, rolled to reduce the solvent content and then pressed into a block in a heated hydraulic press. The block is cut into sheets of the desired thickness, which after the removal of the residual solvent by stoving may be polished between metal plates by the application of heat and pressure.

Celluloid rods and tubes are manufactured by hydraulic extrusion. Celluloid films are made by dissolving the cellulose nitrate in a suitable solvent such as a mixture of 70 parts of ether and 30 parts of alcohol. The film is made by the casting process already described in Chapter 10 (Section 10.3).

Ethyl Cellulose

The ethyl ether of cellulose is made by heating together cellulose, alkali and ethyl chloride at 100°C.

$$ROH + C_2H_5Cl + NaOH \longrightarrow ROC_2H_5 \quad + \quad NaCl + H_2O \qquad (14.9)$$
Cellulose   Ethyl cellulose

A suitable material contains, on the average, 2.5 ethoxy groups per anhydro glucose unit.

Ethyl cellulose is a tough flexible material with high impact strength. It has got low inflammability. Its defects are its low moisture resistance and low softening temperature.

Q.14.3　Write equations for the formation of the following cellulose derivatives: (i) alkali cellulose, (ii) viscose, (iii) ethyl cellulose, (iv) cellulose nitrate.

14.5. CROSS-LINKING REACTIONS

In the previous chapters, a number of cross-linking reactions have been described. These included the preparation of phenol-formaldehyde and urea-formaldehyde resins (Sections 2.5.2 and 2.5.3), the cross-linking of unsaturated poly(ester) resins (Section 2.5.1), the vulcanization of natural rubber (Section 8.2.3), neoprene rubber (Section 8.3.5) and cross-linking of silcone resins (Section 2.6.1).

14.5.1. Peroxide Cross-linking

Peroxide cross-linking is resorted to in the case of those polymers in which cross-linking cannot be introduced by vulcanization with sulphur. In the case of poly(ethylene) strength and workability at higher temperatures is achieved by peroxidation:

$$\underset{\text{Peroxide}}{\text{ROOR}} \longrightarrow \underset{\text{Free-radicals}}{2RO^{\cdot}}$$

$$\underset{\text{Poly (ethylene)}}{RO^{\cdot} + \text{-}\!\!\backslash\!\backslash\!\!\text{-}CH_2-CH_2-CH_2-\text{-}\!\!\backslash\!\backslash\!\!\text{-}} \longrightarrow \underset{\text{(Alcohol) Poly (ethylene radical)}}{ROH + \text{-}\!\!\backslash\!\backslash\!\!\text{-}CH_2-\overset{\cdot}{C}H-CH_2-\text{-}\!\!\backslash\!\backslash\!\!\text{-}}$$

$$\begin{array}{c} -CH_2-\overset{\cdot}{C}H-CH_2- \\ \\ -CH_2-\overset{\cdot}{C}H-CH_2- \end{array} \longrightarrow \underset{\text{Cross - linked poly (ethylene)}}{\begin{array}{c} \text{-}\!\!\backslash\!\backslash\!\!\text{-}CH_2-CH-CH_2-\text{-}\!\!\backslash\!\backslash\!\!\text{-} \\ | \\ \text{-}\!\!\backslash\!\backslash\!\!\text{-}CH_2-CH-CH_2-\text{-}\!\!\backslash\!\backslash\!\!\text{-} \end{array}} \qquad (14.10)$$

However, only one cross-link is introduced between two polymer chains per molecule of the peroxide.

The introduction of a vinyl group in a siloxane increases the efficiency of cross-linking, thus:

$$\underset{\substack{\text{(Vinyl methyl} \\ \text{silanol)}}}{HO-\overset{\overset{\displaystyle CH_3}{|}}{\underset{\underset{\overset{\displaystyle CH_2}{\|}}{\overset{\displaystyle CH}{|}}}{Si}}-OH} + \underset{\substack{\text{(Dimethyl} \\ \text{silanol)}}}{HO-\overset{\overset{\displaystyle CH_3}{|}}{\underset{\underset{\displaystyle CH_3}{|}}{Si}}-OH} \longrightarrow \text{-}\!\!\backslash\!\backslash\!\!\text{-}O-\overset{\overset{\displaystyle CH_3}{|}}{\underset{\underset{\overset{\displaystyle CH_2}{\|}}{\overset{\displaystyle CH}{|}}}{Si}}-O-\overset{\overset{\displaystyle CH_3}{|}}{\underset{\underset{\displaystyle CH_3}{|}}{C}}-O-\text{-}\!\!\backslash\!\backslash\!\!\text{-} \qquad (14.11)$$

The peroxide cross-linking of the co-polymer is much more efficient than that of the corresponding homopolymer of dimethyl silanol. The process becomes a chain reaction, involving pendant vinyl groups on the poly(siloxane) chains.

14.5.2. Cross-linking in Alkyds

Linseed oil contains both oleic acid [CH_3 $(CH_2)_7$ $CH{=}CH$ $(CH_2)_7$ $COOH$] and linoleic acid [CH_3 $(CH_2)_7$ $CH{=}CH - CH{=}CH$ $(CH_2)_5$ $COOH$].

Oleic acid contains only one double bond whereas linoleic acid contains conjugated double bonds. Both are utilized in the preparation of unsaturated poly(ester) resins:

CH$_3$(CH$_2$)$_7$ CH = CH (CH$_2$)$_7$ COOH + O = C   C = O + HOH$_2$C – CH – CH$_2$OH
(Oleic acid) |
                                                                                OH
                                        (Phthalic   (Glycerol)
                                        anhydride)   (14.12a)

CH$_3$(CH$_2$)$_7$ CH = CH (CH$_2$)$_7$ – COOC H$_2$ — CHCH$_2$-O-C   C-O-

(Alkyd resin based on oleic acid)

The alkyds, used in varnishes and paints, dry in air by a process of cross-linking by the initial formation of an allylic hydroperoxide,

$$\text{CH}_2\text{CH}=\text{CH} - + \text{O}_2 \longrightarrow \text{CH-CH}=\text{CH-}$$
$$|$$
$$\text{OOH}$$

The hydroperoxide can decompose as follows, leading to cross–linking

CH – CH = CH   CH – CH=CH + ·OH
| |
OOH   O·

CH – CH = CH
|
O
|
O
|
CH – CH = CH   (14.12b)

The decomposition of the hydroperoxide can lead to other secondary reactions, which also contribute to the cross-linking between two alkyd chains by the formation of carbon-carbon bonds or ether linkages.

For the drying of unsaturated alkyds, driers are used which can be salts of Co, Pb and Mn (usually octanoates, naphthenates and linoleates).

The reaction of oxygen with alkyds containing conjugated double bonds occurs as follows:

CH= CH — CH= CH   + O$_2$ →
                                              CH
                                        CH   CH
                                        | |
                                        O   CH →   (14.12c)
                                          O

CH-CH=CH-CH
| ·
OO

A diradical is formed and cross-linking can take place by 1,4-polymerization of the polymer molecules.

14.5.3. Cross-linking of 1,3-Dienes

Sulphur Vulcanization

Goodyear in 1839 found that if natural rubber is heated with sulphur, its mechanical properties improve. Although a century and a half has elapsed since then, the mechanism of sulphur vulcanization has not been elucidated unambiguously. At the present, of the many mechanisms proposed, the ionic mechanism is the preferred one.

$$S_8 \longrightarrow \overset{\delta^+}{S_X} - \overset{\delta^-}{S_Y} \longrightarrow S_X^+ + S_Y^-$$

$$\text{(Sulphur)} \qquad \downarrow \quad \text{—}CH_2 - CH = CH - CH_2 \text{—}$$

$$1,4 \text{ poly (butadiene)} \qquad (14.13a)$$

$$\text{—}CH_2 - CH \overset{+}{\underset{S_X}{-}} CH - CH_2 \text{—} \qquad + S_Y^-$$

The sulphonium ion formed (Eq. 14.13a) can then react with a polymer molecule (polybutadiene) either by a hydride ion abstraction or a proton transfer,

$$\text{—}CH_2 - CH \underset{S_X}{-} CH - CH_2 \text{—} \xrightarrow[\text{Poly (butadiene)}]{\text{Polymer}} CH_2 - CH_2 - CH - CH_2 \text{—}$$

$$\begin{array}{c} | \\ S_X \text{(hydride ion} \\ \text{abstraction)} \end{array} \qquad (14.13b)$$

$$\text{Polymer} \mid \text{Poly (butadiene)} \qquad + \text{—}\overset{+}{C}H - CH = CH - CH_2 \text{—}$$

$$\text{—}CH_2 - CH - CH = CH \text{—} + \text{—}CH_2 - CH_2 - \overset{+}{C}H - CH_2 \text{—}$$
$$\qquad | \\ \qquad S_X$$

$$\text{(Proton transfer)}$$

The cation can then react with sulphur and cross-link with a polymer, thus:

$$\text{—}\overset{+}{C}H - CH = CH - CH_2 \text{—} \xrightarrow{S_8} \text{—}CH - CH = CH - CH_2 \text{—}$$

$$\begin{array}{cc} & | \\ & + S_X \qquad \downarrow \text{Polymer} \end{array}$$

$$\text{—}CH - CH = CH - CH_2 \text{—} \qquad \text{—}CH - CH = CH - CH_2 \text{—}$$
$$\quad | \qquad\qquad\qquad\qquad\qquad | $$
$$\quad S_X \qquad\qquad\qquad\qquad\qquad S_X \qquad (14.13c)$$
$$\text{—}CH_2 - C = CH - CH_2 \text{—} \xleftarrow{\text{Polymer}} \text{—}CH_2 - CH - \overset{+}{C}H - CH_2 \text{—}$$

$$+ \text{—}CH_2 - CH_2 - \overset{+}{C}H - CH_2 \text{—} \qquad\qquad \downarrow \text{Polymer}$$

$$\text{—}CH - CH = CH - CH_2 \text{—}$$
$$\quad | $$
$$\quad S_X $$
$$\text{—}CH_2 - CH - CH_2 - CH_2 \text{—}$$
$$\quad + $$
$$\text{—}\overset{+}{C}H - CH = CH - CH_2 \text{—}$$

Table 14.1. Recipe for Vulcanization of Rubber

Rubber	100 parts
Sulphur	0.25–1.5 parts
Accelerator	0.25–1.5 parts
Activator (zinc oxide)	1–10 parts
Soap (stearic acid)	1–5 parts
Antioxidant	0–1.5 parts

Vulcanization with Sulphur and Accelerators

Sulphur when used alone leads to a very slow vulcanization process and gets wasted due to the formation of a large number of sulphur linkages. Sulphur is usually used in conjunction with accelerators and activators. A typical recipe for the vulcanization of rubber is given in Table 14.1.

A large number of elastomers such as poly(isoprene), poly(butadiene), co-polymers of butadiene with styrene (SBR) and acrylonitrile (Buna-N, and isoprene-butylene co-polymer can be vulcanized in this way. The structures of a few accelerators have been shown in Chapter 8 (Section 8.2.3). The detailed mechanism of the vulcanization of elastomers in the presence of an accelerator and activator is obscure.

Q. 14.4 By what reactions can the following polymers be cross-linked:

   (i)   Poly(ester) obtained from ethylene glycol and maleic anhydride.

   (ii) Poly(ester) obtained from oleic acid, ethylene glycol and phthalic acid.

   (iii) Poly(ethylene).(iv) Poly(butadiene).

14.6. SOLID PHASE SYNTHESIS OF PEPTIDES

Proteins consist of poly(peptide) chains, which in turn are made up of amino acids joined together by peptide linkages $\{CONH\}$.

The utilization of solid support in poly(peptide) synthesis was developed nearly simultaneously by R.B. Merrifield at Rockefeller University and Robert Letsinger at Northwestern University. Merrifield immobilized an amino acid to the backbone of a poly(styrene) derivative and carried out sequential peptide bond formation with the original amino acid serving as an anchor to the poly(styrene) resin.

A difficulty arises due to the fact that every amino acid has both an amino residue and a carboxyl residue. Unless one or the other function is converted to a derivative which cannot react in the sequence, several different products are likely to be obtained. To circumvent this problem, *blocking groups* are used, which prevent the functional group which one does not wish to react, from participating in the sequence.

In the synthesis involving polymer support two additional steps are necessary. The first step involves coupling of the organic amino acid to the polymer support. This is followed by sequential peptide forming reactions. At the end of the reaction, the grown poly(peptide) must be removed from the polymer support backbone.

Blocking of the amino group by *t*-butoxy carbonyl group can be achieved by reacting the amino acid with *t*-butoxy carbonylazide, $(CH_3)_3 C-O-CO-N_3$, thus:

$$(CH_3)_3 C - O - CON_3 + H_2N - CHR - COOH \longrightarrow$$

$$(CH_3)_3 C - O - CONH. CHR - COOH \qquad (14.14)$$

The sequence for the solid phase synthesis of a poly(peptide) is shown in Fig. 14.1.

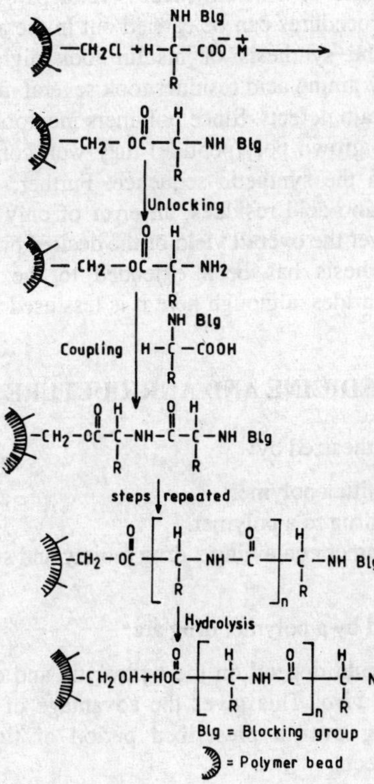

Fig. 14.1 Solid phase synthesis of a poly(peptide).

The unblocking can be done by treating the blocked amino group with trifluoroacetic acid (CF_3COOH). The final hydrolysis to free the poly(peptide) from the polymer support is achieved by treating the product with hydrofluoric acid (HF). Coupling of the second amino acid to the amino group terminum of the first poly(styrene) anchored amino acid is accomplished either with dehydrating agents like dicyclohexyl carbodiimide (DCC) or with activated acid derivatives like acid chlorides.

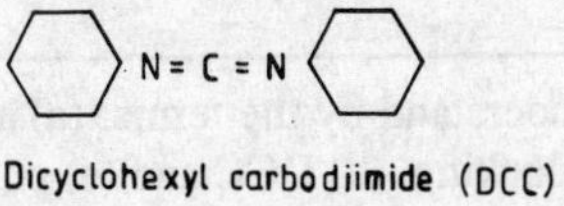

Dicyclohexyl carbodiimide (DCC)

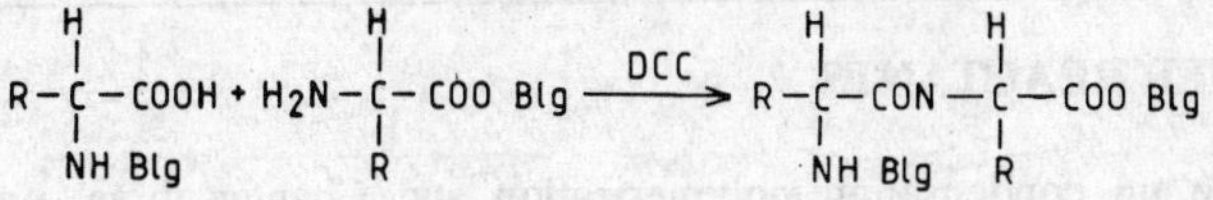

$$(14.15)$$

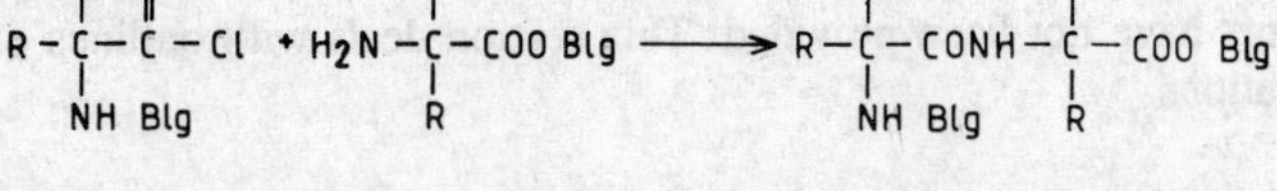

$$(14.16)$$

The success of resin bound poly(peptide) synthesis is most striking. The technique has been automated so that the successive steps of deblocking, coupling and the associated wash procedures can be carried out in the absence of an operator. Under these conditions the synthesis of insulin took eight days while that of ribonuclease-A with its 127 amino acid residues took several months.

The technique has certain defects. Since polymers incorporate small quantities of reagents and products, the grown poly(peptide) may well contain small amounts of impurities picked up from the synthetic sequence. Further, for the synthesis of a protein containing 100 amino acid residues, an error of only 1% in the addition of each amino acid would lower the overall yield of the desired product by 36%.

The solid phase synthesis has been extended to the preparation of oligo-nucleotides and oligosaccharides, although here it is less used than that in the case of poly(peptides).

14.7. POLYMERS IN MEDICINE AND AGRICULTURE

A polymer drug can be synthesized by:

(i)   Covalent bonding with a polymer.
(ii) Complexing of the drug to a polymer.
(iii) Synthesis of a monomer containing a drug moiety and subsequent polymerization.

The advantages offered by a polymer drug are:

(i)   The drug action would depend on the hydrolytic and enzymatic cleavage of the drug moiety *in vivo*. This gives the advantage of a slow and sustained release of the drug over a prescribed period of time which avoids the undesirable side effects.
(ii) It is possible to make a polymer drug with a specific purpose and desirable properties.
(iii) Where local absorption is undesirable, it is possible to synthesize a polymer drug with low absorptivity, e.g., for skin affections.
(iv) It is possible to couple two or more drugs on the same polymer to enhance the synergic activity.
(v)   The physical encapasulation of a drug by a polymer may have advantage over polymer drug.

In agriculture, the use of polymeric pesticides, fertilizers and herbicides can lead to slow release of these chemicals over a period of time.

Q. 14.5 What do you understand by the terms: (a) a blocking reagent, (b) a polymer support, (c) unblocking, (d) DCC and (e) polymer drug.

14.8. POLYMER REACTANTS

In Chapter two on condensation polymerization and Chapter three, on addition polymerization, many reactants used in the synthesis of polymers have been described. However, in several cases, the synthesis or other modes of preparation of these reactants have not been provided. This section deals with outlines of some of these preparations.

14.8.1. Non Ethylenic Polymer Reactants

Acids

The preparation of adipic acid starting from phenol has already been given (Section 2.4.2). There is another method for the preparation of adipic acid from tetrahydrofuran.

$$
\begin{array}{c}
\text{H}_2\text{C} \underline{\quad\quad} \text{CH}_2 \\
\mid \quad\quad\quad \mid \\
\text{H}_2\text{C} \underset{\text{O}}{\diagdown} \diagup \text{CH}_2
\end{array}
\; + \; 2\text{CO} + \text{H}_2\text{O} \; \xrightarrow[\text{Ni(CO)}_4,\text{NiI}_2\,(\text{Adipic acid})]{270°} \; \text{HOOC}\,(\text{CH}_2)_4\,\text{COOH}
$$

(Tetrahydrofuran) (High pressure) (14.17)

Sebacic acid is obtained by the alkaline hydrolysis of castor oil, which is glyceryl ester of ricinoleic acid.

$$
\text{Castor oil} \xrightarrow{\text{Na}^+ \text{OH}^-} \quad \underset{\underset{\text{OH}}{\mid}}{\overset{\overset{\text{H}}{\mid}}{\text{CH}_3\,(\text{CH}_2)_5 - \text{C} - \text{CH}_2 - \text{CH} = \text{CH}\,(\text{CH}_2)_7\,\text{COOH}}} \xrightarrow[\text{H}_2\text{O}]{\text{Na}^+\text{OH}^-}
$$

$$
\underset{\underset{\text{OH}}{\mid}}{\overset{\overset{\text{H}}{\mid}}{\text{CH}_3\,(\text{CH}_2)_5 - \text{C} - \text{CH}_3}} \; + \; \text{Na}^{+-}\,\text{OOC} - (\text{CH}_2)_8\,\text{COO}^-\,\text{Na}^+
$$

Capryl alcohol   Sebacic acid
                        (Na salt)   (14.18)

Orthophthalic acid can be obtained by the oxidation of naphthalene in presence of vanadium pentoxide.

(Nephthalene) $\xrightarrow[\text{V}_2\text{O}_5]{[\text{O}],\,360°}$ (Phthalic anhydride) $\xrightarrow{+\text{H}_2\text{O}}$ (Pthalic acid) (14.19)

Terephthalic acid is obtained by the oxidation of *p*-xylene:

(p–Xylene) $\xrightarrow{[\text{O}]\ 360°}$ (Terephalic acid) (14.20)

Similarly, isophthalic acid (metaisomer) is obtained by the oxidation of *m*-xylene. Pyromellitic acid can be similarly obtained by the vapour phase oxidation of durene either by nitric acid or in the presence of vanadium pentoxide:

$$\underset{\text{(Durene)}}{\overset{\text{H}_3\text{C}}{\text{H}_3\text{C}}\boxed{}\overset{\text{CH}_3}{\text{CH}_3}} \xrightarrow[\text{V}_2\text{O}_5]{[\text{O}],500°} \underset{\substack{\text{(Pyromellitic}\\ \text{acid)}}}{\overset{\text{HOOC}}{\text{HOOC}}\boxed{}\overset{\text{COOH}}{\text{COOH}}} \longrightarrow \underset{\substack{\text{(Pyromellitic}\\ \text{anhydride)}}}{\text{Pyromellitic anhydride}} \quad (14.21)$$

Maleic acid is obtained commercially by the oxidation of benzene in the presence of V_2O_5 as catalyst.:

$$C_6H_6 \xrightarrow[\text{V}_2\text{O}_5]{[\text{O}],450°} \underset{\text{(Maleic acid)}}{\overset{\text{CH.COOH}}{\underset{\text{CH.COOH}}{\|}}} + 2CO + 2H_2O \longrightarrow \underset{\text{(Maleic anhydride)}}{\overset{\text{CH.CO}}{\underset{\text{CH.CO}}{\|}}\diagdown_{\text{O}}} \quad (14.22)$$

Phenols

Phenol can be prepared by the sulphonation process as follows:

$$\underset{\text{Benzene}}{C_6H_6} + H_2SO_4 \longrightarrow \underset{\substack{\text{Benzene sul-}\\ \text{phonic acid}}}{C_6H_5SO_3H} \xrightarrow[\text{fusion}]{\text{NaOH}} C_6H_5ONa \xrightarrow{SO_2} \underset{\text{Phenol}}{C_6H_5OH} \quad (14.23)$$

In the Reschig process, chlorobenzene is hydrolysed in the presence of a catalyst:

$$\underset{\text{Benzene}}{C_6H_6} + HCl + \frac{1}{2}O_2 \longrightarrow \underset{\text{Chlorobenzene}}{C_6H_5Cl} \xrightarrow[\text{catalyst}]{\text{H}_2\text{O}, \Delta} \underset{\text{Phenol}}{C_5H_5OH} \quad (14.24)$$

Phenol can also be obtained from Cumene:

$$\underset{\text{(Benzene)}}{\boxed{}} \xrightarrow[\text{CH}_2=\overset{\text{H}}{\underset{}{\text{C}}}-\text{CH}_3]{} \boxed{}-\overset{\text{CH}_3}{\underset{\text{CH}_3}{\text{C}}}-\text{H} \xrightarrow{[\text{O}]} \underset{\text{(Cumene hydroperoxide)}}{\boxed{}-\overset{\text{CH}_3}{\underset{\text{CH}_3}{\text{C}}}-\text{OOH}} \quad (14.25)$$

$$\Big\downarrow \text{H}_2\text{SO}_4$$

$$\underset{\text{(Acetone)}}{\overset{}{\text{CH}_3-\underset{\overset{\|}{\text{O}}}{\text{C}}-\text{CH}_3}} + \underset{\text{(Phenol)}}{\boxed{}-\text{OH}}$$

Bisphenol-A is obtained by the condensation of phenol with acetone:

$$\underset{\text{(Phenol)}}{\text{HO}-\boxed{}-\text{H}} + \underset{\substack{\text{(Acetone)}}}{\overset{\text{CH}_3}{\underset{\text{CH}_3}{\text{C}}}=\text{O}} + \underset{\text{(Phenol)}}{\text{H}-\boxed{}-\text{OH}} \xrightarrow{\text{H}_2\text{SO}_4} \underset{\text{(Bisphenol-A)}}{\text{HO}-\boxed{}-\overset{\text{CH}_3}{\underset{\text{CH}_3}{\text{C}}}-\boxed{}-\text{OH}} + H_2O \quad (14.26)$$

Reactants containing amino or cyanato groups

Melamine is obtained from calcium cyanamide by the following series of reactions:

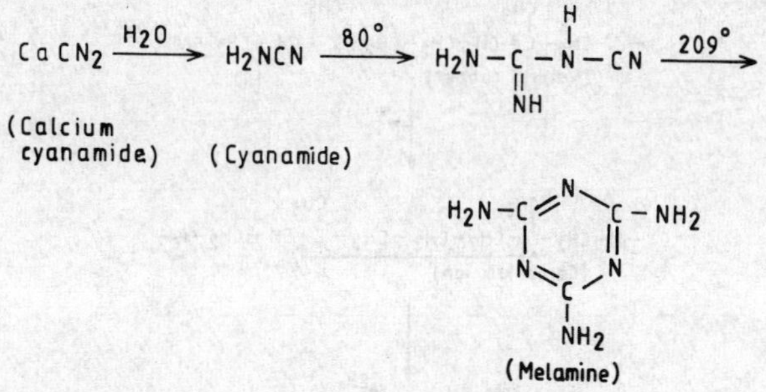

$$CaCN_2 \xrightarrow{H_2O} H_2NCN \xrightarrow{80°} H_2N-\overset{\overset{H}{|}}{\underset{\overset{\|}{NH}}{C}}-N-CN \xrightarrow{209°}$$

(Calcium cyanamide)   (Cyanamide)

$$H_2N-C\begin{smallmatrix}N\\ \\N\end{smallmatrix}C-NH_2$$

(14.27)

(Melamine)

Toluene di-isocyanate is obtained by the action of phosgene on 2, 4-diamino toluene:

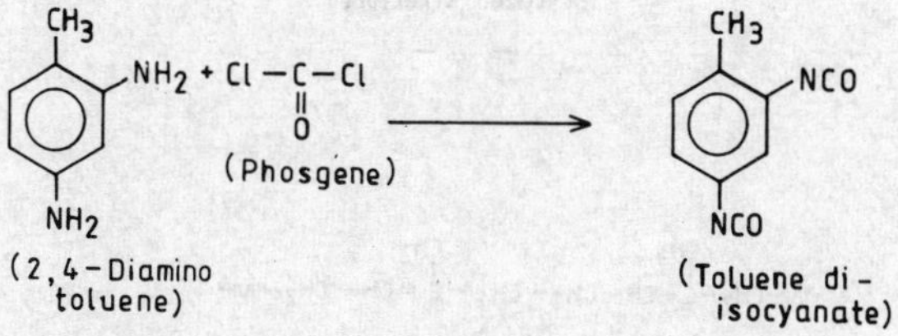

(2, 4-Diamino toluene) + Cl−C−Cl (Phosgene) ⟶ (Toluene di-isocyanate)

(14.28)

Hexamethylene tetramine is obtained by the action of ammonia on formaldehyde. Formaldehyde is obtained by the oxidation of methanol:

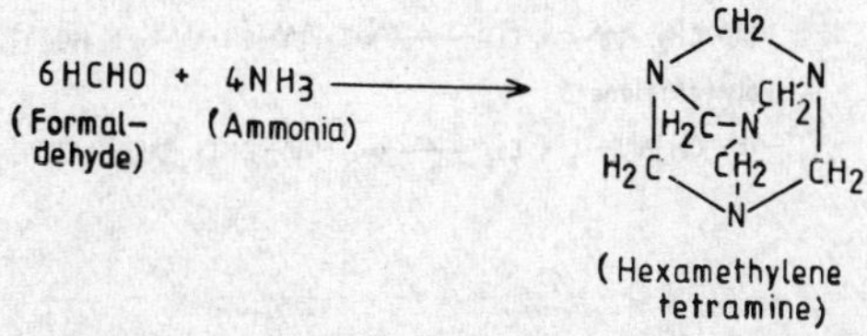

$$6\,HCHO + 4\,NH_3 \longrightarrow$$
(Formaldehyde)   (Ammonia)

(14.29)

(Hexamethylene tetramine)

The preparation of a large number of polymer reactants has been discussed in Chapters two and three.

14.8.2. Ethylenic Reactants

The preparation of a large number of vinyl monomers and other reactants derived from ethylene has been described in Chapter three (3.11).

ANSWERS TO QUESTIONS

14.1

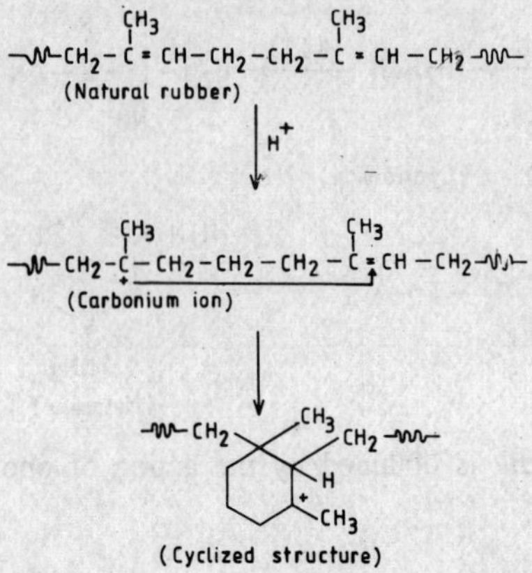

$$-\text{W}-CH_2-\underset{\underset{CH_3}{|}}{C} = CH-CH_2-CH_2-\underset{\underset{CH_3}{|}}{C} = CH-CH_2-\text{W}-$$
(Natural rubber)

$$\Big\downarrow H^+$$

$$-\text{W}-CH_2-\underset{\underset{CH_3}{|}}{\overset{+}{C}}-CH_2-CH_2-CH_2-\underset{\underset{CH_3}{|}}{C}\overset{+}{=}CH-CH_2-\text{W}-$$
(Carbonium ion)

$$\downarrow$$

(Cyclized structure)

14.2

(i)

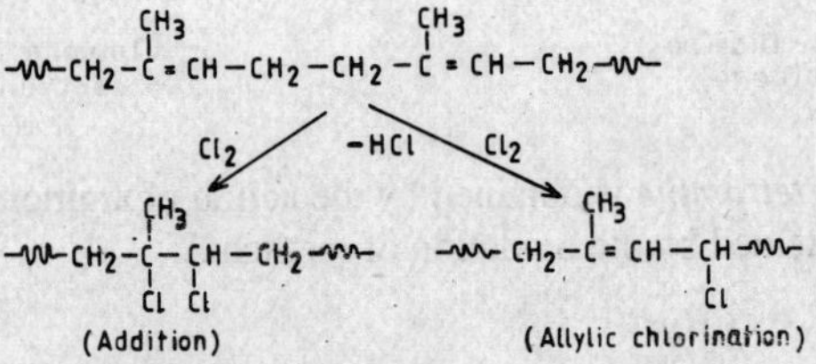

$$-\text{W}-CH_2-\underset{\underset{CH_3}{|}}{C}=CH-CH_2-CH_2-\underset{\underset{CH_3}{|}}{C}=CH-CH_2-\text{W}-$$

$$Cl_2 \swarrow \quad -HCl \quad \searrow Cl_2$$

$$-\text{W}-CH_2-\underset{\underset{Cl}{|}}{\overset{\overset{CH_3}{|}}{C}}-\underset{\underset{Cl}{|}}{CH}-CH_2-\text{W}- \qquad -\text{W}-CH_2-\underset{\underset{CH_3}{|}}{C}=CH-\underset{\underset{Cl}{|}}{CH}-\text{W}-$$
(Addition)   (Allylic chlorination)

(ii)

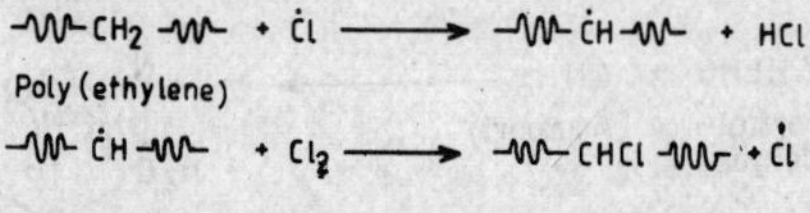

$$-\text{W}-CH_2-\text{W}- + \dot{C}l \longrightarrow -\text{W}-\dot{C}H-\text{W}- + HCl$$
Poly (ethylene)

$$-\text{W}-\dot{C}H-\text{W}- + Cl_2 \longrightarrow -\text{W}-CHCl-\text{W}- + \dot{C}l$$

14.3

(i) For answer please see scheme 2.10.

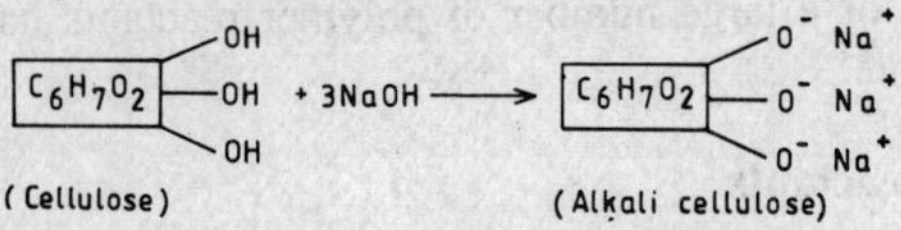

$$C_6H_7O_2\!\!\left[\begin{array}{l}OH\\OH\\OH\end{array}\right. + 3NaOH \longrightarrow C_6H_7O_2\!\!\left[\begin{array}{l}O^-\,Na^+\\O^-\,Na^+\\O^-\,Na^+\end{array}\right.$$
(Cellulose)   (Alkali cellulose)

(ii) $R-OH + CS_2 + NaOH \longrightarrow R-O-\underset{\underset{S}{\|}}{C}-SNa$
   Cellulose   S   Viscose

$$R-O-\underset{\underset{S}{\|}}{C}-SNa \xrightarrow{\;H^+\;} ROH + CS_2 + Na^+$$
                          Cellulose

(iii) $R-OH + C_2H_5Cl + NaOH \longrightarrow R-OC_2H_5 + NaCl + H_2O$
    Cellulose   Ethyl Cellulose

(iv) $R - OH + HNO_3 + H_2SO_4 \longrightarrow R - ONO_2 \xrightarrow[\text{hydrolysis}]{\text{Partial}}$ (2.5 OH groups out of 3 nitrated)

Cellulose

Cellulose nitrate

14.4

(i)

-ᴡᴡ- OCH₂CH₂OCO CH = CH CO -ᴡᴡ-

$+$ $\xrightarrow[\text{vinyl monomer}]{nCH_2=CHX}$

-ᴡᴡ-OCH₂CH₂OCO CH = CH CO -ᴡᴡ- -ᴡᴡ-OCH₂CH₂OCOCH-CH CO -ᴡᴡ-
$\quad$ |
$\quad$ (CH₂
$\quad$ |
polyester $\qquad\qquad\qquad$ CHX)n
$\qquad\qquad\qquad\qquad\qquad$ |
$\qquad\qquad$ -ᴡᴡ- OCH₂CH₂OCOCH-CH -ᴡᴡ-

cross-linked polyester

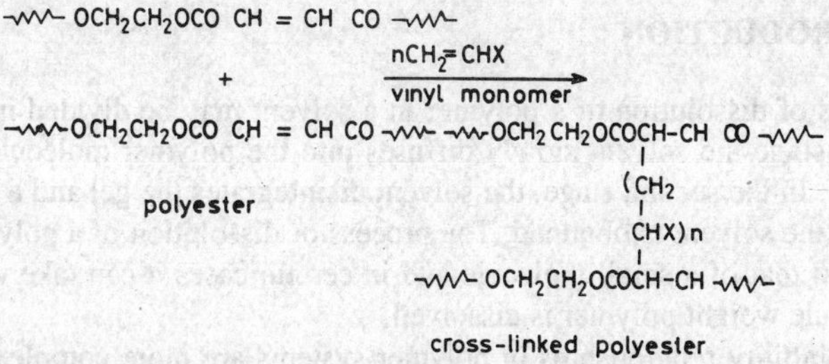

(ii) Please see equations (14.12a) and (14.12b).

(iii) See equation (14.10)

(iv) See equations (14.13a), (14.13b) and (14.13c)

14.5.

(a) *Blocking agent*--In the poly(peptide) synthesis one of the two reactive groups of an amino acid namely $-NH_2$ or $-COOH$ has to be blocked by the use of a blocking reagent to prevent it from taking part in the sequential reactions. A group such as *t*-butoxy carbonyl azide (I) is used as a blocking agent

$Me_3C–O–CON_3 + H_2NCHR – COOH \longrightarrow Me_3C – O – CONHCHR – COOH$
(I) $\qquad\qquad$ Amino acid

(b) *Polymer support*—It is usually a cross-linked poly(styrene) bead on which a group like $-CH_2Cl$ is introduced. This reacts with the first amino acid to create a polymer support on which further sequential poly(peptide) synthesis is done.

(c) *Unblocking*—A blocked group such as $-NHBlg$ is unblocked or set free to $-NH_2$ by the action of an agent such as trifluoroacetic acid ($F_3C.COOH$).

(d) DCC or dicyclohexyl carbodiimide ($C_6H_{11} – N = C = N – C_6H_{11}$) is a reagent used in coupling a/second amino acid to the first anchored on a polymer support.

(e) *Polymer drugs* — See Section 14.6.

CHAPTER 15

POLYMER SOLUBILITY

15.1. INTRODUCTION

The process of dissolution of a polymer in a solvent may be divided into two stages. In the first stage the solvent slowly diffuses into the polymer molecule to produce a swollen gel. In the second stage, the solvent disintegrates the gel and a solution of the polymer in the solvent is obtained. The process of dissolution of a polymer is slow as compared to that of a small molecule and in certain cases it can take weeks before a high molecule weight polymer is dissolved.

The solubility relationships in polymer systems are more complex than those in the case of low molecular weight compounds. These depend on the size difference between the polymer and the solvent molecules, the viscosity of the solution and the texture and molecular weight of the polymer. Cross-linked polymers do not dissolve but swell only if they interact at all with the solvent. This is illustrated by considering the example of rubber. Raw rubber is soluble in a number of solvents, moderately vulcanized rubber swells in contact with a solvent and hard rubbers do not appreciably swell when treated with a solvent, as is also the case with cross-linked thermosetting resins.

This insolubility is also exhibited by several crystalline polymers as well, particularly by non-polar ones. However, the crystallinity decreases as the melting point is approached and the melting point is depressed by the presence of the solvent. Thus, linear poly (ethylene) ($T_m = 135°$) is soluble in many solvents at $100°$, which is below the crystalline melting point (T_m is defined in chapter 7). Even poly(tetrafluoroethylene) which has a crystalline melting point (T_m) of $325°$ dissolves in some solvents which boil above $300°C$. Several polar polymers like Nylon-6, 6 which has a crystalline melting point (T_m) of $265°$ dissolves in several common solvents even at room temperature. These solvents interact strongly with Nylon to form hydrogen bonds.

15.2. SOLUBILITY PARAMETERS

The free enthalpy of mixing of a polymer solution, ΔG, is given by the thermodynamic equation:

$$\Delta G = \Delta H - T\Delta S \qquad (15.1)$$

where ΔH is the enthalpy of mixing, ΔS is the entropy of mixing and T is the temperature. For polymer solution, ΔS is very small and to obtain negative value of ΔG, ΔH is required to be small or negligible. J.H. Hildebrand and R.L. Scott have shown that to a first approximation, ΔH can be written as,

$$\Delta H = v_1 v_2 (\delta_1 - \delta_2)^2 \qquad (15.2)$$

where v_1 is the volume fraction of the solvent and v_2 that of the polymer. δ_1 is the solubility parameter of the solvent and δ_2 that of the polymer. δ_1^2 and δ_2^2 are the cohesive energy densities of the substances and δ_1^2 can be approximated by the energy of vaporization per unit volume of the solvent.

Cellulose nitrate is insoluble in both ethyl alcohol and diethyl ether, but it is soluble in a 50:50 mixture of the two solvents. This can be explained on the basis that the solubility parameter of a mixture of solvents is equal to the sum of the products of the mole fractions and solubility parameters of the two components, provided the two solvents are not too dissimilar in molecular structure.

Solubility parameters of volatile solvents can be calculated from the latent heat of vaporization (ΔH) but this technique is not applicable in the estimation of the solubility parameters of resinous products. The solubility parameter (δ_1) of a volatile liquid can be calculated by the application of the following formula:

$$\delta_1 = \left[\frac{\Delta H - RT}{M/D} \right]^{1/2} \tag{15.3}$$

where R is the gas constant, M is the molecular weight and D the density of the liquid. M/D is thus the molar volume V.

The relationship (15.3) can be used to estimate the solubility parameter of n-heptane.

$$\delta_1 = \left[\frac{\Delta H - RT}{M/D} \right]^{1/2} = \left[\frac{87(100) - 2(298)}{100/0.68} \right]^{1/2} = [7.4 \text{ cal/cm}^3]^{1/2}$$

The solubility parameter of any substance can be obtained by the application of P.A. Small's formula (Eq. 15.4),

$$\delta_2 = \frac{D \Sigma G}{M} \tag{15.4}$$

where G is the molar attraction constant, D is the density and M the molecular weight of the segment. The use of the formula is illustrated by considering the case of poly(propylene), which has a segment ($- CH (CH_3) CH_2 -$) with a formula weight of 42.

Typical G values in Small's formula are as follows:

$$- CH_3 = 303; - CH_2 - = 269; CH - = 176 \text{ (J} - cm^3)^{1/2} \text{ mol}$$

Hence, for poly(propylene),

$$\delta_2 = \frac{D \Sigma G}{M} = \frac{(0.905)(303 + 269 + 176)}{42} = 16.1 \text{ (J} - cm^3)^{1/2}$$

The value of the solubility-parameter approach is that it can be calculated for both polymer and solvent. In the absence of strong interactions such as hydrogen bonding, solubility can be expected if $\delta_1 - \delta_2$ is less than $3.5 - 4$ (approximately). A few typical values of δ_1 and δ_2 are given in Table 15.1. For polymers, they are the square roots of the values of cohesive energy density. Values of G are given in Table 15.2.

Table 15.1. Solubility Parameters of a Few Solvents and Polymers

Solvent	δ_1 $(J-cm^3)^{1/2}$	Polymer	δ_2 $(J-cm^3)^{1/2}$
n-Hexane	14.8	Poly(tetrafluoroethylene)	12.7
Carbon tetrachloride	17.6	Poly(dimethyl siloxane)	14.9
Toluene	18.3	Poly(ethylene)	16.2
2-Butanone	18.5	Poly(propylene)	16.6
Benzene	18.7	Poly(butadiene)	17.6
Cyclohexanone	19.0	Poly(styrene)	17.6
Styrene	19.0	Poly(methylmethacrylate)	18.6
Chlorobenzene	19.4	Poly(vinyl chloride)	19.4
Acetone	19.9	Poly(vinyl acetate)	21.7
Tetrahydrofuran	20.3	Poly(ethylene terephthalate)	21.9
Methanol	29.7	Nylon 6, 6	27.8
Water	47.9	Poly(acrylonitrile)	31.5

Table 15.2. Values of Molar Attraction Constant g

Group	G $(J-cm^3)^{1/2}/mole$	Group	G $(J-cm^3)^{1/2}/mole$
$-CH_3$	303	NH_2	463
$-CH_2$	269	$-NH-$	368
$-CH{<}$	176	$-N-$	125
${>}C{<}$	65	$C \equiv N$	725
$CH_2 =$	259	NCO	733
$-CH =$	249	$-S-$	439.
${>}C =$	173	Cl_2	701
$-CH =$ (aromatic)	239	Cl (primary)	419
${>}C =$ (aromatic)	200	Cl (secondary)	425
$-O-$ (ether, acetal)	235	Cl (aromatic)	329
$-O-$ (epoxide)	360	F	84
$-COO-$	668	Conjugation	17
${>}C = O$	538	$-cis$	-14
$-CHO$	599	$-trans$	-28
$(CO)_2O$	1159	Six membered ring	-48
$OH-$	462	ortho	-19
$OH-$ (aromatic)	350	meta	-13
H (acidic dimer)	-103	para	-82

Cohesive energy density

In the Hildebrand expression (Eq. 15.2) δ_1^2 and δ_2^2 have been stated to be cohesive energy densities of the substances or for small molecules, the energy of vaporization per unit volume. Cohesive energy is the total, energy required to remove a molecule from a liquid or solid to a position far from its neighbours. The cohesive energy per unit volumes is the cohesive energy density. The cohesive energy can be calculated from thermodynamic data.

15.3. SOLUTION VISCOSITY

15.3.1. Theta Temperature

It has been found that $[\eta]M$ product is directly proportional to the average conformational size of a dissolved flexible polymer. Assuming a situation where polymer/solvent/temperature interactions are so adjusted that the polymer assumes its `unperturbed', average shape, P.J. Flory showed that,

$$[\eta]_M \propto (\overline{\gamma_0^2})^{3/2} \tag{15.5}$$

where $(\overline{\gamma_0^2})$ is defined as the near square end-to-end dimension of linear chains. The special `unperturbed' state is referred to as θ (theta) state.

According to Flory, θ temperature represents the lowest temperature at which a polymer with infinite molecular weight would be completely miscible with a specific solvent. Thus, θ is the critical miscibility temperature at which the molecule assumes an `unperturbed' conformation. In this conformation there are no long range or short range interactions. The solution is said to be pseudo-ideal and there is free-rotation around the bonds.

An infinite molecular weight polymer would be on the verge of precipitation in this θ situation. Also the second virial coefficient would be approaching zero as the θ point is being achieved (see equation 6.4). At the other extreme, better solution of the polymer would cause expansion of the characteristic average chain dimension as solvent/polymer contacts lower the free energy of the system. A multiplicative chain expansion factor $\alpha \geq 1$ is introduced in the model in order to express average molecular size.

$$[\eta]_M = \phi \, (\overline{\gamma_0^2})^{3/2} \, \alpha^3 \tag{15.6}$$

$[\eta]_M$ is referred to as `hydrodynamic volume' in accordance with its dimensions, and ϕ is a constant. Eq. (15.6) can be written as:

$$[\eta] = \phi \left(\frac{\overline{\gamma_0^2}}{M} \right)^{3/2} M^{1/2} \alpha^3 \tag{15.7}$$

The term $\phi \left(\dfrac{\overline{\gamma_0^2}}{M} \right)^{3/2}$ is a constant for a polymer and is independent of solvent and temperature and can be denoted by K.

$$[\eta]_\theta = K \, M^{1/2} \, \alpha^3 \tag{15.8}$$

At Flory temperature θ, α becomes unity, and the Eq. 15.8 is reduced to,

$$[\eta]_\theta = K \, M^{1/2} \tag{15.9}$$

where $[\eta]_\theta$ is the intinsic viscosity at the temperature θ in a θ solvent. K is around 1×10^{-3} for a number of polymer systems.

Table 15.3. Theta Temperatures for Some Polymer-solvent Systems

Polymer	Solvent	Theta temp. (°C)
Poly(ethylene)	Diphenyl ether	161
Poly(propylene)	Isoamyl acetate	34
Poly(styrene)	Decalin	31
Poly(vinyl chloride)	Benzyl alcohol	155
Poly(vinyl acetate)	3-Heptanone	29
Poly(methyl methacrylate)	Acetone	−50
Poly(acrylic acid)	Dioxane	30
Poly(isobutylene)	Benzene	24
Nylon−6,6	90% Formic acid + 0.3 M K Cl	25

15.3.2. Viscosity of Dilute Solutions

The Mark-Howink-Sakurada equation (Eq. 15.10), which relates the viscosity with the molecular weight, can be derived from Eq. (15.6),

$$[\eta] = K' M^{\alpha} \tag{15.10}$$

H. Staudinger proposed on empirical grounds that $[\eta]$ is proportional to the molecular weight of a given polymer-solvent combination. The determination of viscosity average molecular weight has been described in detail in Chapter 6 (see Section 6.5.3), where the terms intrinsic viscosity, specific viscosity, relative viscosity and concentration have been defined and explained. In Fig. 15.1, the correlation for viscosity of dilute solutions has been illustrated.

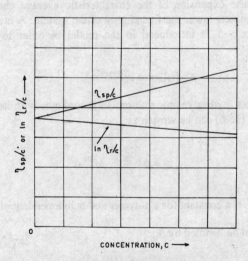

Fig. 15.1. Correlations for viscosity of dilute solutions.

It would be seen from the upper part of Fig. 15.1 that both η_{sp}/c and $\ln \eta/c$ vessus concentration (c) curves converge at the same point on the ordinate.

It is found that Eq. (15.10) becomes inaccurate for molecular weights below around 50,000. This is because deviations from linear relationships set in. For theoretical reasons and for better results in this region, the following relationship is preferred,

$$[\eta] = KM^{1/2} + K''M \tag{15.11}$$

Here the first term is determined by short range interactions and the second term by long range interactions.

Q. 15.1 Calculate the solubility parameter for poly(vinyl acetate), using the data in Table (15.2). Molecular weight for $C_4H_6O_2$ is 86 and density of the polymer is 1.2.

Q. 15.2 How can you make the best guess at M from measurements of $[\eta]$ for an entirely new kind of polymer.

Q. 15.3 A polymer with $M = 100,000$ obeys the Mark-Howink-Sakurada equation with $K' = 1 \times 10^{-4}$ and $\alpha = 0.80$. Huggin's constant is 0.33. Calculate the relative viscosity at $c = 0.30$ g/dl.

ANSWERS TO QUESTIONS

15.1 Molecular weight for $C_4H_6O_2 = 86$, density $= 1.2$

1 CH$_3$	303
1 CH$_2$	269
1 CH	176
1 ester COO	668
G	1416

$$\delta = 1.2 \times \frac{1416}{86} = 19.75$$

15.2 The Mark-Howink-Sakurada equation is:

$$[\eta] = K' M^{\alpha}_t$$

For many systems α lies between 0.6 and 0.8, and K' values are between 0.5 and 5×10^{-4}

$$\log [\eta] = \log K' + \alpha \log M$$

Hence,
$$\log M = \frac{\log[\eta] - \log K'}{\alpha}$$

By using the two values of α and the two values of K', different values of molecular weight can be computed.

15.3 Huggins equation is;

$$\frac{\eta_{sp}}{c} = [\eta] + k'[\eta]^2 c$$

Mark-Howink Sakurada relationship is:

$$[\eta] = K' M^{\alpha}$$

$K' = 1 \times 10^{-4}$ and $\alpha = 0.80$ and $M = 100,000$

Hence, $[\eta] = 1 \times 10^{-4} \times (100,000)^{0.80}$

$$= \frac{(100,000)^{0.80}}{10^4} = \frac{(10^5)^{0.80}}{10^4} = 1.0$$

$$\frac{\eta_{sp}}{c} = [\eta] + K'' [\eta]^2 c = 1 + 0.33 \times 0.30 = 1.099$$

$$\eta_{sp} = 1.099 \times 0.30 = 0.3297 \text{ or } \eta_{rd} = 1 + 0.3297 = 1.33$$

CHAPTER 16

FLOW PROPERTIES OF POLYMERS

16.1. INTRODUCTION

The branch of science dealing with the study of deformation and flow of materials is known as *rheology*. The prefix rheo is derived from the Greek word rheos, which means current or flow. The subject of rheology includes fluid and solid mechanics. In polymer chemistry one deals with viscoelastic materials which have both a solid and a fluid characteristic.

16.2. VISCOELASTICITY

It is found that low molecular weight solids and liquids follow Hooke's and Newton's laws in their characteristic flow behaviour.

16.2.1. Spring Model (Hooke Model)

Hooke's law states that the applied stress (S) is proportional to the resultant strain (γ), but is independent of the rate change of the strain $\dfrac{(d\gamma)}{dt}$,

$$S = E\gamma \tag{16.1}$$

Stress is equal to force per unit area, and strain or elongation is the extension per unit length. When the strain in simple, tension is measured as the change in length of the test specimen; this stress/strain ratio is called *Young's modulus of elasticity* (E). Stress applied tangentially to an object is called *shear*. The shear modulus of elasticity (G) is the ratio of shear stress (S) to shear strain (γ).

$$G = \frac{S}{\gamma} \tag{16.2}$$

or $$S = G\gamma \tag{16.2a}$$

Differentiation of Eq. (16.2a) gives,

$$\frac{dS}{dt} = G\frac{d\gamma}{dt} \tag{16.3}$$

Now, elongation is given by the stress divided by the modulus,

$$\gamma = \frac{S}{G} \tag{16.4}$$

The behaviour of an ideal solid can be compared to that of a steel spring which elongates readily when stress is applied and attains its original length upon release of stress (unloading) instantaneously. The behaviour of an ideal solid is called elastic deformation and is time-independent. The stress-strain plot of such a deformation is represented in part (a) of Fig. 16.1. It is seen from this figure that the applied stress (S) is proportional to the resultant strain (γ) and is independent of the rate of strain ($\dfrac{d\gamma}{dt}$).

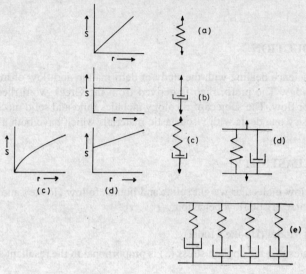

Fig. 16.1. Plots showing stress (S)-strain (γ) behaviour of solids and liquids in terms of different models: (a) Hooke model, (b) Newton model, (c) Maxwell model, (d) Voigt model, and (e) Maxwell-Weichert Model.

16.2.2. Dash Pot Model (Newton Model)

The viscous component is dominant in liquids, hence, their flow properties may be described in terms of Newton's law (Eq. 16.5).

$$S = \eta \frac{d\gamma}{dt} \tag{16.5}$$

Newton model states that the applied stress (S) is proportional to the rate of strain ($d\gamma/dt$), but is independent of the strain (γ) or applied velocity gradient. η is the coefficient of viscosity of the liquid.

When a force is applied to a viscous liquid, it starts to elongate and continues to do so until the force is released when the flow stops. However, the material never recovers its original shape. The deformation in this case is time dependent and is permanent. This phenomenon of viscous flow is illustrated in part (*b*) of Figure (16.1) which also depicts the stress-strain curve. Elongation in this case is given by the equation,

$$\gamma = \frac{S}{\eta} t \tag{16.6}$$

The elongation is dependent on the magnitude of the stress and time and is inversely proportional to the viscosity coefficient of the liquid. This model can be considered to be akin to a 'dash pot' or a barrel containing a viscous liquid fitted with a piston. When a force is applied, the piston moves slowly and when the stress is released, the piston does not move back to its original position. Thus, there is a permanent displacement of the piston.

16.2.3. Maxwell Model

The model in which a spring and a 'dash pot' are connected in series is called the Maxwell model of 'visco-elasticity'. This is illustrated in part (c) of Fig. (16.1) which also shows the stress-strain behaviour. According to this model the material under test shows the combined effect of a spring (with modulus G) and a viscous liquid (with viscosity coefficient η). Thus, these two effects are additive and can be represented as shown below:

$$\gamma_{Maxwell} = \gamma_{elastic} + \gamma_{viscous} = \frac{S}{G} + \frac{S}{\eta}t \tag{16.7}$$
$$\text{or}$$
$$\text{total}$$

On differentiation, Eq. (16.7) gives the following equation:

$$\frac{d\gamma}{dt} = \frac{dS}{dt}\frac{1}{G} + \frac{S}{\eta} \tag{16.8}$$

The rate of change of strain ($d\gamma/dt$) is equal to zero under conditions of constant stress (S); hence,

$$\frac{dS}{dt}\frac{1}{G} + \frac{S}{\eta} = 0 \tag{16.9}$$

With the assumption $S = S_0$ at zero time, integration gives,

$$S = S_0 e^{-tG/\eta} \tag{16.10}$$

and since the relaxation time $\tau = \eta/G$,

$$S = S_0 e^{-t/\tau} \tag{16.11}$$

Thus, according to Eq. (16.11) for the Maxwell model under conditions of constant strain, the stress (or stresses) will decrease exponentially with time and at the relaxation time (τ) will be equal to $1/e = 1/2.7$ or 0.37 of the original value (S_0).

Maxwell was studying the behaviour of metals like copper and lead. A copper wire on loading underwent an immediate elongation, and continued to elongate slowly to reach a certain length. On releasing the load, it contracted immediately just as much as it had elongated originally. After this, it remained set permanently at this length. Thus, the total strain is an additive function of elastic strain and viscous strains as shown in Eq. (16.7).

16.2.4. Voigt and Other Models

In the Voigt model the spring and dash pot are parallel as shown in part (d) of Fig. (16.1). In this model, the stress is shared between the spring and the dash pot. The elastic response is thus retarded by the viscous resistance of the liquid in the dash pot. In this model the vertical movement of the spring is essentially equal to that of the piston in the dash pot. Thus, if G is much larger than η, the retardation time (η/G) or τ is small, but if η is larger than G, τ is large. In this model, the total stress is equal to the sum of elastic and viscous stresses, i.e.,

$$S_{Voigt} = S_{elastic} + S_{viscous} \tag{16.12}$$

or
$$S = G\gamma + \eta \frac{d\gamma}{dt} \qquad (16.13)$$

On integration of Eq. (16.13) one obtains,

$$\gamma = \frac{S}{G}(1 - e^{-tG/\eta}) = \frac{S}{G}(1 - e^{-t/\tau}) \qquad (16.14)$$

The retardation time τ is the time for the model to decrease to $1 - (1/e)$ or $1 - (1/2.7)$ = 0.63 of the original value.

The stress-strain curve for the Voigt model is shown in part (*d*) of Fig. (16.1). Voigt proposed this model on the basis of the behaviour of rubber band under strain. On applying a small load, the rubber band underwent a small but continuous elongation. On releasing the load, the recovery was not instantaneous. It was slow and continuous but complete.

It is easy to understand that although Maxwell and Voigt models explain viscoelastic behaviour these simple models do not possess sufficient flexibility to mirror the mechanical properties of polymers with high fidelity.

This situation may be remedied by considering combinations of models. Although many such combinations are possible, discussion will be limited only to Weichert model illustrated in part (*e*) of Fig. (16.1). The Maxwell-Weichert model, a parallel combination of Maxwell elements, is very helpful in analysing stress relaxation behaviour. The spring modulii as well as viscosity coefficients or relaxation time for each element may be chosen independently. It is easy to show that the time-dependent modulii for each element sum up to give total modulus for the model.

In Fig. (16.2) are represented the strain-time relationships at a constant stress for simple models of Hooke, Newton (dash pot), Maxwell and Voigt.

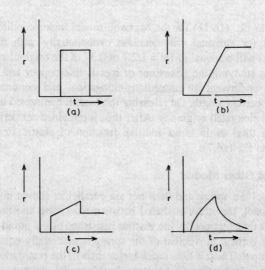

Fig. 16.2. Plots showing elongation-time behaviour for (a) Hooke, (b) Newton (c) Maxwell and (d) Voigt models.

16.3. DEFORMATION BEHAVIOUR OF POLYMER MATERIALS

While polymer melts and elastomers flow readily when stress is applied, structural plastics resist irreversible deformation and behave as elastic solids, when small stresses are applied. These plastics are called *ideal* or *Bingham* plastics. The plastics with shear thinning are called *pseudoplastics* and those with shear thickening are known as *dilatants*. Further if the shear rate does not increase as rapidly as the applied stress, the system is a dilatant and if it increases more rapidly, the system is *pseudoplastic*. A block with zero inertia, resting on a plane surface, called *St. venant* body may also be used to represent some properties of polymers.

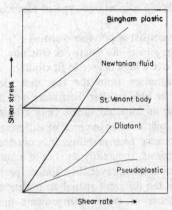

Fig. 16.3. Types of flow.

Liquids which undergo a decrease in viscosity with time are called *thixotropic* and those which undergo increase in viscosity with time are known as *rheopectic* (shear thickening). The term *creep* is used to describe the slipping of polymer chains over a long period of time.

The Bingham plastics differ from Newtonian fluids on the basis of the yield value (S_0). As shown in Fig. (16.3), Bingham plastics exhibit Newtonian flow above the stress yield value (S_0). The curves for shear thickening (dilatant) and shear thinning (pseudoplastics) materials are also shown in this figure.

The Herschel-Buckley equation (Eq. 16.15) is a general equation which reduces to the Bingham equation (Eq. 16.15) when $\eta = 1$ and to the Newtonian equation when $\eta = 1$ and $S_0 = 0$.

$$(S - S_0) = \phi \, \frac{d\gamma}{dt} \tag{16.15}$$

The stress-strain curves of polymeric materials, vary depending on factors like crystallinity, glass transition temperature (T_g) and the degree of cross-linking. Taking these factors into consideration, polymeric materials can be classified into the following categories:

   (1) Hard and brittle, e.g., poly(styrene) (below T_g)

   (2) Hard and strong, e.g., nylon (below T_g)

   (3) Hard and tough, e.g., poly(ethylene) (above T_g)

   (4) Soft and weak, e.g., uncross-linked poly(isobutylene) elastomer (above T_g)

   (5) Soft and tough, e.g., vulcanized rubber (above T_g)

16.4. RUBBER ELASTICITY

Rubber elasticity may be defined as very large deformability accompanied with essentially complete recoverability. Rubber elasticity is not a special behaviour of a few polymeric compounds such as rubber or poly(isobutylene), but it is a property which all macromolecular compounds have above their softening temperature, as long as they have a softening temperature and do not decompose first (softening temperature = freezing point = glass temperature, T_g).

16.4.1. Requirements for Elasticity

In order that a material should exhibit this type of elasticity, it should satisfy three molecular conditions:

 (1) The material must consist of polymeric chains.
 (2) The chains must be joined into a network structure.
 (3) The chains must have a high degree of flexibility.

The first requirement arises from the fact that molecules in a rubber or elastomeric material must be able to alter their arrangement and extension in space dramatically in response to an imposed stress. Only a long chain molecule has the required large number of spatial arrangements of different extensions. The second requirement ensures elastomeric recoverability. It is obtained by joining together or 'cross-linking' pairs of segments, apparently one out of a hundred, thereby preventing the extended polymer chains from reversibly sliding past one another. The third characteristic specifies that the different spatial arrangements be accessible, i.e., the change from one to another of these arrangements must not be hindered by constraints resulting from rigidity of the chains.

16.4.2. Stress-Strain Relationship

It can be shown that the stress (S) is related to the strain (γ) by Eq. (16.16),

$$S = 2\, mk\, Tb^2\left(\gamma - \frac{1}{\gamma^2}\right) \tag{16.16}$$

and the modulus of elasticity (G) is given by,

$$G = \frac{ds}{d\gamma} = 2\, mk\, Tb^2\left(1 + \frac{2}{\gamma^3}\right) \tag{16.17}$$

where $b^2 = \frac{3}{2}\, xl^2$, x being the number of links with length l; k is the Boltzman's constant, m is the number of chains in the network of average length $(\gamma^2)^{1/2}$ per unit volume directed along each of the three perpendicular axes and T is the absolute temperature.

Fig. (16.4) depicts a stress-strain curve for a typical elastomer. As the strain increases from zero ($\gamma = 1$) to large values of γ, the slope decreases to one-third its initial value as required by the experimental facts. Eq. (16.16) predicts the stress-strain relation of actual elastomers very well upto elongations of 300% or more. Fig. (16.5) shows in a schematic manner the coil deformation occurring during stretching.

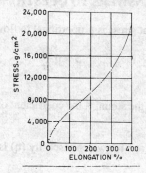

Fig. 16.4. Elongation of an elastomer as a function of load.

At elongations greater than 300%, the stress-strain curves of elastomers have a higher slope than that predicted by Eq. (16.16). This is far below the ultimate elongation of 1000% for a good elastomer.

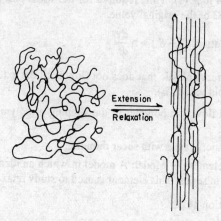

Fig. 16.5. Schematic representation of the transition from the randomly coiled to the orientated state on elongation of a cross-linked rubber.

Q. 16.1 State Newton's law. What is a Newtonian fluid?

Q. 16.2 Define the terms (i) relaxation time and (ii) retardation time.

Q. 16.3 State Bingham's equation. What do you understand by a Bingham plastic?

Q. 16.4 Define (i) creep, (ii) pseudoplastic, (iii) Maxwell element.

Q. 16.5 What conditions must be fulfilled by a material to show rubber elasticity?

ANSWERS TO QUESTIONS

16.1. Newton's law states that stress is proportional to flow,

$$S = \eta \frac{d\gamma}{dt}$$

where S is the stress, γ the strain, η the coefficient of viscosity and $\frac{d\gamma}{dt}$ the rate of strain.

A newtonian fluid is one for which viscosity (η) is proportional to the applied viscosity gradient $\frac{d\gamma}{dt}$.

16.2. (i) *Relaxation time* (τ): Time required for the stress of a polymer under constant strain to decrease to $1/e$ or 0.37 of its original value.

(ii) *Retardation time* (τ): Time required for the stress of a deformed polymer to decrease to 0.63 of the original value.

16.3. Bingham's equation: $(S - S_0) = \eta \frac{d\gamma}{dt}$

Bingham plastic: A plastic that does not flow until the external stress exceeds a critical threshold value (S_0).

16.4. (i) *Creep*: cold flow of a polymer or the slipping of polymer chains over a long period of time.

(ii) Pseudoplastic: Plastic with shear thinning.

(iii) Maxwell element or Model: A model in which an ideal spring and dash pot are connected in series. This element is used to study relaxation of polymers.

16.5 See under Section 16.3.1.

CHAPTER 17

SOME NATURALLY OCCURRING POLYMERS

17.1. INTRODUCTION

All living matter, animal or vegetable, contains polymers as major constituents. One of the major constituents of vegetable matter is carbohydrate. Carbohydrates are usually made up of cellulose, starch and lignin. Both animal and vegetable kingdom contain a complex nitrogenous material known as protein. Besides these, the nucleic acids form part of all living cells, both animal and vegetable. Natural rubber, which is obtained from *Hevea Brasiliensis* species of trees, is an important ingredient for the manufacture of many industrial products. The structure and uses of rubber have already been considered in detail in an earlier chapter (Chapter 8).

17.2. POLYSACCHARIDES AND LIGNINS

17.2.1. Cellulose

Cellulose is a polyanhydroglucose. Its empirical formula is $C_6H_{10}O_5$. On hydrolysis it yields eventually D–glucose. The linkage between the glucose units is a β–1:4 glucosidic-linkage. The α–1:4 linkage, distinct from the β–1:4 bonding, is present in amyose fraction of starch. The β- and α-1:4 linkages are illustrated in the structures given below:

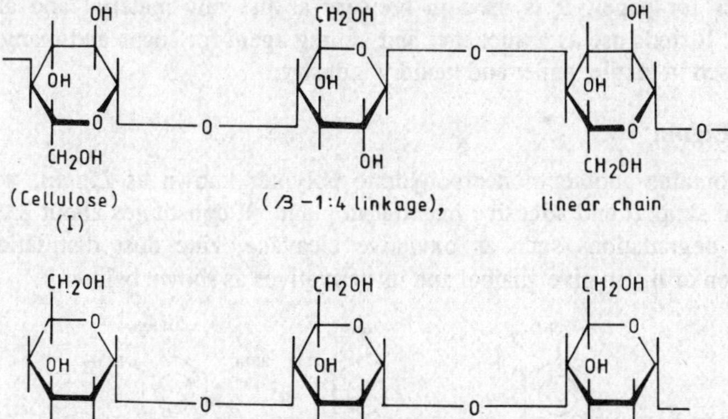

(Cellulose) (I) (β –1:4 linkage), linear chain

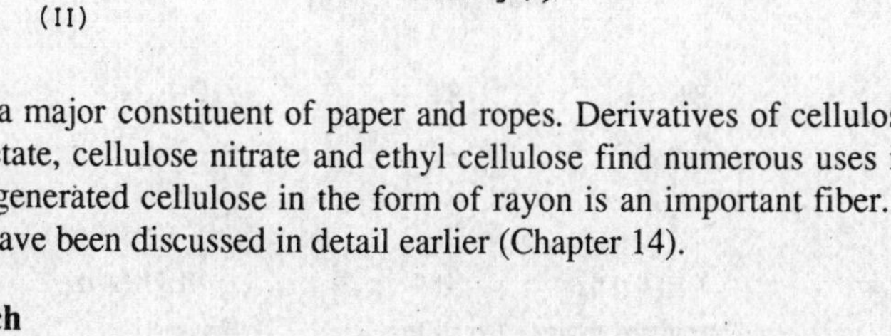

(Amylose) (II) (α – 1:4 linkage), linear chain

Cellulose is a major constituent of paper and ropes. Derivatives of cellulose such as cellulose acetate, cellulose nitrate and ethyl cellulose find numerous uses in plastics industry. Regenerated cellulose in the form of rayon is an important fiber. Cellulose derivatives have been discussed in detail earlier (Chapter 14).

17.2.2. Starch

Starch is the reserve carbohydrate material of plants. The main sources of starch are the grain and potato. Starch contains two kinds of polymers which differ in their structures; and molecular weights; one component is *amylose*, a linear polyanhydroglucose in which the glucose units are joined by α–1:4 linkages (already illustrated). This linkage gives greater flexibility to the chain, which can adopt a coiled configuration in contrast to the β–1:4 linkage in cellulose, in which the chains are stiff. The amylose fraction can have molecular weights ranging from 10,000 to 400,000. The structure of amylose has already been given.

The other fraction, *amylopectin*, consists of a chain of linear α–1:4 linked glucose units with branches of poly(anhydroglucose), joined to the main linear chain by both α–1:4 and α–1:6 linkages. The molecular weight of the latter can be about 10,000. The structure of amylopectin is shown below:

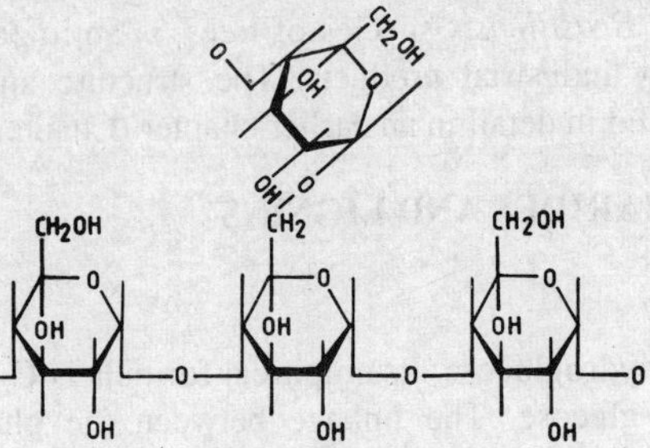

(Amylopectin) (branched)
(III)

Starch finds numerous uses in industry for the manufacture of paper boards and adhesives for paper. It is used in brewery as the raw material and also in food industry. It finds use as a thickener and gelling agent for soaps and many desserts. It is also used in textile, paper and laundry industry.

17.2.3. Lignin

Wood contains another noncarbohydrate polymer known as *Lignin*, which is the structural support and adhesive material in plants. It constitutes about 25% of wood. Various degradations such as oxidative cleavage, zinc dust distillation and dry distillation of lignin give guaicol and its derivatives as shown below:

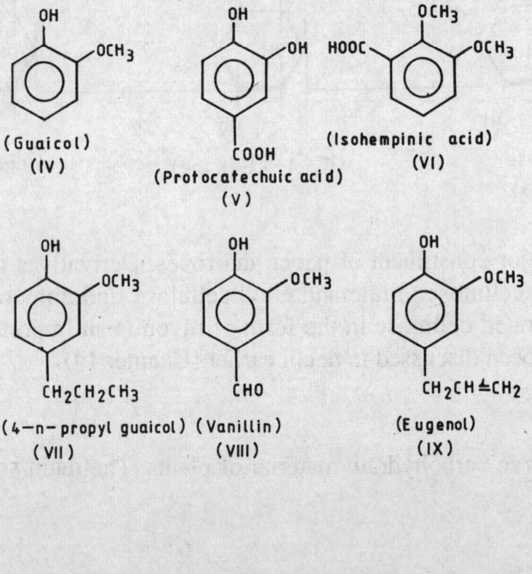

(Guaicol)
(IV)

(Protocatechuic acid)
(V)

(Isohempinic acid)
(VI)

(4–n–propyl guaicol)
(VII)

(Vanillin)
(VIII)

(Eugenol)
(IX)

The structure of lignin has been elucidated, but it is too complex to be discussed here. Lignin finds numerous uses; as a dispersant and wetting agent used in oil well digging muds, in adhesives, road binders, industrial cleaners and leather tanning. It is also used in the industrial preparation of vanillin.

Glycogen is a highly branched poly(anhydroglucose) with a structure somewhat similar to that of amylopectin. It has a chain length of about 10–14 glucose units. It occurs in almost all animal cells and some plants and is the main carbohydrate food reservoir of animals.

Q. 17.1.(a) Why is cellulose stronger than amylose?

(b) How many hydroxyl groups are present in the anhydroglucose units of cellulose?

(c) Why is starch digestible by human beings whereas cellulose is not?

(d) From which wood product is vanillin obtained industrially?

17.3. PROTEINS

Proteins are the naturally occurring condensation polymers of α–amino acids. Hydrolysis of proteins with acids, bases or enzymes produces a mixture of α–amino acids. In protein the amino acid units are linked by peptide linkages (–CO–NH–).

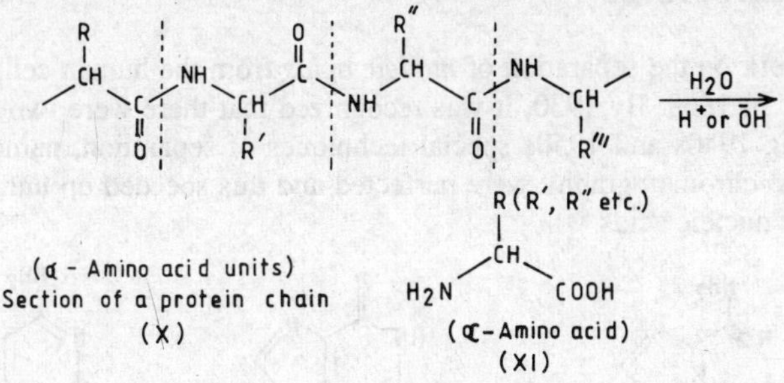

Scheme 17.1

where R, R′, R″, etc., can be either hydrogen or an organic group. Molecules which contain only short sequences of α–amino acids are known as *poly(peptides)*. There are about 20 amino acids present in naturally occurring proteins. The amino acids obtained from the hydrolysis of proteins can be separated by paper chromatography and identified by spraying them with ninhydrin reagent. The sequence of amino acids in a protein can be obtained by carrying out reactions with the amino or carboxyl end group, followed by controlled hydrolysis.

The α–amino acids are dipolar ions and do not migrate to the poles at their isoelectric points. As they migrate to the poles at different pH values, the amino acids and proteins can be separated by electrophoresis.

The simple proteins can be divided into *fibrous* and *globular* proteins. Proteins with high intermolecular hydrogen bonding are fibrous and those with intramolecular hydrogen bonding are globular. Furthermore, the globular proteins become insoluble or denatured on heating. An example is the egg albumin. Fibrous proteins include those obtained from silk, wool, hides and hooves. Globular proteins include the albumins, globulins and glutelins.

The *primary structure* of a protein describes the sequence of α–amino acids in a poly(peptide) chain or its configuration.

The *secondary structure* of a protein describes its conformation or molecular shape. It has been shown that the poly(peptide) chains may have helical conformations in their crystal lattices. These helices are very stable even in the dissolved state, being stabilized by intramolecular hydrogen bonds involving −NHCO−groups. This arrangement may be depicted by a right handed α−helix conformation of a poly(peptide).

The term *tertiary structure* is used to describe the shape or folding resulting from the presence of sulphur-sulphur cross-links between the polymer chains.

The poly(peptides) can be synthesized by the classical methods of Emil Fischer and other workers. In these syntheses one of the groups, −NH$_2$ or −COOH, an amino acid is protected before condensing it with another amino acid. In the recent Merifield solid phase synthesis (see Chapter 14, Section 14.5) the reaction takes place on the surface of a cross-linked poly(styrene) bead.

Q. 17.2. (a) Which linkages are present in protein molecules?

(b) Which group is responsible for the solubility of proteins in water?

(c) Define isoelectric point.

(d) What is the primary structure of a protein?

17.4. NUCLEIC ACIDS

The early work on the separation of *nucleic acids* from the human cells was initiated by Miescher in 1868. By 1930, it was recognized that there were two types of these acids. During 1940s and 1950s special techniques of separation, namely, paper and ion-exchange chromatography were perfected and this speeded up unravelling of the chemistry of nucleic acids.

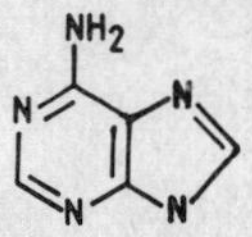

(Adenine) (A)
(XII)

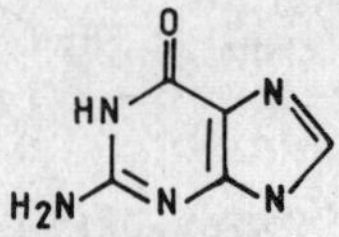

(Guanine) (G)
(XIII)

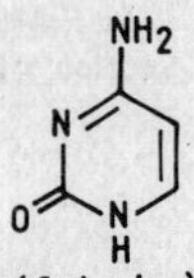

(Cytosine) (C)
(XIV)

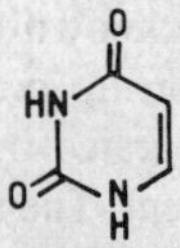

(Uracil) (U)
(XV)

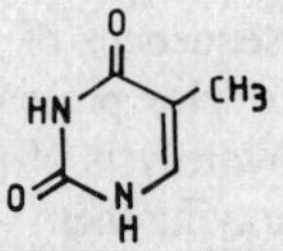

(Thymine) (T)
(XVI)

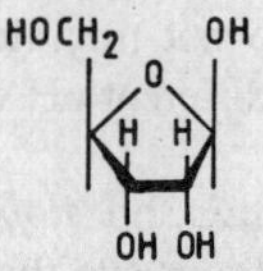

(D−Ribose)
(XVII)

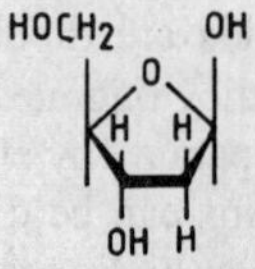

(Deoxyribose)
(XVIII)

The nucleic acids are the condensation products of nucleoside triphosphates and contain heterocyclic bases. In deoxyribose nucleic acid (DNA), the heterocyclic bases are adenine, guanine, thymine, and cytosine. In ribose nucloic acid (RNA), these are adenine, guanine, uracil and cytosine. The names of the two nucleic acids are based on the sugar moiety present in them. Thus, DNA and RNA differ in that one contains the carbohydrate D–2–deoxyribose and the other contains D–ribose. Also one of the heterocylic bases is different: DNA contains thymine and RNA contains uracil.

Nucleosides

On enzymatic hydrolysis, RNA yields four nucleosides, namely, adenosine, guanosine, uridine and cytidine, which on acid hydrolysis give D–ribose and the four bases adenine, guanine, uracil and cytosine, respectively. DNA, on a similar reaction with an enzyme, yields the four nucleosides deoxyadenosine, deoxyguanosine, deoxycytidine and deoxythymidine. Acid hydrolysis of the nucleosides yields besides deoxy D–2–ribose, the four heterocyclic bases adenine, guanine, cytosine and thymine.

Nucleotides

Nucleic acids on hydrolysis by specific enzymes or very mild chemical hydrolysis yield a mixture of *nucleotides*. These were shown to be phosphate esters of nucleosides, which had been obtained under more drastic hydrolysis.

(Adenosine)
(a· nucleoside)
(XIX)

(Cytosine deoxyribonucleotide)
(a nucleotide)
(XX)

17.4.1. Structures of RNA and DNA

Segments of RNA and DNA polymers are represented on the next page

17.4.1.1. *Conformation of the Nucleic Acids*

By 1950, the chemical structure of nucleic acids was settled. The accepted conformation of DNA was put forward by F.H.C. Crick and J.D. Watson. They suggested that DNA consisted of two poly(nucleotide) chains wound helically in opposite directions around a central axis. They postulated that the nitrogen bases in each linear chain of the helix were paired across the axis of the helix with the nitrogen bases in the other linear chain. In each case the pairing was made by hydrogen bonding between the bases. Since only a purine paired with a pyrimidine would conveniently fit the dimensions across the axis of the helix, this model also explained the regular ratios of the base compositions.

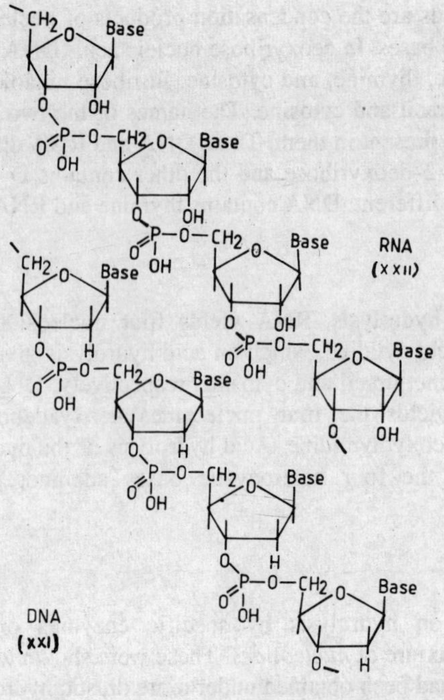

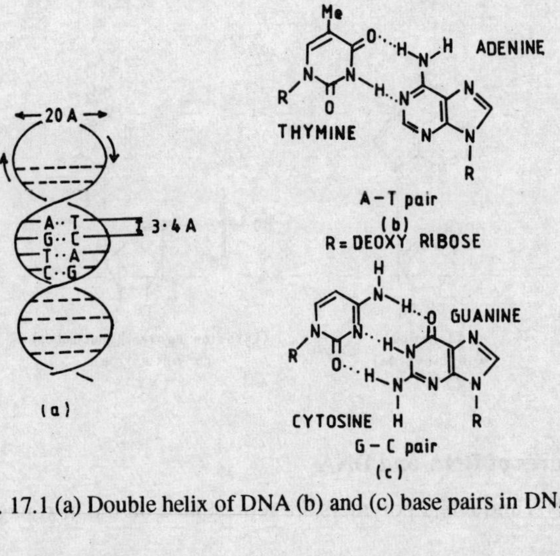

Fig. 17.1 (a) Double helix of DNA (b) and (c) base pairs in DNA.

Q. 17.3. Define the terms (a) nucleoside and (b) nucleotide. Which of the two is more acidic? What is the repeating unit in DNA and what is the difference in the structures of DNA and RNA?

Q. 17.4. If the sequence in one of the chains of a double helix is CTATCGGATCG, what is the sequence in the other, adjacent chain?

ANSWERS TO QUESTIONS

17.1. (a) Cellulose is stronger than amylose because of the presence of intermolecular hydrogen bonding in the former.
   (b) Three hydroxyl groups are present in the anhydroglucose unit of cellulose.
   (c) This is because of the presence of enzymes in the digestive system of humans, which can hydrolyse the α–linkage of starch, but not the β-linkages of cellulose.
   (d) Lignin.

17.2. (a) Peptide $-$CONH $-$ linkages are present in proteins.
   (b) The $-$COOH group is responsible for the solubility of proteins in water.
   (c) This is the pH at which the protein does not migrate to either of the two poles.
   (d) The primary structure of a protein shows the simple sequences of α–amino acids present in the protein.

17.3. (a) The structure of a nucleoside is $\overset{\beta}{S}$ where B = the base and S = sugar
   (b) Nucleotides are the phosphate esters of nucleoside, $HO-P-O-S$

$$\underset{HO \quad\quad O \quad\quad B}{\diagup \quad \| \quad\quad |}$$

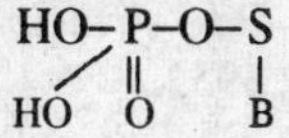

Nucleotides are more acidic than the nucleosides.
The repeat unit in DNA is $-(O-P-O-B)-$
$$\underset{S.}{|}$$
where S = deoxyribose and B = base
The difference between RNA and DNA is twofold:
   (i) The carbohydrate present in RNA is D–ribose whereas in DNA it is D–2–deoxyribose.
   (ii) The heterocyclic bases present in both RNA and DNA are four in number. Three are common. The fourth base is different. In RNA it is uracil and in DNA it is thymine.

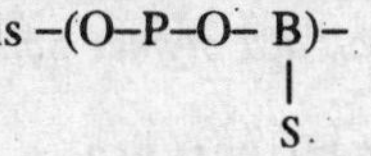

17.4.   GATAGCCTAGC

SUGGESTIONS FOR FURTHER READING

The following books, articles and encyclopaedia are suggested for further reading:

CHAPTER 1

1. Mark, H.F., *Giant Molecules* (Time-Life Books, New York), 1966.
2. Mark, H.F., *J. Chem. Educ.*, **50**, (1973) 757.
3. Flory, P.J., *J. Chem. Educ.* **50** (1973) 732.

CHAPTER 2

1. Bawn, C.E.H., *The Chemistry of High Polymers*, (Butterworth, London), 1948.
2. Flory, P.J., *Principles of Polymer Chemistry*, (Cornell University Press, New York), 1953.
3. Odian, G., *Principles of Polymerization*, (John Wiley, New York), 2nd Edition, 1981.
4. Billmeyer, Jr., F.W. *Textbook of Polymer Science*, (John Wiley, New York) 3rd Edition, 1984.
5. Stille, J.K., *J. Chem. Educ.* **58**, (1981) 862.

CHAPTER 3

1. Flory, P.J., *Principles of Polymer Chemistry*, (Cornell University Press, New York), 1953.
2. Allen, P.E.M. and Patrick, C.M., *Kinetics and Mechanism of Polymerization Reactions*, John Wiley, New York), 1974.
3. Bamford, C.H., Jenkins, A.D., Barb, W.G. and Onyon, P.F., *The Kinetics of Vinyl Polymerization by Radical Mechanisms*, (Butterworth, London), 1958.
4. Bevington, J.C., *Radical Polymerization*, (Academic Press, New York), 1961.
5. Ham, G.E., *Vinyl Polymerization*, **I**, Part I, (Marcel Dekker, New York), 1967.
6. Odian, G., *Principles of Polymerization*, (John Wiley, New York), 2nd Edition, 1981.
7. Billneyer, Jr. F.W., *Text book of Polymer Science*, John Wiley, New York, 3rd Edition.
8. Natta, G. and Giannini, U., "Coordination Polymerization" in *Encyclopaedia of Polymer Science & Technology*, **4**, (wiley-Interscience, New York), 1966, pp. 137-150.
9. McGrath, J.E., *J. Chem. Educ.* **58**, (1981) 844.

CHAPTER 4

1. Alfrey, Jr., T. Bohrer, J.J. and Mark, H., *Copolymerization* (Interscience, New York), 1952.

2. Ceresa, R.J., "Block and Graft Copolymers" in *Encyclopaedia of Polymer Science and Technology*, **2**, (Wiley-Interscience, New York), 1965, pp. 485-528.
3. Ceresa, R.J., *Block and Graft Copolymers* (Butterworth, Lonon), 1962.
4. Stille, J.K., *Introduction of Polymer Chemistry*, (John Wiley, New York), 1962.

CHAPTER 5

1. Natta, G. and Dannisso, F., *Stereoregular Polymers and Sterospecific Polymerization*, (Pergamon, Oxford), 1967.
2. Bovey, F.A., *Polymer Conformation and Configuration*, (Academic Press, New York), 1969.
3. Price, C.C., *J. Chem. Educ.* **36**, (1959) 160-163; **42**, (1965) 13-17.
4. Odian, G., *Principles of Polymerization*, (John Wiley, New York), 2nd Edition, 1981.
5. Jenkins, A.D. and Ledewith, A., *Reactivity, Mechanism and Structure in Polymer Chemistry*, Chapter 12, (Wiley-Interscience, London), 1974.

CHAPTER 6

1. Vollmert, B., *Polymer Chemistry*, (Springer-Verlag, New York), 1973.
2. Billmeyer, Jr., F.W., *Text Book of Polymer Science*, (John Wiley, New York), 3rd Edition, 1984.
3. Flory, P.J., *Principles of Polymer Chemistry*, (Cornell University Press, New York), 1953.
4. Allen, P.W. (Ed.)., *Techniques of Polymer Characterization*, (Butterworth, London), 1959.
5. Bonner, R.V., Dimbat, M. and Stross, F.H., *Number-average Molecular Weights*, (Wiley-Interscience, New York), 1950.
6. Chien Jen-Yuen, *Determination of Molecular Weight of High Polymers*, (Oldbourne, London), 1963.
7. D'Alelio, G.F., *Fundamental Principles of Polymerization* "Rubbers, Plastics and Fibers" (John Wiley, New York), 1952.
8. Ward, T.C., *J. Chem. Educ.*, **58** (1981), 867.

CHAPTER 7

1. Harris, F.W., *J. Chem. Educ.*, **58**, (1981) 837.
2. Rodriguez, F., *Principles of Polymer systems*, (McGraw Hill, New York), 1970.
3. Ravve, A., *Organic Chemistry of Macromolecules*, (Marcel Dekker, New York), 3rd Edition, 1967.
4. Billmeyer, Jr., F.W., *Text Book of Polymer Science*, (John Wiley, New York), 3rd Edition, 1984.
5. Tager, A., *Physical Chemistry of Polymers*, (Mir Publishers, Moscow), 1972.

CHAPTER 8

1. Schildknecht, C.E. (Ed.), *Polymer Processes* (Interscience, New York), 1956.
2. Morton, M. (Ed.)., *Rubber Technology*, 2nd Edition, (Van Nostrand-Reinhold, New York), 1981.
3. Braun. D. *et al.*, *Encyclopaedia of Polymer Science & Technology*, "Rubber Natural", **12**, (Wiley-Interscience, New York), 1970, pp. 178-256.

4. Cooper W. in *Encyclopaedia of Polymer Science & Technology*, "Elastomer, Synthetic", **5**, (Wiley-Interscience, New York), 1966, pp. 406-482.
5. Blow, C.M., *Rubber Technology and Manufacture*, (Butterworth, London), 1971.

CHAPTER 9

1. Mark, H.F. *et al.*, *Man Made Fibers, Science and Technology*, **1**, **2** and **3**, (Wiley-Interscience, New York), 1967-68.
2. Melville, H.W., *Big Molecules* (G. Bell, London), 1958.
3. Billmeyer, Jr., F.W., *Text Book of Polymer Science*, (John Wiley, New York) 3rd Edition, 1984.
4. Moncrieff, R.W., *Man Made Fibers* (Halsted, New York), 1975.

CHAPTER 10

1. Dubois, J.H. and John, F.W., *Plastics* (Reinhold, New York), 1981.
2. Seymour, W.B., *Modern Plastics Technology* (Reston Publishing Co., Reston), 1975).

CHAPTER 11

1. Grassie, N., and Scott, G., *Polymer Degradation and Stabilization*, (Cambridge University Press, Cambridge), 1985.
2. Grassie, N. in *Encyclopaedia of Polymer Science & Technology*, "Degradation of Polymers", (Wiley-Interscience), 1966, pp. 647-716.
3. Jellinek, H.H.G. (Ed.) *Aspects of Degradation and Stabilization of Polymers*, (Elsevier, Amsterdam), 1978.
4. Segal, C.L., *High Temperature Polymers* (Marcel-Dekker, New York), 1967.

CHAPTER 12

1. Frissell, W.J. in *Encyclopaedia of Polymer Science and Technology*, "Fillers." **6**, (Wiley-Interscience, New York) 1967.
2. Buttery, D.N., *Plasticizers*, (Franklin Publishing, Pallsides, New Jersey), 2nd Edition, 1960.
3. Ritchie, P.D. (Ed.) *Plasticizers, Stabilizers and Fillers*, (Butterworth, London), 1972.

CHAPTER 13

1. Rodriguez, F., *Principles of Polymer Systems*, (McGraw Hill, New York), 1970.
2. Nyquist, R.A., *Infrared Spectra of Plastics and Resins*, (The Dow Chemical Co., Midland), 1961, 2nd Edition.
3. Bovey, F.A., In *Encyclopaedia of Polymer Science and Technology*, "Nuclear Magnetic Resonance", **9** (Wiley Interscience, New York), 1968.
4. Bovey, F.A., *High Resolution NMR of Macromolecules*, (Academic Press, New York), 1972.
5. Kline, G.M.E., *Analytical Chemistry of Polymers*, Part I, (Interscience, New York), 1959.
6. Cloutier, H. and Pud'homme, R.E., *J. Chem. Educ.* **62**, (1985) 815.

CHAPTER 14

1. Moor, J.A. (Ed.), *Reactions of Polymers*, (Reidel, Dordrecht, (1973).
2. Odian, G., *Principles of Polymerization*, (John Wiley, New York), 2nd Edition 1981.
3. Neckers, D.C., *J. Chem. Educ.* **52**, (1975) 695.

CHAPTER 15

1. Seymour, R.B. and Carraher, C.E., *Polymer Chemistry*: An *Introduction*, (Marcel Dekker, New York), Second Editions, 1988.
2. Billmeyer, Jr., F.W., *A Text Book of Polymer Science*, (John-Wiley, New York), 3rd Edition, 1984.
3. Morawetz, H., *Macromolecules in Solution*, (Wiley-Interscience, New York), 2nd Edition, 1975.

CHPATER 16

1. Aklonis, J.J., Mcknight, W.J. and Sten, M., *Introduction to Polymer Viscoelasticity*, (Wiley-Interscience, New York), 1982.
2. Seymour, R.B. and Carraher, C.E., *Polymer Chemistry: An Introduction* (Marcel Dekker, New York) second Edition, 1988.
3. Mark, J.E., *J. Chem. Educ.* **58**, (1981) 898.
4. Aklonis, J.J., *J. Chem. Educ.* **58**, (1981) 892.
5. Rosen, S.L., *Fundamental Principles of Polymeric Materials*, (Wiley-Interscience, New York), 1982.

CHAPTER 17

1. Seymour, R.B. and Carraher, Jr., C.E., *Polymer Chemistry: An Introduction*, (Marcel Dekker, New York), 1988, 2nd Edition.
2. Hendrickson, J.E., Cram, D.J. and Hammond, G.S., *Organic Chemistry*, (McGraw Hill, New York), 1970, 3rd Edition.
3. Walker, B.J., Organophosphorus Chemistry, (Penguin, Hammondsworth, Middlesex, England), 1972.
4. Grant, J., *Cellulose, Pulp and Allied Products*, (Interscience, New York), 1959.
5. Doty, P., *Proteins in Organic Chemistry of Life*, (Freeman, New York), 1973, pp. 137-146.

LABORATORY MANUALS ON POLYMER CHEMISTRY

1. D'Alelio, G.F., *Laboratory Manual of Plastics and Synthetic Resins*, (John Wiley, New York), 1943.
2. Redfarn, C.A. and Bedford, J., *Experimental Plastics*, (Iliffe, London), 1960.
3. Pinner, S.H., *A Practical Course in Polymer Chemistry*, (Pergamon, New York), 1961.
4. McCaffery, E.L., *Laboratory Preparation for Macromolecular Chemistry* (McGraw Hill, New York), 1970.
5. Collins, E.A., Bares, J. and Billmeyer, Jr., F.W., *Experiments in Polymer Science*, (Wiley-Interscience, New York), 1973.
6. Sorensen, W. and Campbell, T.W., *Preparative Methods of Polymer Chemistry*, (Wiley-Interscience, New York), 2nd Edition 1968.

LABORATORY EXERCISES

A course in polymer chemistry is incomplete if it does not include suitable experiments to illustrate the basic principles. There are some excellent manuals on laboratory experiments in polymer science and technology some of which are listed in Appendix I.

The present introductory text only purports to be a guide book on polymer chemistry. Therefore, detailed experimental details of all the processes involved in polymer synthesis and their characterization cannot be provided here. However, to evoke the interest of the users of this book, a few laboratory exercises are provided in this appendix. The exercises are calculated to stimulate interest in polymer chemistry which is in its formative stages in this country. A set of ten exercises has been selected to illustrate step and chain polymerization including interfacial polymerization, modification of polymers and depolymerization. These do not require any special equipment or chemicals and can be performed in ordinary organic chemistry laboratories of Indian universities, colleges and technological institutions.

EXPERIMENT 1

One-stage Synthesis of Phenol-Formaldehyde Resin

Materials and Equipment

(i) Phenol
(ii) Formalin (40% w/v)
(iii) Ammonia (sp. g. 0.880)
(iv) One liter bolt head flask and reflux condenser
(v) One liter distilling flask and receiver
(vi) Vacuum pump

Procedure

Phenol (250 g) formalin (300 ml) and ammonia (25 ml, sp. g. 0.880) are mixed together in a bolt-head flask with cooling. The mixture is then warmed cautiously under the reflux condenser with a Bunsen burner over a wire gauge until the reaction commences. As soon as the bubbles start rising, showing that heat is being generated by the reaction, the burner is removed. The heat of the reaction brings the reaction mixture to boil. A container of cold water is kept ready and the flask is cooled by immersing in it, if the reaction becomes unduly violent.

When the reaction slows down, heating is resumed to bring the contents to boil. Heating is continued until the mixture clouds and separates into two layers and ten minutes thereafter. Water is then removed by distillation under vacuum, heating on a waterbath at about 75°. Distillation is continued until the resin is clear and a sample yields a brittle bead on cooling with water. The resin is then poured out into a shallow tray where it hardens to a solid.

EXPERIMENT 1(a)

Thermosetting of Phenol-formaldehyde Resin

Materials and Equipment

- (i) Phenolic resin from experiment 1
- (ii) Boiling tube
- (iii) Clamp and Clamp stand
- (iv) Oil bath
- (v) Glass rod with rounded ends
- (vi) Thermometer reading upto 350° C

Procedure

About 5 g of the resin obtained from experiment 1 is roughly powdered and put into a boiling tube. The boiling tube is heated on the oil-bath maintained at 120–130°C. The temperature is monitored using the thermometer and the resin is kept stirred with the glass rod.

It may be noted that the resin first melts (Bakelite 'A' or Resol). On further heating, it becomes a jelly (Bakelite 'B' or Resitol) and on still further heating it becomes hard and infusible (Bakelite 'C' or Resit).

This is the process of thermosetting.

EXPERIMENT 2

Urea-Formaldehyde Adhesive

Materials and Equipment

- (i) Urea
- (ii) Formalin (40%, w/v)
- (iii) Sodium hydroxide (4%, w/v) solution
- (iv) Lactic acid solution (7.2%, v/v)
- (v) B.D.H. Universal indicator paper
- (vi) Wood-flour (80-120 mesh)
- (vii) Benzyl alcohol or furfural
- (viii) Ammonium chloride
- (ix) One liter bolt-head flask
- (x) Vacuum distillation assembly
- (xi) Thermometer
- (xii) Hot water bath
- (xiii) Pieces of wood
- (xiv) G-clamps

Procedure

(i) *Resin preparation*

Urea (100 g), formalin (270 ml) (mol ratio 1:2.2) and sodium hydroxide solution (5 ml) are mixed in the flask and boiled under reflux for five minutes. The liquor is adjusted to pH–5.5 with lactic acid using the B.D.H. indicator paper and boiling is continued for a further period of 30 minutes under reflux. The liquor is adjusted to pH 7.5 with sodium hydroxide solution and vacuum distilled on a water bath at 60-65°C to get a thick syrup.

(ii) *Production of gap filling*

Wood flour (10 g) is mixed with benzyl alcohol (20 ml) or furfural.

(iii) *Production and application*

Resin (100 g) all of the gap filler type and ammonium chloride (1 g) are thoroughly mixed. Then within the course of an hour or so, the adhesive is applied to about one inch of one surface of each of the piece of the wood.

The pieces are placed together in pairs with end to end overlapping contact, clamped in position and allowed to stand overnight.

The adhesion is tested by putting one end of the assembly in a vice and breaking the joint open with a hammer blow. A good joint should show wood failure when breaking, that is there should be splinters of wood torn away with the adhesive.

EXPERIMENT 3

Alkyd Resins

Materials and Equipment

    (i)   Phthalic anhydride
   (ii)   Glycerol
  (iii) Ethylene glycol
  (iv) Two pyrex boiling tubes
   (v) 360° C thermometer
  (vi) Bunsen burner

Procedure

The following mixtures are heated in separate tubes for 90 min. at 160-180°C. (The experiment should be carried out in a fume cup-board) (a) 4.8 g (1.5 mol) phthalic anhydride + 6.1 g (4.9 ml) (1 mol) glycerol; (b) 14.8 g (1 mol) phthalic anhydride + 6.2 g (5.6 ml) (1 mol) ethylene glycol.

After this treatment, a drop of each resin should yield a brittle bead on cooling with water. Approximately half of each resin is cooled and retained for solubility tests.

The remainders of the mixtures are heated to 240° and maintained at this temperature under observation. The glycerol resin gels indicating thermosetting properties, while the ethylene glycol resin remains mobile.

Small amounts of the samples should be tested for hot and cold solubility in acetone, ethyl alcohol, linseed oil and white spirit and the results recorded.

Alkyd resins are used in plastics moulding compositions on account of their good surface wetting properties. The oil modified types are extensively used in enamels, varnishes and paints.

EXPERIMENT 4

Preparation of 6-10 Nylon Poly(hexamethylene sebacamide) by Interfacial Polymerization

Materials and Equipment

    (i)   Sebacoyl chloride
   (ii)   Hexamethylene diamine
  (iii) Tetrachloroethylene
  (iv) A tall beaker
   (v) Pair of tweezers

(vi)   Acetone or ethanol
(vii) *m*-Cresol
(viii) Ostwald viscometer
(ix)   Oven

Procedure

In a tall beaker of 200 ml capacity is placed a solution of 3.0 ml sebacoyl chloride in 100 ml redistilled tetrachloroethylene. Over the acid chloride solution, a solution of 4.4 g of hexamethylene diamine in 50 ml distilled water is carefully poured. The polymer film is grasped with a tweezer as it is formed at the interface of the two solutions and raised from the beaker as a continuously forming rope. The polymer is washed several times with 50% aqueous ethanol or acetone and dried in an oven at 60°C. The product has inherent viscosity of 0.4–1.8 in *m*-cresol, 0.5%, concentration at 25°C, depending on the reaction conditions. The polymer melting temperature is 215°C.

Report

The report should include: (i) m.p. of the nylon, and (ii) viscosity number.

The techniques useful in determining the rate of polymerization in chain polymerization reactions are based on the fact that from the very beginning *only monomer and polymer molecules* are present. The main techniques are:

(i)   Isolation of the polymer (followed by drying and weighing) from the reaction mixture through precipitation by addition of non-solvent.
(ii) The rate of chain polymerization of vinyl monomers can be followed by addition of bromine to the unreacted double bonds and those of the monomer.
(iii) Use of dilatometry. This is based on the large difference in densities of the monomer and the polymer. The conversion is followed in a dilatometer (Fig. AII.1) whose volume has been calibrated. Purified monomer is added to the dilatometer so that the level comes into the region of the capillary. After removing the dissolved oxygen, the dilatometer is placed in a thermostat and the volume change is followed by means of a scale or a cathetometer. This volume change is quantitatively related to the percentage conversion of monomer.

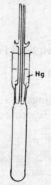

Fig. AII. 1. A dilatometer.

EXPERIMENT 5

Polymerization of Styrene in Bulk Initiated by Benzoyl Peroxide (effect of initiator concentration)

Materials and Equipment

(i)   Styrene
(ii) Sodium hydroxide solution (10%)
(iii) Anhydrous calcium chloride
(iv) Methanol
(v)   Pyrex ampoules
(vi) Thermostat

Procedure

(i) *Purification of monomer*—Styrene is shaken with 10% sodium hydroxide solution to free it from inhibitors, followed by several washings with water. It is then allowed to stand overnight over anhydrous calcium chloride. The dried styrene is then distilled under reduced pressure and nitrogen atmosphere (b.p. 82°C/100 mm; 46°C/20 mm). It is then stored in a refrigerator under nitrogen until required for use.

(ii) *Polymerization*—Four pyrex ampoules are cleaned with chromic acid, washed with water and steamed. 4.7 mg, (0.019 mmol), 9.3 mg (0.038 mmol), 46.5 mg (0.19 mmol) and 93 mg (0.38 mmol) of benzoyl peroxide are weighed separately into the four ampoules followed by the addition of 4.0 g (38.4 mmol) of purified styrene. The initiator is dissolved by thoroughly shaking the contents of the ampoules and the dissolved air displaced by oxygen-free nitrogen. The ampoules are then sealed and held in a thermostat at $60 \pm 1°C$ for four hours. They are cooled and the contents poured into excess of ice-cooled methanol under stirring. The contents are filtered. The solid obtained is dissolved in benzene, reprecipitated from methanol, filtered in a sintered glass crucible and weighed.

Report

The rate of reaction (% conversion per hour) is plotted against the initiator concentration (moles per litre) on a logarithmic paper. The plot should be a straight line. The slope of the line is also recorded.

EXPERIMENT 6

Determination of the Molecular Weight of Poly(styrene)

Materials and Equipment

(i)   Ostwald viscometer with fittings
(ii) Poly(styrene) samples from experiment 5
(iii) Benzene
(iv) Thermostat
(v)   Bellows
(vi) Stoppered bottle
(vii) Stop watch

Procedure

(i) *Fitting of a viscometer*—Fittings of a viscometer are shown in Fig. A II. 2.

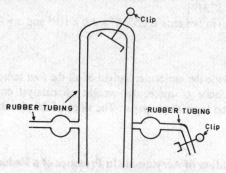

Fig. A II.2 Fitting of a viscometer.

The fittings shown in the above figure are convenient for closing the top of the U-tube of the Ostwald viscometer and for raising the liquid into the upper bulb. It consists of two glass T-pieces each of which is provided with a bulb (1–1.5 cm diameter) filled with cotton wool. The two pieces are attached to the two arms of the U–tube by rubber tubing and are connected above by another short length of similar rubber tubing, provided with a spring pinch cock. After attaching calcium chloride tubes the liquid is blown up to fill the upper viscometer bulb and its level adjusted. A pinchcock is used to close the tubing through which the air has been blown and at the required moment the pinchcock on the tubing joining the two T's is opened. The liquid then flows down without drawing in air from the room.

(ii) *Determination of specific viscosity*—A small sample of poly(styrene) (0.1–0.2 g) is weighed into a stoppered bottle and dissolved in a measured volume of pure benzene to furnish a 1% solution (approx). The solvent (10 ml) is introduced into the reservoir of a thoroughly cleaned viscometer, which is held in an erect position inside a constant temperature water bath maintained at $25 \pm 0.1°C$ until it attains the temperature of the water bath. The benzene is then blown up into the upper viscometer bulb by means of bellows and the time of efflux of the solvent noted with the help of a stop watch. This is repeated several times. The bulb of the viscometer is then emptied and it is thoroughly dried. A solution (10 ml) of poly(styrene) is then introduced into the viscometer reservoir and the time of efflux of the solution recorded in the same way as for the solvent. The specific viscosity is then given by the following expression:

$$\text{Specific viscosity } (\eta_{sp}) = \left[\frac{\text{Time of efflux of the polymer solution at } 25°}{\text{Time of efflux of the solvent at } 25°} \right] - 1$$

The viscosity number or (η_{sp}/c) is obtained by dividing the specific viscosity by the concentration of the polymer solution (given in g/100 ml).

(iii) *Calculation of molecular weight*—The limiting viscosity number $[\eta]$ can be calculated by the use of the following equation,

$$[\eta] = \frac{\eta_{sp}/c}{1 + K_\eta \eta_{sp}}$$

where k_η has a value of 0.28 for many polymer/solvent systems.

The viscosity average molecular weight $\bar{M}_v$ of the poly(styrene) sample may be calculated from the limiting viscosity number, using the appropriate value of K' and α in the equation, $[\eta = K' M^\alpha]$.

For poly(styrene) in benzene at 25°, $K' = 1.0 \times 10^{-4}$ and $\alpha = 0.74$.

Report

(i) It should include the molecular weight of all the four samples.

(ii) The dependence of molecular weight on catalyst concentration may be checked by plotting on logarithmic paper. The slope of the plot should be (−0.5).

EXPERIMENT 7

Solution Polymerization of Acrylamide in Presence of a Redox Initiator

Materials and Equipment

(i)   A pyrex three-necked reaction flask
(ii)   A thermostat
(iii) Nitrogen gas (oxygen-free)
(iv)   Iodine flasks (5)
(v)   Conductivity water
(vi)   Acrylamide
(vii) Potassium persulphate
(viii) Sodium metabisulphite
(ix)   Sodium thiosulphate solution (approx. N/50)
(x)   Brominating mixture (a mixture of KBr and potassium bromate in the molar ratio 5:1)
(xi)   Potassium iodide
(xii) Potassium bromate

Procedure

Known quantities of acrylamide and sodium metabisulphite together with requisite quantity of water are taken in a pyrex reaction flask, protected from light. The flask is immersed upto the neck in a thermostat maintained at $35 \pm 0.2°C$ and a controlled stream of oxygen-free dry nitrogen is passed through the contents of the flask. The nitrogen is passed initially at a fast rate and then at the rate of one bubble per second, to displace dissolved oxygen. When thermal equilibrium is attained, polymerization is initiated by the addition of a calculated amount of $K_2S_2O_8$ solution in water. The reaction mixture is shaken thoroughly. An aliquot is quickly drawn and introduced into ice-cooled brominating reagent, known amounts of which are previously taken in iodine flasks. Several aliquots are drawn at desired intervals of time. These are directly introduced into ice-cooled, standard brominating reagent in order to short stop the polymerization reaction. Iodine flasks are quickly stoppered and shaken well and kept in dark to complete the bromination of monomer double bonds. Remaining bromine is titrated back against potassium iodide and thiosulphate to starch end point. Percentage conversion is calculated by the following formula,

$$\% \text{ Conversion} = \left[\frac{C (V_2 - V_1) M}{20W} \right]$$

where C = normality of thiosulphate solution against KIO_3, V_1 = volume of thiosulphate required at zero time, V_2 = volume of thiosulphate at different intervals, M = molecular weight of monomer (acrylamide = 71.08) and W = weight of acrylamide present in the aliquot withdrawn at different intervals.

Short, variable induction periods may be observed due to the presence of different impurities.

Reactions

$$KBrO_3 + 5KBr + 3H_2SO_4 = 3Br_2 + 3K_2SO_4 + 3H_2O$$
$$CH_2 = CH - CONH_2 + Br_2 = CH_2Br - CH\ BrCONH_2$$
$$Br_2 + 2KI = 2KBr + I_2$$
(excess)
$$I_2 + 2Na_2S_2O_3 = 2NaI + 2Na_2S_4O_6$$

Sodium thiosulphate is standardized by using potassium iodate and potassium iodide in acid solution

$$IO_3^- + I^- + 6H^+ = 3I_2 + 3H_2O$$
$$I_2 + 2Na_2S_2O_3 = 2NaI + Na_2S_4O_6$$

Determination of molecular weight

A desired volume of the reaction mixture is poured into purified methanol. The polymer is allowed to settle down, filtered and purified by dissolving in water and reprecipitating from methanol. Samples after precipitation are dried at 40°C. The viscosity of the dilute solution (0.5%) is determined in aqueous medium at 30°C. Intrinsic viscosity is calculated by using the relationship given in Experiment 6. Molecular weights are obtained by the use of the following equation.

$$[\eta]_{30°C} = K' M^\alpha = 6.8 \times 10^{-4}\ M^{0.66}$$

Report

(i) A conversion curve may be constructed by plotting % conversion against time and initial rate of polymerization obtained in % conversion per minute from the initial slope of the curve.

(ii) By changing the amounts of oxidant (potassium persulphate) at constant monomer (acrylamide) and reductant (sodium metabisulphite) concentrations, several conversion curves can be plotted and the initial rate of polymerization obtained therefrom.

EXPERIMENT 8

Suspension Polymerization of Methyl Methacrylate

Materials and Equipment

(i) Three-necked 500 ml flask with stirrer and reflux condenser and dropping funnel

(ii) Thermometer

(iii) Hot water-bath

(iv) Buchner funnel with filtration flask

(v) Water-pump

   (vi)   Drying oven
   (vii) Methyl methacrylate monomer
   (viii) Benzoyl peroxide
   (ix)   Ammonia (sp. gr. 0.880)
   (x)   Sodium phosphate solution (10%, w/v)
   (xi)   Calcium chloride solution (2%, w/v)
   (xii) Sodium hydroxide solution (2%, w/v)
   (xiii) Hydrochloric acid solution (2%, w/v)

Procedure

Methyl methacrylate is freed from the inhibitor by washing it three times with a 2% solution of sodium hydroxide followed by three washings with distilled water.

A 500 ml three-necked flask, fitted with an efficient stirrer, a reflux condenser and a dropping funnel is placed on a water bath. The suspending agent is first prepared by placing hydrated sodium phosphate (0.36 g), ammonia (5 ml, sp. gr. 0.880) and 125 ml of distilled water in the flask. With slow stirring a solution of hydrated calcium chloride (1.5 g) in 70 ml water is added through the dropping funnel over a period of 30 min. The water bath is now heated and the temperature maintained around 70°C.

While stirring at 70° and keeping a gentle stream of nitrogen flowing, a solution of benzoyl peroxide (1.0 g) in 100 ml methyl methacrylate is run dropwise into the flask over a period of 15 min. so that a stable suspension of monomer droplets results. After replacing the dropping funnel by a thermometer, stirring speed and temperature are held constant until a sample drawn from the flask yields a hard bead on cooling with water (approx. 2 hr.). There is little tendency for the monomer droplets to coalesces until they become viscous as polymer forms. Eventually, hard beads will form. If the stirring is stopped when the beads are still tacky, agglomeration may occur although suspension aids tend to prevent this.

The apparatus is now dismantled and the polymer beads are filtered at the pump. The beads are then washed with HCl and then with water and eventually dried in air or an oven at 50°C.

The experiment can be repeated using 0.5 g, 0.2 g and 1.5 g of benzoyl peroxide per 100 ml of the monomer and noting the minimum time required for the production of hard beads; the yield of the polymer is noted in each case.

Report

It should include; (i) yield, (ii) minimum time required to get a hard bead, and (iii) the size of beads.

EXPERIMENT 9

Preparation of Poly(vinyl alcohol) from Poly(vinyl acetate) (polymer modification)

Materials and Equipment

   (i)   Round bottom flask (500 ml) with reflux condenser and $CaCl_2$ tube
   (ii) One litre beaker
   (iii) Mechanical stirrer
   (iv) Poly(vinyl acetate), low viscosity grade
   (v)   Anhydrous methanol

(vi) Potassium hydroxide

(vii) Acetone (water free)

Procedure

Into a 500 ml round bottom flask fitted with a reflux condenser and protected from moisture by an anhydrous calcium chloride guard tube are placed poly(vinyl acetate) (8.6 g, 0.1 mol) and anhydrous methanol (200 ml). The contents are heated on a water bath from about 20 min to dissolve poly(vinyl acetate) and then potassium hydroxide (0.25 g) is added and the refluxing continued for another hour. Poly(vinyl alcohol) separates out as a fine powder.

To purify poly(vinyl alcohol), it is first filtered using the pump and an alkali resistant filter paper. It is finally washed several times with anhydrous methanol. It is air-dried using the pump, dissolved in 150 ml distilled water by heating at 80°. The solution is then poured dropwise into vigorously stirred anhydrous acetone (500 ml). The precipitated poly(vinyl alcohol) is then filtered at the pump and dried in a desiccator.

Report

The report should include yield.

EXPERIMENT 10

Depolymerization of Poly(methyl methacrylate)

Materials and Equipment

(i)   Poly(methyl methacrylate)

(ii)   A vacuum distillation assembly

(iii) Two traps

(iv) Dewar flasks

(v)   Solid CO_2 or a freezing mixture

(vi) Acetone

(vii) Sand bath or metal alloy bath

(viii) Thermometer, reading upto 450°

Procedure

A vacuum distillation assembly is set up and in the distilling flask 20–30 g of poly(methyl methacrylate) is placed. The receiver is attached to a vacuum system of 0.5 to 2 mm capacity through a series of two traps cooled with dry ice-acetone cooling mixture or a freezing mixture. The flask is heated by means of a suitable sand or metal alloy bath to 300-400°C and the distillate is collected under vacuum for a period of 3-4 hrs. The combined distillate is purified in a suitable distilling apparatus and identified.

Report

(i) The purified distillate is identified by determination of boiling point, refractive index and specific gravity.

(ii) The yield of the distillate is noted.

COATINGS AND ADHESIVES

COATINGS

The use of gums and natural resins in painting has been in vogue from times immemorial. Lac, which is a natural resin extruded by the lac insect, has been known in India and China for over 3000 years. The refined form of lac resin, known as shellac, when dissolved in spirit gives a varnish, which has been used for preservation and decoration of wood all over the world for several centuries now. It is only recently that synthetic resins have replaced shellac and other natural resin in the decorative coatings industry.

The first major advance in this field was the discovery of nitrocellulose or cellulose nitrate. The preparation of this resin has been described earlier (Section 14.3). Nitrocellulose, containing 2–2.5 nitro groups per anhydroglucose unit of cellulose, is used as lacque when dissolved in a suitable solvent and mixed with other ingredients.

With the discovery of alkyds (derived from alcohols and acids), the synthetic organic coating industry received a fillip. Alkyds based on a glycol such as ethylene glycol or glycerol and phthalic acid were commercially marketed around 1920 and after extensive chemical modification and improvement they remain among the most widely used coating components.

$$O=C \underset{\underset{(\text{Phthalic anhydride})}{}}{\overset{O}{\diagdown}} C=O \quad + \quad HOCH_2-\underset{\underset{OH}{|}}{CH}-CH_2 OH \quad \longrightarrow \quad O=C \underset{\underset{(\text{Polymer ester})}{}}{\overset{O}{\diagdown}} COCH_2-CH-CH_2\text{-}\!\!\!\text{w-}$$

(Phthalic anhydride) (Glycerol) (Polymer ester) (I)

Drying oil modifications, which employ mostly linseed oil and a drier such as cobalt naphthenate, are used to a large extent in paint industry (see Sections 2.5.1 and 14.4.2). The development of silicone alkyds is another advance in this field.

Coatings based on vinyls

Poly(ethylene) is usually not used in coatings due to difficulty in applying it, but its two derivatives, poly(vinyl chloride) and poly(methyl acrylate) find a wide use in the form of water-based or latex paints in houses for external and internal coatings.

These acrylic paints are only second in importance to the alkyds in their widespread use.

Coatings based on Epoxy resins and poly(urethanes):

The introduction of coatings based on epoxy resins (II) and poly(urethanes) (III) has revolutionized modern paint technology. Their high cost is, however, a disadvantage in their wide-spread use.

$$
\begin{array}{c}
\text{O} \\
/ \ \backslash \\
-\text{CH} - \text{CH}_2 \ +\text{NH}_2\text{CH}_2- \ \longrightarrow \ \text{\textasciitilde}-\text{CH}-\text{CH}_2\text{NH}-\text{CH}_2\text{\textasciitilde} \\
\end{array}
$$

(epoxy group) (amine) (epoxy resin) (II)
OH

$$
-\text{N}=\text{C}=\text{O} \ + \ \text{HO}-\text{CH}_2- \ \longrightarrow \ \text{\textasciitilde}-\text{N}-\overset{\text{O}}{\underset{\text{H}}{\overset{\|}{\text{C}}}}-\text{OCH}_2-\text{\textasciitilde}
$$

(isocyanate group)   (alcohol)   poly(urethane)

(III)

The coating systems are so designed that the polymers which form on the surface are substantially large in molecular dimensions, forming so called infinite networks. These huge molecules make the epoxies and poly(urethanes) exceptionally tough and durable.

Special coatings

Among the special organic coatings mention may be made of fluorocarbons, two notable examples of which are poly(vinyl fluoride) and poly(tetrafluoroethylene) (Section 3.11.5).

$$
\text{\textasciitilde}\text{CH}_2-\underset{\text{F}}{\text{CH}}-\text{CH}_2-\underset{\text{F}}{\text{CH}}-\text{\textasciitilde}
$$

Poly (Vinyl fluoride)
(IV)

$$
\text{\textasciitilde}\underset{\text{F}}{\overset{\text{F}}{\text{C}}}-\underset{\text{F}}{\overset{\text{F}}{\text{C}}}-\underset{\text{F}}{\overset{\text{F}}{\text{C}}}-\underset{\text{F}}{\overset{\text{F}}{\text{C}}}-\text{\textasciitilde}
$$

Poly (Tetrafluoroethylene)
(V)

The substitution of hydrogens in poly(ethylene) by fluoride atoms makes poly(fluorocarbons) some of the most environmentally stable coatings. They are very resistant to water, ultraviolet light and micro-organisms. They are extremely heat resistant and are used in the manufacture of cooking utensils. Their high cost and difficulties in applying to surfaces are factors against their wide use.

Inorganic coatings

The poly(siloxanes) have an inorganic backbone and as such are very resistant to heat and are water repellants (see Section 2.6.1).

$$
\text{\textasciitilde}\underset{\text{CH}_3}{\overset{\text{CH}_3}{\text{Si}}}-\text{O}-\underset{\text{CH}_3}{\overset{\text{CH}_3}{\text{Si}}}-\text{O}-\underset{\text{CH}_3}{\overset{\text{CH}_3}{\text{Si}}}-\text{O}-\underset{\text{CH}_3}{\overset{\text{CH}_3}{\text{Si}}}-\text{O}-\text{\textasciitilde}
$$

Poly (dimethyl siloxane)
(VI)

Another 'fire-proof' inorganic polymer is poly(phosphazene) (see Section 2.6.2).

$$\begin{array}{ccccccc} & \text{R} & & \text{R} & \vdots & \text{R} & \\ & | & & | & & | & \\ -\text{\small WW}-\!\!\!&\text{P}&\!\!=\text{N}-\!\!\!&\text{P}&\!\!=\text{N}-\!\!\!&\text{P}&\!\!=\text{N}-\text{\small WW}- \\ & | & & | & & | & \\ & \text{R} & & \text{R} & & \text{R} & \end{array}$$

Poly (phosphazene)

(VII)

This polymer is in the process of development.

Classification of coatings

The industrially important coatings are classified into the following five categories:

(1) *Spirit varnishes*: Coatings compounded by dissolving in a suitable solvent or mixture of solvents. The most important example of this type is shellac varnish.

(2) *Lacquers*: Coatings based on cellulose derivatives as the main constituent. They may or may not contain a drying oil as a basic constituent.

(3) *Oil varnishes*: This group of coatings is prepared by the joint use of an oil and a resin. It is sometimes referred to as oleo resinous varnish. If a thermosetting resin has been used, heat must be applied to cause curing or hardening of the resin. If a thermoplastic resin has been used, heat may be applied to accelerate the drying of the oil.

(4) *Enamels*: These are highly pigmented varnishes used to produce coloured or opaque films.

(5) *Paints*: These comprise a pigment dispersed in an oil vehicle. Hardening occurs as a result of oxidation and polymerization of the oil film.

To take an example, the more important components of a lacquer are: Cellulose derivate, modifying resin, plasticizer, oil, solvents (active and latent), diluent, pigment or dye, and stabilizer.

The coatings are applied by spraying or manually by a brush.

ADHESIVES

Adhesion is defined as a process that occurs when (i) a solid and a liquid are brought together to form an interface and (ii) the surface energies of the two substances are transformed into the energy of the interface and heat is evolved. The forces between the adhesive and the adherent may be small but, as in the case of other molecular interactions, these forces are additive. Adhesives must be fluid at some stage during application and must wet the surfaces of the adherent when in the fluid stage. Adhesion is favoured when the solubility parameter of the adhesive has a high numerical value. The adhesion is advanced by the presence of polar groups. Adhesives may be applied as melts, solutions or aqueous dispersions. The adhesive film must be stabilized by setting or curing.

The two most used adhesives are based on phenol-formaldehyde and urea-formaldehyde condensation products. Their most important use is in the bonding of wood veneers into plywood.

A cold setting phenol-formaldehyde adhesive is prepared by condensing together phenol and formaldehyde in the presence of an alkaline catalyst (NaOH solution) in the molar ratio (1:2.3). The resin is vacuum-distilled and adjusted to pH 7-7.5 by addition of an organic acid (e.g. lactic acid). A hardening catalyst such as a solution of hydrochloric acid is mixed before the use of the adhesive. The preparation of a gap filling cold setting urea-formaldehyde adhesive has already been described in Appendix II (Experiment 2). Other formulations for the preparation of heat setting and cold setting phenolics and ureas have been developed.

The heat setting adhesive becomes effective after the addition of an acidic hardening catalyst. Approximately 3-5% of the catalyst may be added to the resin and curing done at 125-135° C at a pressure of 1000-2000 lbs/sq. inch. The curing time may vary between 2 and 10 min. Phenolic adhesives are used where the strength of the bond is of importance, and urea based adhesives are preferred where colourability is important as in the case of decorative surfaces.

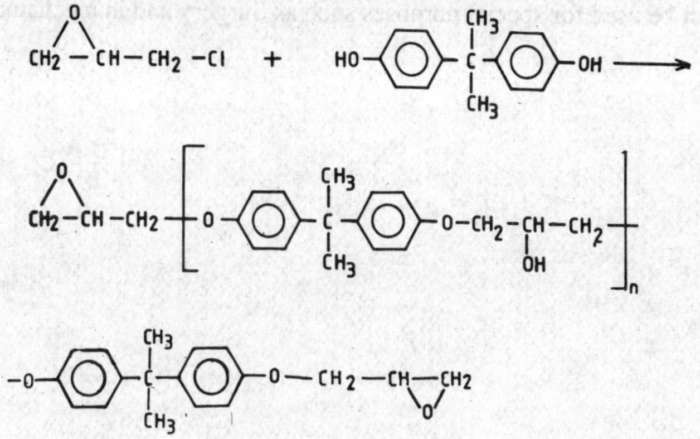

(Resin A)

(VIII)

Scheme 1

Another important adhesive based on epoxy resin is prepared as described below. A mixture of two resins, one obtained from bisphenol-A and epichlorohydrin (mol. wt., ~ 370) (resin A) and the other obtained from the condensation of glycerol and epichlorohydrin (mol. wt., ~ 324) (resin B), is reacted with a hardening reagent, (triethylamine). The resulting mixture can be applied to the surfaces which are to be joined. They are then clamped and kept at a temperature of 40-45° C for about 6 days to give a strong bond.

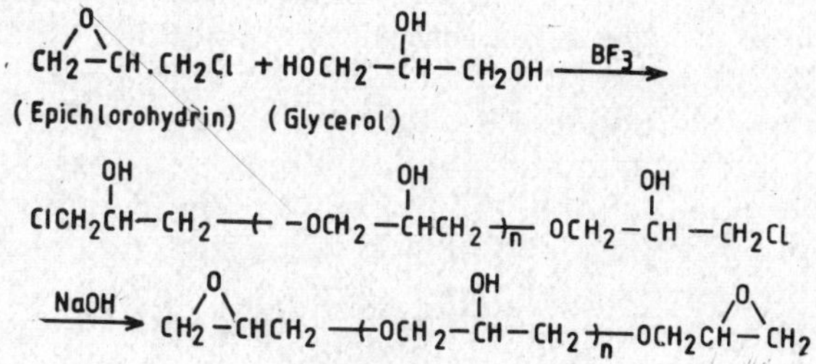

(Resin B)

(IX)

Scheme 2

One recent addition to adhesive family are the Cyanoacrylates. Thus butyl-α-cyanoacrylate polymerizes in the presence of moist air, to produce an excellent

adhesive (x)

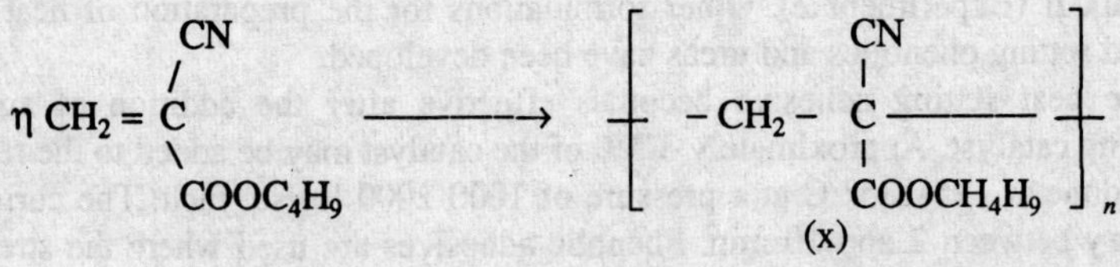

$$\eta \; CH_2 = C \begin{array}{c} CN \\ | \\ \\ \backslash \\ COOC_4H_9 \end{array} \longrightarrow \left[\begin{array}{c} CN \\ | \\ -CH_2 - C \\ | \\ COOCH_4H_9 \end{array} \right]_n$$

(x)

Scheme 3

These can be used for special purposes such as Surgery and in mechanical appliances.

RECOMMENDED ACS (AMERICAN CHEMICAL SOCIETY) SYLLABUS FOR AN INTRODUCTORY COURSE IN POLYMER CHEMISTRY

The following topics have been recommended for an introductory course in polymer chemistry by a syllabus committee appointed by the American Chemical Society under the chairmanship of Prof. R.B. Seymour. These recommendations may be found useful by teachers while planning their courses and selecting materials. [Raymond B. Seymour, *J. Chem. Edn.*, **59** (1982) 652]

Major Topics*	Percent of course time
1. Macromolecules—An Introduction	5
2. Molecular Weights	10
Average Molecular Weight	
Fractionation of Polydisperse Systems	
Characterization Techniques	
3. Step Reaction Polymerization	10
Kinetics	
Polymers Produced by Step Polymerizations	
4. Chain Reaction Polymerization	
Kinetics of Ionic Chain Reaction Polymerization	
Kinetics of Free Radical Chain Reaction	
Polymerization	
5. Copolymerization	10
Kinetics	
Types of Copolymers	
Polymer Blends	
Principal copolymers	
6. Morphology	10
Stereochemistry of Polymers	
Molecular Interactions	
Crystallinity	
7. Testing and Characterization of Polymers	10
Structure-Property Relationships	
Physical Tests	
Instrumental Characterizations	
Additional Topics	35
8. Flow Properties	
Viscoelasticity	
Rubber Elasticity	
9. Solubility	
10. Natural and Biomedical Polymers	
11. Additives	
Fillers	
Plasticizers	
Stabilizers	
Flame Retardants	
Colourants	
12. Reactions of Polymers	
13. Synthesis of Polymers Reactants	

Contd·

*The order of presentation may be changed by the Instructor.

Major Topics*	*Percent of course time*

14.   Polymer Technology
     Plastics
     Elastomers
     Fibers
     Coatings
     Adhesives	

The time allowed to some additional topics may be shortened or some of the topics may be outlined in one quarter courses. Other topics and literature assignments should be considered for two semester courses.

INDEX

A B S resins 80
Acrylonitrile
  copolymers 83
  fibers 143
  preparation 64
Addition polymerization
  kinetics 36-37
  mechanism 36-37
  vs condensation polymerization 66
Additives 170-176
Adhesives 244-246
Adipic acid 15, 201
Ageing 151
Alfrey-Price equation 75
Alkyd-resins
  cross-linking of 195-196
  in coatings 242
  preparation 19
  unsaturated 19
Amylopectin 222
Anionic polymerization
  kinetics 50
  mechanism 50
Antioxidant 2246 160
Antioxidants 160
Asbestos filler 171
Atactic polymers 90-91
Autoxidation 157
Azobis isobutyronitrile 40

Bakelite 20, 233
Balata 131
Benzimidazole 164
Benzophenone
  2-hydroxy-4-octoxy 163
Benzotriazole
  2-hydroxy-3, 5-dialkyl 163
Bimetallic mechanism
  of tacticity 52
Bingham plastics 217
Bisphenol-A 16
Block copolymers 78-79
Bulk polymerization 47
Buna rubbers
  Buna-N 135
  Buna-S 135
  Buna-85 135
  Buna-115 135
Butadiene
  preparation 133
Butylated hydroxy toluene (BHT) 160
Butyl rubber 136

Calendering 149
Camphor 174
Captax 132
Caprolactam 15
Carbon black 162
Carother's equation 18
Casein fibers 141
Cationic polymerization
  kinetics 49
  mechanism 49
Cellophane 193
Celluloid 194
Cellulose 221
  acetate 193
  acetate butyrate 194
  alkali 138
  nitrate 194
  regenerated 140, 193
  xanthate 38
Chain-breaking
  antioxidants 160-161
  mechanisms 160-162
Chain-flexibility 125-126
Chain polymerization 33-70
Chain-transfer
  with initiator 44
  with monomer 44
  with modifier 44
  with solvents 42-44
Chain-transfer agents 44
Characterization of polymers 180-187
Chromatographic
  methods of analysis 152
Chloroprene 135
Coatings 244-246
Cohesive energy density 208
Cold drawing 145
Colourants 175
Compression moulding 146-147
Configuration 86
Conformation 86
Coordination polymerization 51-55
Copolymerization 71-85
  azeotropic 74
  reactivity ratio 83
Copolymer
  alternating 75
  composition equation 72
  feed composition 74
  ideal 74-75
  random 74-75
Cotton fabrics as filler 171
Creep 216

RELATED BOOKS

POLYMER SCIENCE

Gowarikar, V.R., N.V. Vishwanathan and J. Sreedhar

Contents: The Genesis of Polymers. Chemistry of Polymerisation. Molecular Weight and Size. Kinetics of Polymerisation. Chemical and Geometrical Structure of Polymer Molecules. Glass Transition Temperature. Crystallinity in Polymers. Copolymerisation. Individual Polymerrs. Polymer Degradation. Polymer Reactions. Polymer Solutions. Experimental Methods. Elastomeric. Fibre-Forming and Plastic Materials. Polymer Processing.

| 0852263074 | 1988 | 50 pp | Paper | Rs. 95 |

ADVANCES IN MATERIALS AND THEIR APPLICATIONS

P. Rama Rao

The book covers a range of special topics pertainig to advanced materials by authors of longstanding experience in the field. Apart from high temperature and other engineering materials and semiconductor materials which have acquired prominence in high technology applications, nanoscale and composition gradient materials, which also figure in this compilation, are bound to yield new products. With reference to engineering materials, a topic of current interest is mechanics in impact situations while microstructure is being studied with enhanced attention to detail where semiconductor materials are concerned. Processing and application of materials in modern high technology systems generate, often unforeseen, specific challenges. A flavour of these aspects is contained in the mix of papers that deal with selected materials as well as actual systems, from a materials perspective, like automobiles, aeroengines, naval systems, nuclear reactors and modern missiles.

Contents: Preface. Materials Development to Product Deployment: Some Reflections on Strategies for Success. Fundamental Materials Science Approach in Advanced Materials Manufacturing and Structural Integrity Assessment. Advanced Materials and Processes for the Next Two Decades. Nanometer Scale Materials: A New Frontier. Nanophase Materials Assembled from Clusters. The Role of Surfaces in Atomic Ordering and Phase Separation in Ternary and Quaternary III-V Semiconductors. Metastable Phase Formation in Thin Films. Composition Gradient Materials. Material Properties for Structural Impact Problems. Material Deformation Mechanisms in Crashworthiness Applications. Dislocation Mechanics Description of Dynamic Plasticity and Fracturing Properties. Grain Boundaries and High-Temperature Deformation. Deformation and Damage Behaviour of Superalloys: A Mechanism Based Analysis. Advanced Surface Modification of Steels. Hydrogen Induced Cold Cracking in Multi-Pass Steel Weld Metal. Nickel-Containing Materials in Applications. Materials for Aerogas Turbine Engines. Materials for Automobile Industry. Materials for Naval Systems. Advanced Materials for Nuclear Systems. Materials in Indian Missile Programme. Benjamin Robins: A Neglected Mid-18th Century Military Engineer-Scientist.

PUBLISHING FOR ONE WORLD

WILEY EASTERN LIMITED

NEW DELHI • BANGALORE • BOMBAY • CALCUTTA • GUWAHATI
HYDERABAD • LUCKNOW • MADRAS • PUNE